HYPERSPECTRAL REMOTE
OF VEGETATION
SECOND EDITION
VOLUME II

Hyperspectral Indices and Image Classifications for Agriculture and Vegetation

Hyperspectral Remote Sensing of Vegetation
Second Edition

Volume I: Fundamentals, Sensor Systems, Spectral Libraries, and Data Mining for Vegetation

Volume II: Hyperspectral Indices and Image Classifications for Agriculture and Vegetation

Volume III: Biophysical and Biochemical Characterization and Plant Species Studies

Volume IV: Advanced Applications in Remote Sensing of Agricultural Crops and Natural Vegetation

HYPERSPECTRAL REMOTE SENSING OF VEGETATION
SECOND EDITION
VOLUME II

Hyperspectral Indices and Image Classifications for Agriculture and Vegetation

Edited by
Prasad S. Thenkabail
John G. Lyon
Alfredo Huete

CRC Press
Taylor & Francis Group
Boca Raton London New York

CRC Press is an imprint of the
Taylor & Francis Group, an **informa** business

CRC Press
Taylor & Francis Group
6000 Broken Sound Parkway NW, Suite 300
Boca Raton, FL 33487-2742

First issued in paperback 2022

© 2019 by Taylor & Francis Group, LLC
CRC Press is an imprint of Taylor & Francis Group, an Informa business

No claim to original U.S. Government works

ISBN 13: 978-1-03-247585-1 (pbk)
ISBN 13: 978-1-138-06603-8 (hbk)

DOI: 10.1201/9781315159331

Visit the Taylor & Francis Web site at
http://www.taylorandfrancis.com

and the CRC Press Web site at
http://www.crcpress.com

Dr. Prasad S. Thenkabail, *Editor-in-Chief of these four volumes would like to dedicate the four volumes to three of his professors at the Ohio State University during his PhD days:*

1. Late Prof. Andrew D. Ward, *former professor of The Department of Food, Agricultural, and Biological Engineering (FABE) at The Ohio State University,*

2. Prof. John G. Lyon, *former professor of the Department of Civil, Environmental and Geodetic Engineering at the Ohio State University, and*

3. Late Prof. Carolyn Merry, *former Professor Emerita and former Chair of the Department of Civil, Environmental and Geodetic Engineering at the Ohio State University.*

Contents

SECTION I Hyperspectral Vegetation Indices

SECTION II Hyperspectral Image Classification Methods and Approaches

SECTION III Hyperspectral Vegetation Index Applications to Agriculture and Vegetation

SECTION IV   Conclusions

Foreword to the First Edition

The publication of this book, *Hyperspectral Remote Sensing of Vegetation*, marks a milestone in the application of imaging spectrometry to study 70% of the Earth's landmass which is vegetated. This book shows not only the breadth of international involvement in the use of hyperspectral data but also in the breadth of innovative application of mathematical techniques to extract information from the image data.

Imaging spectrometry evolved from the combination of insights from the vast heterogeneity of reflectance signatures from the Earth's surface seen in the ERTS-1 (Landsat-1) 4-band images and the field spectra that were acquired to help fully understand the causes of the signatures. It wasn't until 1979 when the first hybrid area-array detectors, mercury-cadmium-telluride on silicon CCD's, became available that it was possible to build an imaging spectrometer capable of operating at wavelengths beyond 1.0 μm. The AIS (airborne imaging spectrometer), developed at NASA/JPL, had only 32 cross-track pixels but that was enough for the geologists clamoring for this development to see *between* the bushes to determine the mineralogy of the substrate. In those early years, vegetation cover was just a nuisance!

In the early 1980s, spectroscopic analysis was driven by the interest to identify mineralogical composition by exploiting absorptions found in the SWIR region from overtone and combination bands of fundamental vibrations found in the mid-IR region beyond 3 μm and the electronic transitions in transition elements appearing, primarily, short of 1.0 μm. The interests of the geologists had been incorporated in the Landsat TM sensor in the form of the add-on, band 7 in the 2.2 μm region based on field spectroscopic measurements. However, one band, even in combination with the other six, did not meet the needs for mineral identification. A summary of mineralogical analyses is presented by Vaughan et al. in this volume. A summary of the historical development of hyperspectral imaging can be found in Goetz (2009).

At the time of the first major publication of the AIS results (Goetz et al., 1985), very little work on vegetation analysis using imaging spectroscopy had been undertaken. The primary interest was in identifying the relationship of the chlorophyll absorption red-edge to stress and substrate composition that had been seen in airborne profiling and in field spectral reflectance measurements. Most of the published literature concerned analyzing NDVI, which only required two spectral bands.

In the time leading up to the 1985 publication, we had only an inkling of the potential information content in the hundreds of contiguous spectral bands that would be available to us with the advent of AVIRIS (airborne visible and infrared imaging spectrometer). One of the authors, Jerry Solomon, presciently added the term "hyperspectral" to the text of the paper to describe the "…multidimensional character of the spectral data set," or, in other words, the mathematically, over-determined nature of hyperspectral data sets. The term hyperspectral as opposed to multispectral data moved into the remote sensing vernacular and was additionally popularized by the military and intelligence community.

In the early 1990s, as higher quality AVIRIS data became available, and the first analyses of vegetation using statistical techniques borrowed from chemometrics, also known as NIRS analysis used in the food and grain industry, were undertaken by John Aber and Mary Martin of the University of New Hampshire. Here, nitrogen contents of tree canopies were predicted from reflectance spectra by regression techniques using reference measurements from laboratory wet chemical analyses of needle and leaf samples acquired by shooting down branches. At the same time, the remote sensing community began to recognize the value of "too many" spectral bands and the concomitant wealth of spatial information that was amenable to information extraction by statistical techniques. One of them was Eyal Ben-Dor who pioneered soil analyses using hyperspectral imaging and who is one of the contributors to this volume.

As the quality of AVIRIS data grew, manifested in increasing SNR, an ever-increasing amount of information could be extracted from the data. This quality was reflected in the increasing number of nearly noiseless principal components that could be obtained from the data or, in other words, its dimensionality. The explosive advances in desktop computing made possible the application of image processing and statistical analyses that revolutionized the uses of hyperspectral imaging. Joe Boardman and others at the University of Colorado developed what has become the ENVI software package to make possible the routine analysis of hyperspectral image data using "unmixing techniques" to derive the relative abundance of surface materials on a pixel-by-pixel basis.

Many of the analysis techniques discussed in this volume, such as band selection and various indices, are rooted in principal components analysis. The eigenvector loadings or factors indicate which spectral bands are the most heavily weighted allowing others to be discarded to reduce the noise contribution. As sensors become better, more information will be extractable and fewer bands will be discarded. This is the beauty of hyperspectral imaging, allowing the choice of the number of eigenvectors to be used for a particular problem. Computing power has reached such a high level that it is no longer necessary to choose a subset of bands just to minimize the computational time.

As regression techniques such as PLS (partial least squares) become increasingly adopted to relate a particular vegetation parameter to reflectance spectra, it must be remembered that the quality of the calibration model is a function of both the spectra and the reference measurement. With spectral measurements of organic and inorganic compounds under laboratory conditions, we have found that a poor model with a low coefficient of determination (r^2) is most often associated with inaccurate reference measurements, leading to the previously intuitive conclusion that "spectra don't lie."

Up to this point, AVIRIS has provided the bulk of high-quality hyperspectral image data but on an infrequent basis. Although Hyperion has provided some time series data, there is no hyperspectral imager yet in orbit that is capable of providing routine, high-quality images of the whole Earth on a consistent basis. The hope is that in the next decade, HyspIRI will be providing VNIR and SWIR hyperspectral images every 3 weeks and multispectral thermal data every week. This resource will revolutionize the field of vegetation remote sensing since so much of the useful information is bound up in the seasonal growth cycle. The combination of the spectral, spatial, and temporal dimensions will be ripe for the application of statistical techniques and the results will be extraordinary.

Dr. Alexander F. H. Goetz PhD
Former Chairman and Chief Scientist
ASD Inc.
2555 55th St. #100
Boulder, CO 80301, USA
303-444-6522 ext. 108
Fax 303-444-6825
www.asdi.com

REFERENCES

Goetz, A. F. H., 2009, Three decades of hyperspectral imaging of the Earth: A personal view, *Remote Sensing of Environment*, 113, S5–S16.

Goetz, A.F.H., G. Vane, J. Solomon and B.N. Rock, 1985, Imaging spectrometry for Earth remote sensing, *Science*, 228, 1147–1153.

BIOGRAPHICAL SKETCH

Dr. Goetz is one of the pioneers in hyperspectral remote sensing and certainly needs no introduction. Dr. Goetz started his career working on spectroscopic reflectance and emittance studies of the Moon and Mars. He was a principal investigator of Apollo-8 and Apollo-12 multispectral photography studies. Later, he turned his attention to remote sensing of Planet Earth working in collaboration with Dr. Gene Shoemaker to map geology of Coconino County (Arizona) using Landsat-1 data and went on to be an investigator in further Landsat, Skylab, Shuttle, and EO-1 missions. At NASA/JPL he pioneered field spectral measurements and initiated the development of hyperspectral imaging. He spent 21 years on the faculty of the University of Colorado, Boulder, and retired in 2006 as an Emeritus Professor of Geological Sciences and an Emeritus Director of Center for the Study of Earth from Space. Since then, he has been Chairman and Chief Scientist of ASD Inc. a company that has provided more than 850 research laboratories in over 60 countries with field spectrometers. Dr. Goetz is now retired. His foreword was written for the first edition and I have retained it in consultation with him to get a good perspective on the development of hyperspectral remote sensing.

Foreword to the Second Edition

The publication of the four-volume set, *Hyperspectral Remote Sensing of Vegetation*, second edition, is a landmark effort in providing an important, valuable, and timely contribution that summarizes the state of spectroscopy-based understanding of the Earth's terrestrial and near shore environments. Imaging spectroscopy has had 35 years of development in data processing and analysis methods. Today's researchers are eager to use data produced by hyperspectral imagers and address important scientific issues from agricultural management to global environmental stewardship. The field started with development of the Jet Propulsion Lab's Airborne Imaging Spectrometer in 1983 that measured across the reflected solar infrared spectrum with 128 spectral bands. This technology was quickly followed in 1987 by the more capable Advanced Visible Infrared Imaging Spectrometer (AVIRIS), which has flown continuously since this time (albeit with multiple upgrades). It has 224 spectral bands covering the 400–2500 nm range with 10 nm wavelength bands and represents the "gold standard" of this technology. In the years since then, progress toward a hyperspectral satellite has been disappointingly slow. Nonetheless, important and significant progress in understanding how to analyze and understand spectral data has been achieved, with researchers focused on developing the concepts, analytical methods, and spectroscopic understanding, as described throughout these four volumes. Much of the work up to the present has been based on theoretical analysis or from experimental studies at the leaf level from spectrometer measurements and at the canopy level from airborne hyperspectral imagers.

Although a few hyperspectral satellites have operated over various periods in the 2000s, none have provided systematic continuous coverage required for global mapping and time series analysis. An EnMap document compiled the past and near-term future hyperspectral satellites and those on International Space Station missions (EnMap and GRSS Technical Committee 2017). Of the hyperspectral imagers that have been flown, the European Space Agency's CHRIS (Compact High Resolution Imaging Spectrometer) instrument on the PROBA-1 (Project for On-Board Autonomy) satellite and the Hyperion sensor on the NASA technology demonstrator, Earth Observing-1 platform (terminated in 2017). Each has operated for 17 years and have received the most attention from the science community. Both collect a limited number of images per day, and have low data quality relative to today's capability, but both have open data availability. Other hyperspectral satellites with more limited access and duration include missions from China, Russia, India, and the United States.

We are at a threshold in the availability of hyperspectral imagery. There are many hyperspectral missions planned for launch in the next 5 years from China, Italy, Germany, India, Japan, Israel, and the United States, some with open data access. The analysis of the data volumes from this proliferation of hyperspectral imagers requires a comprehensive reference resource for professionals and students to turn to in order to understand and correctly and efficiently use these data. This four-volume set is unique in compiling in-depth understanding of calibration, visualization, and analysis of data from hyperspectral sensors. The interest in this technology is now widespread, thus, applications of hyperspectral imaging cross many disciplines, which are truly international, as is evident by the list of authors of the chapters in these volumes, and the number of countries planning to operate a hyperspectral satellite. At least some of the hyperspectral satellites announced and expected to be launched in this decade (such as the HyspIRI-like satellite approved for development by NASA with a launch in the 2023 period) will provide high-fidelity narrow-wavelength bands, covering the full reflected solar spectrum, at moderate (30 m pixels) to high spatial resolution. These instruments will have greater radiometric range, better SNR, pointing accuracy, and reflectance calibration than past instruments, and will collect data from many countries and parts of the world that have not previously been available. Together, these satellites will produce an unprecedented flow of information about the physiological functioning (net primary production, evapotranspiration, and even direct measurements related to respiration), biochemical characteristics (from spectral indices and from radiative transfer first principle methods), and direct measurements of the distributions of plant and soil biodiversity of the terrestrial and coastal environments of the Earth.

This four-volume set presents an unprecedented range and scope of information on hyperspectral data analysis and applications written by leading authors in the field. Topics range from sensor characteristics from ground-based platforms to satellites, methods of data analysis to characterize plant functional properties related to exchange of gases CO_2, H_2O, O_2, and biochemistry for pigments, N cycle, and other molecules. How these data are used in applications range from precision agriculture to global change research. Because the hundreds of bands in the full spectrum includes information to drive detection of these properties, the data is useful at scales from field applications to global studies.

Volume I has three sections and starts with an introduction to hyperspectral sensor systems. Section II focuses on sensor characteristics from ground-based platforms to satellites, and how these data are used in global change research, particularly in relation to agricultural crop monitoring and health of natural vegetation. Section III provides five chapters that deal with the concept of spectral libraries to identify crops and spectral traits, and for phenotyping for plant breeding. It addresses the development of spectral libraries, especially for agricultural crops and one for soils.

Volume II expands on the first volume, focusing on use of hyperspectral indices and image classification. The volume begins with an explanation of how narrow-band hyperspectral indices are determined, often from individual spectral absorption bands but also from correlation matrices and from derivative spectra. These are followed by chapters on statistical approaches to image classification and a chapter on methods for dealing with "big data." The last half of this volume provides five chapters focused on use of vegetation indices for quantifying and characterizing photosynthetic pigments, leaf nitrogen concentrations or contents, and foliar water content measurements. These chapters are particularly focused on applications for agriculture, although a chapter addresses more heterogeneous forest conditions and how these patterns relate to monitoring health and production.

The first half of Volume III focuses on biophysical and biochemical characterization of vegetation properties that are derived from hyperspectral data. Topics include ecophysiological functioning and biomass estimates of crops and grasses, indicators of photosynthetic efficiency, and stress detection. The chapter addresses biophysical characteristics across different spatial scales while another chapter examines spectral and spatial methods for retrieving biochemical and biophysical properties of crops. The chapters in the second half of this volume are focused on identification and discrimination of species from hyperspectral data and use of these methods for rapid phenotyping of plant breeding trials. Lastly, two chapters evaluate tree species identification, and another provides examples of mapping invasive species.

Volume IV focuses on six areas of advanced applications in agricultural crops. The first considers detection of plant stressors including nitrogen deficiency and excess heavy metals and crop disease detection in precision farming. The second addresses global patterns of crop water productivity and quantifying litter and invasive species in arid ecosystems. Phenological patterns are examined while others focus on multitemporal data for mapping patterns of phenology. The third area is focused on applications of land cover mapping in different forest, wetland, and urban applications. The fourth topic addresses hyperspectral measurements of wildfires, and the fifth evaluates use of continuity vegetation index data in global change applications. And lastly, the sixth area examines use of hyperspectral data to understand the geologic surfaces of other planets.

Susan L. Ustin
Professor and Vice Chair, Dept. Land, Air and Water Resources
Associate Director, John Muir Institute
University of California
Davis California, USA

REFERENCE

EnMap Ground Segment Team and GSIS GRSS Technical committee, December, 2017. *Spaceborne Imaging Spectroscopy Mission Compilation*. DLR Space Administration and the German Federal Ministry of Economic Affairs and Technology. http://www.enmap.org/sites/default/files/pdf/Hyperspectral_EO_Missions_2017_12_21_FINAL4.pdf

BIOGRAPHICAL SKETCH

Dr. Susan L. Ustin is currently a Distinguished Professor of Environmental and Resource Sciences in the Department of Land, Air, and Water Resources, University of California Davis, Associate Director of the John Muir Institute, and is Head of the Center for Spatial Technologies and Remote Sensing (CSTARS) at the same university. She was trained as a plant physiological ecologist but began working with hyperspectral imagery as a post-doc in 1983 with JPL's AIS program. She became one of the early adopters of hyperspectral remote sensing which has now extended over her entire academic career. She was a pioneer in the development of vegetation analysis using imaging spectrometery, and is an expert on ecological applications of this data. She has served on numerous NASA, NSF, DOE, and the National Research Council committees related to spectroscopy and remote sensing. Among recognitions for her work, she is a Fellow of the American Geophysical Union and received an honorary doctorate from the University of Zurich. She has published more than 200 scientific papers related to ecological remote sensing and has worked with most of the Earth-observing U.S. airborne and spaceborne systems.

Preface

This seminal book on *Hyperspectral Remote Sensing of Vegetation* (Second Edition, 4 Volume Set), published by Taylor and Francis Inc.\CRC Press is an outcome of over 2 years of effort by the editors and authors. In 2011, the first edition of *Hyperspectral Remote Sensing of Vegetation* was published. The book became a standard reference on hyperspectral remote sensing of vegetation amongst the remote sensing community across the world. This need and resulting popularity demanded a second edition with more recent as well as more comprehensive coverage of the subject. Many advances have taken place since the first edition. Further, the first edition was limited in scope in the sense it covered some very important topics and missed equally important topics (e.g., hyperspectral library of agricultural crops, hyperspectral pre-processing steps and algorithms, and many others). As a result, a second edition that brings us up-to-date advances in hyperspectral remote sensing of vegetation was required. Equally important was the need to make the book more comprehensive, covering an array of subjects not covered in the first edition. So, my coeditors and myself did a careful research on what should go into the second edition. Quickly, the scope of the second edition expanded resulting in an increasing number of chapters. All of this led to developing the seminal book: *Hyperspectral Remote Sensing of Vegetation*, Second Edition, 4 Volume Set. The four volumes are:

Volume I: Fundamentals, Sensor Systems, Spectral Libraries, and Data Mining for Vegetation
Volume II: Hyperspectral Indices and Image Classifications for Agriculture and Vegetation
Volume III: Biophysical and Biochemical Characterization and Plant Species Studies
Volume IV: Advanced Applications in Remote Sensing of Agricultural Crops and Natural Vegetation

The goal of the book was to bring in one place collective knowledge of the last 50 years of advances in hyperspectral remote sensing of vegetation with a target audience of wide spectrum of scientific community, students, and professional application practitioners. The book documents knowledge advances made in applying hyperspectral remote sensing technology in the study of terrestrial vegetation that include agricultural crops, forests, rangelands, and wetlands. This is a very practical offering about a complex subject that is rapidly advancing its knowledge-base. In a very practical way, the book demonstrates the experience, utility, methods, and models used in studying terrestrial vegetation using hyperspectral data. The four volumes, with a total of 48 chapters, are divided into distinct themes.

- **Volume I**: There are 14 chapters focusing on hyperspectral instruments, spectral libraries, and methods and approaches of data handling. The chapters extensively address various preprocessing steps and data mining issues such as the Hughes phenomenon and overcoming the "curse of high dimensionality" of hyperspectral data. Developing spectral libraries of crops, vegetation, and soils with data gathered from hyperspectral data from various platforms (ground-based, airborne, spaceborne), study of spectral traits of crops, and proximal sensing at field for phenotyping are extensively discussed. Strengths and limitations of hyperspectral data of agricultural crops and vegetation acquired from different platforms are discussed. It is evident from these chapters that the hyperspectral data provides opportunities for great advances in study of agricultural crops and vegetation. However, it is also clear from these chapters that hyperspectral data should not be treated as panacea to every limitation of multispectral broadband data such as from Landsat or Sentinel series of satellites. The hundreds or thousands of hyperspectral narrowbands (HNBs) as well as carefully selected hyperspectral vegetation indices (HVIs) will help us make significant advances in characterizing, modeling, mapping, and monitoring vegetation biophysical, biochemical, and structural quantities. However, it is also important to properly understand hyperspectral data and eliminate redundant bands that exist for every application and to optimize computing as well as human resources to enable seamless and efficient handling enormous volumes of hyperspectral data. Special emphasis is also put on preprocessing and processing of Earth Observing-1 (EO-1) Hyperion, the first publicly available hyperspectral data from space. These methods, approaches, and algorithms, and protocols set the stage for upcoming satellite hyperspectral sensors such as NASA's HyspIRI and Germany's EnMAP.

- **Volume II**: There are 10 chapters focusing on hyperspectral vegetation indices (HVIs) and image classification methods and techniques. The HVIs are of several types such as: (i) two-band derived, (ii) multi-band-derived, and (iii) derivative indices derived. The strength of the HVIs lies in the fact that specific indices can be derived for specific biophysical, biochemical, and plant structural quantities. For example, you have carotenoid HVI, anthocyanin HVI, moisture or water HVI, lignin HVI, cellulose HVI, biomass or LAI or other biophysical HVIs, red-edge based HVIs, and so on. Further, since these are narrowband indices, they are better targeted and centered at specific sensitive wavelength portions of the spectrum. The strengths and limitations of HVIs in a wide array of applications such as leaf nitrogen content (LNC), vegetation water content, nitrogen content in vegetation, leaf and plant pigments, anthocyanin's, carotenoids, and chlorophyll are thoroughly studied. Image classification using hyperspectral data provides great strengths in deriving more classes (e.g., crop species within a crop as opposed to just crop types) and increasing classification accuracies. In earlier years and decades, hyperspectral data classification and analysis was a challenge due to computing and data handling issues. However, with the availability of machine learning algorithms on cloud computing (e.g., Google Earth Engine) platforms, these challenges have been overcome in the last 2–3 years. Pixel-based supervised machine learning algorithms like the random forest, and support vector machines as well as object-based algorithms like the recursive hierarchical segmentation, and numerous others methods (e.g., unsupervised approaches) are extensively discussed. The ability to process petabyte volume data of the planet takes us to a new level of sophistication and makes use of data such as from hyperspectral sensors feasible over large areas. The cloud computing architecture involved with handling massively large petabyte-scale data volumes are presented and discussed.

- **Volume III**: There are 11 chapters focusing on biophysical and biochemical characterization and plant species studies. A number of chapters in this volume are focused on separating and discriminating agricultural crops and vegetation of various types or species using hyperspectral data. Plant species discrimination and classification to separate them are the focus of study using vegetation such as forests, invasive species in different ecosystems, and agricultural crops. Performance of hyperspectral narrowbands (HNBs) and hyperspectral vegetation indices (HVIs) when compared with multispectral broadbands (MBBs) and multispectral broadband vegetation indices (BVIs) are presented and discussed. The vegetation and agricultural crops are studied at various scales, and their vegetation functional properties diagnosed. The value of digital surface models in study of plant traits as complementary\supplementary to hyperspectral data has been highlighted. Hyperspectral bio-indicators to study photosynthetic efficiency and vegetation stress are presented and discussed. Studies are conducted using hyperspectral data across wavelengths (e.g., visible, near-infrared, shortwave-infrared, mid-infrared, and thermal-infrared).

- **Volume IV**: There are 15 chapters focusing on specific advanced applications of hyperspectral data in study of agricultural crops and natural vegetation. Specific agricultural crop applications include crop management practices, crop stress, crop disease, nitrogen application, and presence of heavy metals in soils and related stress factors. These studies discuss biophysical and biochemical quantities modeled and mapped for precision farming, hyperspectral narrowbands (HNBs), and hyperspectral vegetation indices (HVIs) involved in assessing nitrogen in plants, and the study of the impact of heavy metals on crop health and stress. Vegetation functional studies using hyperspectral data presented and discussed include crop water use (actual evapotranspiration), net primary productivity (NPP), gross primary productivity (GPP), phenological applications, and light use efficiency (LUE). Specific applications discussed under vegetation functional studies using hyperspectral data include agricultural crop classifications, machine learning, forest management studies, pasture studies, and wetland studies. Applications in fire assessment, modeling, and mapping using hyperspectral data in the optical and thermal portions of the spectrum are presented and discussed. Hyperspectral data in global change studies as well as in outer planet studies have also been discussed. Much of the outer planet remote sensing is conducted using imaging spectrometer and hence the data preprocessing and processing methods of Earth and that of outer planets have much in common and needs further examination.

The chapters are written by leading experts in the global arena with each chapter: (a) focusing on specific applications, (b) reviewing existing "state-of-art" knowledge, (c) highlighting the advances made, and (d) providing guidance for appropriate use of hyperspectral data in study of vegetation and its numerous applications such as crop yield modeling, crop biophysical and biochemical property characterization, and crop moisture assessment.

The four-volume book is specifically targeted on hyperspectral remote sensing as applied to terrestrial vegetation applications. This is a big market area that includes agricultural croplands, study of crop moisture, forests, and numerous applications such as droughts, crop stress, crop productivity, and water productivity. To the knowledge of the editors, there is no comparable book, source, and/or organization that can bring this body of knowledge together in one place, making this a "must buy" for professionals. This is clearly a unique contribution whose time is now. The book highlights include:

1. Best global expertise on hyperspectral remote sensing of vegetation, agricultural crops, crop water use, plant species detection, crop productivity and water productivity mapping, and modeling;
2. Clear articulation of methods to conduct the work. Very practical;
3. Comprehensive review of the existing technology and clear guidance on how best to use hyperspectral data for various applications;
4. Case studies from a variety of continents with their own subtle requirements; and
5. Complete solutions from methods to applications inventory and modeling.

Hyperspectral narrowband spectral data, as discussed in various chapters of this book, are fast emerging as practical most advanced solutions in modeling and mapping vegetation. Recent research has demonstrated the advances and great value made by hyperspectral data, as discussed in various chapters in: (a) quantifying agricultural crops as to their biophysical and harvest yield characteristics, (b) modeling forest canopy biochemical properties, (c) establishing plant and soil moisture conditions, (d) detecting crop stress and disease, (e) mapping leaf chlorophyll content as it influences crop production, (f) identifying plants affected by contaminants such as arsenic, and (g) demonstrating sensitivity to plant nitrogen content, and (h) invasive species mapping. The ability to significantly better quantify, model, and map plant chemical, physical, and water properties is well established and has great utility.

Even though these accomplishments and capabilities have been reported in various places, the need for a collective "knowledge bank" that links these various advances in one place is missing. Further, most scientific papers address specific aspects of research, failing to provide a comprehensive assessment of advances that have been made nor how the professional can bring those advances to their work. For example, deep scientific journals report practical applications of hyperspectral narrowbands yet one has to canvass the literature broadly to obtain the pertinent facts. Since several papers report this, there is a need to synthesize these findings so that the reader gets the correct picture of the best wavebands for their practical applications. Also, studies do differ in exact methods most suited for detecting parameters such as crop moisture variability, chlorophyll content, and stress levels. The professional needs this sort of synthesis and detail to adopt best practices for their own work.

In years and decades past, use of hyperspectral data had its challenges especially in handling large data volumes. That limitation is now overcome through cloud-computing, machine learning, deep learning, artificial intelligence, and advances in knowledge in processing and applying hyperspectral data.

This book can be used by anyone interested in hyperspectral remote sensing that includes advanced research and applications, such as graduate students, undergraduates, professors, practicing professionals, policy makers, governments, and research organizations.

Dr. Prasad S. Thenkabail, PhD
Editor-in-Chief
Hyperspectral Remote Sensing of Vegetation, Second Edition, Four Volume Set

Acknowledgments

This four-volume *Hyperspectral Remote Sensing of Vegetation* book (second edition) was made possible by sterling contributions from leading professionals from around the world in the area of hyperspectral remote sensing of vegetation and agricultural crops. As you will see from list of authors and coauthors, we have an assembly of "**who is who**" in hyperspectral remote sensing of vegetation who have contributed to this book. They wrote insightful chapters, that are an outcome of years of careful research and dedication, to make the book appealing to a broad section of readers dealing with remote sensing. My gratitude goes to (mentioned in no particular order; names of lead authors of the chapters are shown in bold): **Drs. Fred Ortenberg** (Technion–Israel Institute of Technology, Israel), **Jiaguo Qi** (Michigan State University, USA), **Angela Lausch** (Helmholtz Centre for Environmental Research, Leipzig, Germany), **Andries B. Potgieter** (University of Queensland, Australia), **Muhammad Al-Amin Hoque** (University of Queensland, Australia), **Andreas Hueni** (University of Zurich, Switzerland), **Eyal Ben-Dor** (Tel Aviv University, Israel), **Itiya Aneece** (United States Geological Survey, USA), **Sreekala Bajwa** (University of Arkansas, USA), **Antonio Plaza** (University of Extremadura, Spain), **Jessica J. Mitchell** (Appalachian State University, USA), **Dar Roberts** (University of California at Santa Barbara, USA), **Quan Wang** (Shizuoka University, Japan), **Edoardo Pasolli** (University of Trento, Italy), (Nanjing University of Science and Technology, China), **Anatoly Gitelson** (University of Nebraska- Lincoln, USA), **Tao Cheng** (Nanjing Agricultural University, China), **Roberto Colombo** (University of Milan-Bicocca, Italy), **Daniela Stroppiana** (Institute for Electromagnetic Sensing of the Environment, Italy), **Yongqin Zhang** (Delta State University, USA), **Yoshio Inoue** (National Institute for Agro-Environmental Sciences, Japan), Yafit Cohen (Institute of Agricultural Engineering, Israel), **Helge Aasen** (Institute of Agricultural Sciences, ETH Zurich), **Elizabeth M. Middleton** (NASA, USA), **Yongqin Zhang** (University of Toronto, Canada), **Yan Zhu** (Nanjing Agricultural University, China), **Lênio Soares Galvão** (Instituto Nacional de Pesquisas Espaciais [INPE], Brazil), **Matthew L. Clark** (Sonoma State University, USA), **Matheus Pinheiro Ferreira** (University of Paraná, Curitiba, Brazil), **Ruiliang Pu** (University of South Florida, USA), **Scott C. Chapman** (CSIRO, Australia), **Haibo Yao** (Mississippi State University, USA), **Jianlong Li** (Nanjing University, China), **Terry Slonecker** (USGS, USA), **Tobias Landmann** (International Centre of Insect Physiology and Ecology, Kenya), **Michael Marshall** (University of Twente, Netherlands), **Pamela Nagler** (USGS, USA), **Alfredo Huete** (University of Technology Sydney, Australia), **Prem Chandra Pandey** (Banaras Hindu University, India), **Valerie Thomas** (Virginia Tech., USA), **Izaya Numata** (South Dakota State University, USA), **Elijah W. Ramsey III** (USGS, USA), **Sander Veraverbeke** (Vrije Universiteit Amsterdam and University of California, Irvine), **Tomoaki Miura** (University of Hawaii, USA), **R. G. Vaughan** (U.S. Geological Survey, USA), Victor Alchanatis (Agricultural research Organization, Volcani Center, Israel), Dr. Narumon Wiangwang (Royal Thai Government, Thailand), Pedro J. Leitão (Humboldt University of Berlin, Department of Geography, Berlin, Germany), James Watson (University of Queensland, Australia), Barbara George-Jaeggli (ETH Zuerich, Switzerland), Gregory McLean (University of Queensland, Australia), Mark Eldridge (University of Queensland, Australia), Scott C. Chapman (University of Queensland, Australia), Kenneth Laws (University of Queensland, Australia), Jack Christopher (University of Queensland, Australia), Karine Chenu (University of Queensland, Australia), Andrew Borrell (University of Queensland, Australia), Graeme L. Hammer (University of Queensland, Australia), David R. Jordan (University of Queensland, Australia), Stuart Phinn (University of Queensland, Australia), Lola Suarez (University of Melbourne, Australia), Laurie A. Chisholm (University of Wollongong, Australia), Alex Held (CSIRO, Australia), S. Chabrillant (GFZ German Research Center for Geosciences, Germany), José A. M. Demattê (University of São Paulo, Brazil), Yu Zhang (North Dakota State University, USA), Ali Shirzadifar (North Dakota State University, USA), Nancy F. Glenn (Boise State University, USA), Kyla M. Dahlin (Michigan State

University, USA), Nayani Ilangakoon (Boise State University, USA), Hamid Dashti (Boise State University, USA), Megan C. Maloney (Appalachian State University, USA), Subodh Kulkarni (University of Arkansas, USA), Javier Plaza (University of Extremadura, Spain), Gabriel Martin (University of Extremadura, Spain), Segio Sánchez (University of Extremadura, Spain), Wei Wang (Nanjing Agricultural University, China), Xia Yao (Nanjing Agricultural University, China), Busetto Lorenzo (Università Milano-Bicocca), Meroni Michele (Università Milano-Bicocca), Rossini Micol (Università Milano-Bicocca), Panigada Cinzia (Università Milano-Bicocca), F. Fava (Università degli Studi di Sassari, Italy), M. Boschetti (Institute for Electromagnetic Sensing of the Environment, Italy), P. A. Brivio (Institute for Electromagnetic Sensing of the Environment, Italy), K. Fred Huemmrich (University of Maryland, Baltimore County, USA), Yen-Ben Cheng (Earth Resources Technology, Inc., USA), Hank A. Margolis (Centre d'Études de la Forêt, Canada), Yafit Cohen (Agricultural research Organization, Volcani Center, Israel), Kelly Roth (University of California at Santa Barbara, USA), Ryan Perroy (University of Wisconsin-La Crosse, USA), Ms. Wei Wang (Nanjing Agricultural University, China), Dr. Xia Yao (Nanjing Agricultural University, China), Keely L. Roth (University of California, Santa Barbara, USA), Erin B. Wetherley (University of California at Santa Barbara, USA), Susan K. Meerdink (University of California at Santa Barbara, USA), Ryan L. Perroy (University of Wisconsin-La Crosse, USA), B. B. Marithi Sridhar (Bowling Green University, USA), Aaryan Dyami Olsson (Northern Arizona University, USA), Willem Van Leeuwen (University of Arizona, USA), Edward Glenn (University of Arizona, USA), José Carlos Neves Epiphanio (Instituto Nacional de Pesquisas Espaciais [INPE], Brazil), Fábio Marcelo Breunig (Instituto Nacional de Pesquisas Espaciais [INPE], Brazil), Antônio Roberto Formaggio (Instituto Nacional de Pesquisas Espaciais [INPE], Brazil), Amina Rangoonwala (IAP World Services, Lafayette, LA), Cheryl Li (Nanjing University, China), Deghua Zhao (Nanjing University, China), Chengcheng Gang (Nanjing University, China), Lie Tang (Mississippi State University, USA), Lei Tian (Mississippi State University, USA), Robert Brown (Mississippi State University, USA), Deepak Bhatnagar (Mississippi State University, USA), Thomas Cleveland (Mississippi State University, USA), Hiroki Yoshioka (Aichi Prefectural University, Japan), T. N. Titus (U.S. Geological Survey, USA), J. R. Johnson (U.S. Geological Survey, USA), J. J. Hagerty (U.S. Geological Survey, USA), L. Gaddis (U.S. Geological Survey, USA), L. A. Soderblom (U.S. Geological Survey, USA), and P. Geissler (U.S. Geological Survey, USA), Jua Jin (Shizuoka University, Japan), Rei Sonobe (Shizuoka University, Japan), Jin Ming Chen (Shizuoka University, Japan), Saurabh Prasad (University of Houston, USA), Melba M. Crawford (Purdue University, USA), James C. Tilton (NASA Goddard Space Flight Center, USA), Jin Sun (Nanjing University of Science and Technology, China), Yi Zhang (Nanjing University of Science and Technology, China), Alexei Solovchenko (Moscow State University, Moscow), Yan Zhu, (Nanjing Agricultural University, China), Dong Li (Nanjing Agricultural University, China), Kai Zhou (Nanjing Agricultural University, China), Roshanak Darvishzadeh (University of Twente, Enschede, The Netherlands), Andrew Skidmore (University of Twente, Enschede, The Netherlands), Victor Alchanatis (Institute of Agricultural Engineering, The Netherlands), Georg Bareth (University of Cologne, Germany), Qingyuan Zhang (Universities Space Research Association, USA), Petya K. E. Campbell (University of Maryland Baltimore County, USA), and David R. Landis (Global Science & Technology, Inc., USA), José Carlos Neves Epiphanio (Instituto Nacional de Pesquisas Espaciais [INPE], Brazil), Fábio Marcelo Breunig (Universidade Federal de Santa Maria [UFSM], Brazil), and Antônio Roberto Formaggio (Instituto Nacional de Pesquisas Espaciais [INPE], Brazil), Cibele Hummel do Amaral (Federal University of Viçosa, in Brazil), Gaia Vaglio Laurin (Tuscia University, Italy), Raymond Kokaly (U.S. Geological Survey, USA), Carlos Roberto de Souza Filho (University of Ouro Preto, Brazil), Yosio Edemir Shimabukuro (Federal Rural University of Rio de Janeiro, Brazil), Bangyou Zheng (CSIRO, Australia), Wei Guo (The University of Tokyo, Japan), Frederic Baret (INRA, France), Shouyang Liu (INRA, France), Simon Madec (INRA, France), Benoit Solan (ARVALIS, France), Barbara George-Jaeggli (University of Queensland, Australia), Graeme L. Hammer (University of Queensland, Australia), David R. Jordan (University of Queensland, Australia), Yanbo Huang (USDA, USA), Lie Tang (Iowa State

University, USA), Lei Tian (University of Illinois. USA), Deepak Bhatnagar (USDA, USA), Thomas E. Cleveland (USDA, USA), Dehua ZHAO (Nanjing University, USA), Hannes Feilhauer (University of Erlangen-Nuremberg, Germany), Miaogen Shen (Institute of Tibetan Plateau Research, Chinese Academy of Sciences, Beijing, China), Jin Chen (College of Remote Sensing Science and Engineering, Faculty of Geographical Science, Beijing Normal University, Beijing, China), Suresh Raina (International Centre of Insect Physiology and Ecology, Kenya and Pollination services, India), Danny Foley (Northern Arizona University, USA), Cai Xueliang (UNESCO-IHE, Netherlands), Trent Biggs (San Diego State University, USA), Werapong Koedsin (Prince of Songkla University, Thailand), Jin Wu (University of Hong Kong, China), Kiril Manevski (Aarhus University, Denmark), Prashant K. Srivastava (Banaras Hindu University, India), George P. Petropoulos (Technical University of Crete, Greece), Philip Dennison (University of Utah, USA), Ioannis Gitas (University of Thessaloniki, Greece), Glynn Hulley (NASA Jet Propulsion Laboratory, California Institute of Technology, USA), Olga Kalashnikova, (NASA Jet Propulsion Laboratory, California Institute of Technology, USA), Thomas Katagis (University of Thessaloniki, Greece), Le Kuai (University of California, USA), Ran Meng (Brookhaven National Laboratory, USA), Natasha Stavros (California Institute of Technology, USA).

Hiroki Yoshioka (Aichi Prefectural University, Japan), My two coeditors, **Professor John G. Lyon** and **Professor Alfredo Huete**, have made outstanding contribution to this four-volume *Hyperspectral Remote Sensing of Vegetation* book (second edition). Their knowledge of hyperspectral remote sensing is enormous. Vastness and depth of their understanding of remote sensing in general and hyperspectral remote sensing in particular made my job that much easier. I have learnt a lot from them and continue to do so. Both of them edited some or all of the 48 chapters of the book and also helped structure chapters for a flawless reading. They also significantly contributed to the synthesis chapter of each volume. I am indebted to their insights, guidance, support, motivation, and encouragement throughout the book project.

My coeditors and myself are grateful to **Dr. Alexander F. H. Goetz** and **Prof. Susan L. Ustin** for writing the foreword for the book. Please refer to their biographical sketch under the respective foreword written by these two leaders of Hyperspectral Remote Sensing.

Both the forewords are a must read to anyone studying this four-volume *Hyperspectral Remote Sensing of Vegetation* book (second edition). They are written by two giants who have made immense contribution to the subject and I highly recommend that the readers read them.

I am blessed to have had the support and encouragement (professional and personal) of my U.S. Geological Survey and other colleagues. In particular, I would like to mention Mr. Edwin Pfeifer (late), Dr. Susan Benjamin, Dr. Dennis Dye, and Mr. Larry Gaffney. Special thanks to Dr. Terrence Slonecker, Dr. Michael Marshall, Dr. Isabella Mariotto, and Dr. Itiya Aneece who have worked closely with me on hyperspectral research over the years. Special thanks are also due to Dr. Pardhasaradhi Teluguntla, Mr. Adam Oliphant, and Dr. Muralikrishna Gumma who have contributed to my various research efforts and have helped me during this book project directly or indirectly. I am grateful to Prof. Ronald B. Smith, professor at Yale University who was instrumental in supporting my early hyperspectral research at the Yale Center for Earth Observation (YCEO), Yale University. Opportunities and guidance I received in my early years of remote sensing from Prof. Andrew D. Ward, professor at the Ohio State University, Prof. John G. Lyon, former professor at the Ohio State University, and Mr. Thiruvengadachari, former Scientist at the National Remote Sensing Center (NRSC), Indian Space Research Organization, India, is gratefully acknowledged.

My wife (Sharmila Prasad) and daughter (Spandana Thenkabail) are two great pillars of my life. I am always indebted to their patience, support, and love.

Finally, kindly bear with me for sharing a personal story. When I started editing the first edition in the year 2010, I was diagnosed with colon cancer. I was not even sure what the future was and how long I would be here. I edited much of the first edition soon after the colon cancer surgery and during and after the 6 months of chemotherapy—one way of keeping my mind off the negative thoughts. When you are hit by such news, there is nothing one can do, but to be positive, trust your doctors, be thankful to support and love of the family, and have firm belief in the higher spiritual being (whatever your beliefs are). I am so very grateful to some extraordinary people who helped me through this difficult life event: Dr. Parvasthu Ramanujam (surgeon), Dr. Paramjeet K. Bangar (Oncologist), Dr. Harnath Sigh (my primary doctor), Dr. Ram Krishna (Orthopedic Surgeon and family friend), three great nurses (Ms. Irene, Becky, Maryam) at Banner Boswell Hospital (Sun City, Arizona, USA), courage-love-patience-prayers from my wife, daughter, and several family members, friends, and colleagues, and support from numerous others that I have not named here. During this phase, I learnt a lot about cancer and it gave me an enlightened perspective of life. My prayers were answered by the higher power. I learnt a great deal about life—good and bad. I pray for all those with cancer and other patients that diseases one day will become history or, in the least, always curable without suffering and pain. Now, after 8 years, I am fully free of colon cancer and was able to edit the four-volume *Hyperspectral Remote Sensing of Vegetation* book (second edition) without the pain and suffering that I went through when editing the first edition. What a blessing. These blessings help us give back in our own little ways. To realize that it is indeed profound to see the beautiful sunrise every day, the day go by with every little event (each with a story of their own), see the beauty of the sunset, look up to the infinite universe and imagine on its many wonders, and just to breathe fresh air every day and enjoy the breeze. These are all many wonders of life that we need to enjoy, cherish, and contemplate.

Dr. Prasad S. Thenkabail, PhD
Editor-in-Chief
Hyperspectral Remote Sensing of Vegetation

Editors

Prasad S. Thenkabail, Research Geographer-15, U.S. Geological Survey (USGS), is a world-recognized expert in remote sensing science with multiple major contributions in the field sustained over more than 30 years. He obtained his PhD from the Ohio State University in 1992 and has over 140+ peer-reviewed scientific publications, mostly in major international journals.

Dr. Thenkabail has conducted pioneering research in the area of hyperspectral remote sensing of vegetation and in that of global croplands and their water use in the context of food security. In hyperspectral remote sensing he has done cutting-edge research with wide implications in advancing remote sensing science in application to agriculture and vegetation. This body of work led to more than ten peer-reviewed research publications with high impact. For example, a single paper [1] has received 1000+ citations as at the time of writing (October 4, 2018). Numerous other papers, book chapters, and books (as we will learn below) are also related to this work, with two other papers [2,3] having 350+ to 425+ citations each.

In studies of global croplands in the context of food and water security, he has led the release of the world's first Landsat 30-m derived global cropland extent product. This work demonstrates a "paradigm shift" in how remote sensing science is conducted. The product can be viewed in full resolution at the web location www.croplands.org. The data is already widely used worldwide and is downloadable from the NASA\USGS LP DAAC site [4]. There are numerous major publication in this area (e.g. [5,6]).

Dr. Thenkabail's contributions to series of leading edited books on remote sensing science places him as a world leader in remote sensing science advances. He edited three-volume *Remote Sensing Handbook* published by Taylor and Francis, with 82 chapters and more than 2000 pages, widely considered a "magnus opus" standard reference for students, scholars, practitioners, and major experts in remote sensing science. Links to these volumes along with endorsements from leading global remote sensing scientists can be found at the location give in note [7]. He has recently completed editing *Hyperspectral Remote Sensing of Vegetation* published by Taylor and Francis in four volumes with 50 chapters. This is the second edition is a follow-up on the earlier single-volume *Hyperspectral Remote Sensing of Vegetation* [8]. He has also edited a book on *Remote Sensing of Global Croplands for Food Security* (Taylor and Francis) [9]. These books are widely used and widely referenced in institutions worldwide.

Dr. Thenkabail's service to remote sensing community is second to none. He is currently an editor-in-chief of the *Remote Sensing* open access journal published by MDPI; an associate editor of the journal *Photogrammetric Engineering and Remote Sensing* (PERS) of the American Society of Photogrammetry and Remote Sensing (ASPRS); and an editorial advisory board member of the International Society of Photogrammetry and Remote Sensing (ISPRS) *Journal of Photogrammetry and Remote Sensing*. Earlier, he served on the editorial board of *Remote Sensing of Environment* for many years (2007–2017). As an editor-in-chief of the open access *Remote Sensing* MDPI journal from 2013 to date he has been instrumental in providing leadership for an online publication that did not even have a impact factor when he took over but is now one of the five leading remote sensing international journals, with an impact factor of 3.244.

Dr. Thenkabail has led remote sensing programs in three international organizations: International Water Management Institute (IWMI), 2003–2008; International Center for Integrated Mountain Development (ICIMOD), 1995–1997; and International Institute of Tropical Agriculture (IITA),

1992–1995. He has worked in more than 25+ countries on several continents, including East Asia (China), S-E Asia (Cambodia, Indonesia, Myanmar, Thailand, Vietnam), Middle East (Israel, Syria), North America (United States, Canada), South America (Brazil), Central Asia (Uzbekistan), South Asia (Bangladesh, India, Nepal, and Sri Lanka), West Africa (Republic of Benin, Burkina Faso, Cameroon, Central African Republic, Cote d'Ivoire, Gambia, Ghana, Mali, Nigeria, Senegal, and Togo), and Southern Africa (Mozambique, South Africa). During this period he has made major contributions and written seminal papers on remote sensing of agriculture, water resources, inland valley wetlands, global irrigated and rain-fed croplands, characterization of African rainforests and savannas, and drought monitoring systems.

The quality of Dr. Thenkabail's research is evidenced in the many awards, which include, in 2015, the American Society of Photogrammetry and Remote Sensing (ASPRS) ERDAS award for best scientific paper in remote sensing (Marshall and Thenkabail); in 2008, the ASPRS President's Award for practical papers, second place (Thenkabail and coauthors); and in 1994, the ASPRS Autometric Award for outstanding paper (Thenkabail and coauthors). His team was recognized by the Environmental System Research Institute (ESRI) for "special achievement in GIS" (SAG award) for their Indian Ocean tsunami work. The USGS and NASA selected him to be on the Landsat Science Team for a period of five years (2007–2011).

Dr. Thenkabail is regularly invited as keynote speaker or invited speaker at major international conferences and at other important national and international forums every year. He has been principal investigator and/or has had lead roles of many pathfinding projects, including the ~5 million over five years (2014–2018) for the global food security support analysis data in the 30-m (GFSAD) project (https://geography.wr.usgs.gov/science/croplands/) funded by NASA MEaSUREs (Making Earth System Data Records for Use in Research Environments), and projects such as Sustain and Manage America's Resources for Tomorrow (waterSMART) and characterization of Eco-Regions in Africa (CERA).

REFERENCES

1. Thenkabail, P.S., Smith, R.B., and De-Pauw, E. 2000b. Hyperspectral vegetation indices for determining agricultural crop characteristics. *Remote Sensing of Environment*, 71:158–182.
2. Thenkabail, P.S., Enclona, E.A., Ashton, M.S., Legg, C., and Jean De Dieu, M. 2004. Hyperion, IKONOS, ALI, and ETM+ sensors in the study of African rainforests. *Remote Sensing of Environment*, 90:23–43.
3. Thenkabail, P.S., Enclona, E.A., Ashton, M.S., and Van Der Meer, V. 2004. Accuracy assessments of hyperspectral waveband performance for vegetation analysis applications. *Remote Sensing of Environment*, 91(2–3):354–376.
4. https://lpdaac.usgs.gov/about/news_archive/release_gfsad_30_meter_cropland_extent_products
5. Thenkabail, P.S. 2012. Guest Editor for Global Croplands Special Issue. *Photogrammetric Engineering and Remote Sensing*, 78(8).
6. Thenkabail, P.S., Knox, J.W., Ozdogan, M., Gumma, M.K., Congalton, R.G., Wu, Z., Milesi, C., Finkral, A., Marshall, M., Mariotto, I., You, S. Giri, C. and Nagler, P. 2012. Assessing future risks to agricultural productivity, water resources and food security: how can remote sensing help? *Photogrammetric Engineering and Remote Sensing*, August 2012 Special Issue on Global Croplands: Highlight Article. 78(8):773–782. IP-035587.
7. https://www.crcpress.com/Remote-Sensing-Handbook---Three-Volume-Set/Thenkabail/p/book/9781482218015
8. https://www.crcpress.com/Hyperspectral-Remote-Sensing-of-Vegetation/Thenkabail-Lyon/p/book/9781439845370
9. https://www.crcpress.com/Remote-Sensing-of-Global-Croplands-for-Food-Security/Thenkabail-Lyon-Turral-Biradar/p/book/9781138116559

John G. Lyon, educated at Reed College in Portland, OR and the University of Michigan in Ann Arbor, has conducted scientific and engineering research and carried out administrative functions throughout his career. He was formerly the Senior Physical Scientist (ST) in the US Environmental Protection Agency's Office of Research and Development (ORD) and Office of the Science Advisor in Washington, DC, where he co-led work on the Group on Earth Observations and the USGEO subcommittee of the Committee on Environment and Natural Resources and research on geospatial issues in the agency. For approximately eight years, he was director of ORD's Environmental Sciences Division, which conducted research on remote sensing and geographical information system (GIS) technologies as applied to environmental issues including landscape characterization and ecology, as well as analytical chemistry of hazardous wastes, sediments, and ground water. He previously served as professor of civil engineering and natural resources at Ohio State University (1981–1999). Professor Lyon's own research has led to authorship or editorship of a number of books on wetlands, watershed, and environmental applications of GIS, and accuracy assessment of remote sensor technologies.

Alfredo Huete leads the Ecosystem Dynamics Health and Resilience research program within the Climate Change Cluster (C3) at the University of Technology Sydney, Australia. His main research interest is in using remote sensing to study and analyze vegetation processes, health, and functioning, and he uses satellite data to observe land surface responses and interactions with climate, land use activities, and extreme events. He has more than 200 peer-reviewed journal articles, including publication in such prestigious journals as *Science* and *Nature*. He has over 25 years' experience working on NASA and JAXA mission teams, including the NASA-EOS MODIS Science Team, the EO-1 Hyperion Team, the JAXA GCOM-SGLI Science Team, and the NPOESS-VIIRS advisory group. Some of his past research involved the development of the soil-adjusted vegetation index (SAVI) and the enhanced vegetation index (EVI), which became operational satellite products on MODIS and VIIRS sensors. He has also studied tropical forest phenology and Amazon forest greening in the dry season, and his work was featured in a *National Geographic* television special entitled "The Big Picture." Currently, he is involved with the Australian Terrestrial Ecosystem Research Network (TERN), helping to produce national operational phenology products; as well as the AusPollen network, which couples satellite sensing to better understand and predict pollen phenology from allergenic grasses and trees.

Contributors

M. Boschetti
Institute for Electromagnetic Sensing of the
  Environment
National Research Council
Milan, Italy

P. A. Brivio
Institute for Electromagnetic Sensing of the
  Environment
National Research Council
Milan, Italy

Jing Ming Chen
Department of Geography and Planning
University of Toronto
Toronto, Canada

Tao Cheng
National Engineering and Technology Center
  for Information Agriculture (NETCIA)
Key Laboratory of Crop System Analysis and
  Decision Making
Ministry of Agriculture and Rural Affairs
Jiangsu Key Laboratory for Information
  Agriculture
Jiangsu Collaborative Innovation Center for
  Modern Crop Production
Nanjing Agricultural University
Nanjing, Jiangsu, China

Panigada Cinzia
Remote Sensing of Environmental
  Dynamics Laboratory
DISAT
University of Milano Bicocca
Milano, Italy

Melba M. Crawford
Department of Agronomy
Schools of Civil and Electrical and
  Computer Engineering
Purdue University
West Lafayette, Indiana

F. Fava
Sustainable Livestock Systems Program
International Livestock Research
  Institute (ILRI)
Nairobi, Kenya

Anatoly Gitelson
Faculty of Civil and Environmental Engineering
Israel Institute of Technology
Haifa, Israel

and

School of Natural Resources
University of Nebraska-Lincoln
Lincoln, Nebraska

Alfredo Huete
School of Life Sciences
University of Technology Sydney
New South Wales, Australia

Jia Jin
Faculty of Agriculture
Shizuoka University
Shizuoka, Japan

and

Xinjiang Institute of Ecology and Geography
Chinese Academy of Sciences
Urumqi, China

Dong Li
National Engineering and Technology Center
  for Information Agriculture (NETCIA)
Key Laboratory of Crop System Analysis and
  Decision Making
Ministry of Agriculture and Rural Affairs
Jiangsu Key Laboratory for Information
  Agriculture
Jiangsu Collaborative Innovation Center for
  Modern Crop Production
Nanjing Agricultural University
Nanjing, Jiangsu, China

Busetto Lorenzo
Institute for Electromagnetic Sensing of
    the Environment
Italian National Research Council
Milan, Italy

John G. Lyon
American Society for Photogrammetry and
    Remote Sensing
Chantilly, Virginia

Susan K. Meerdink
Department of Geography
University of California
Santa Barbara, California

Meroni Michele
European Commission—Joint Research Centre
Directorate D—Sustainable Resources
Food Security Unit—D5
Ispra, Italy

Rossini Micol
Remote Sensing of Environmental
    Dynamics Laboratory
DISAT
University of Milano Bicocca
Milano, Italy

Edoardo Pasolli
Centre for Integrative Biology
University of Trento
Trento, Italy

Ryan L. Perroy
Department of Geography and
    Environmental Science
University of Hawai'I
Hilo, Hawaii

Saurabh Prasad
Electrical and Computer Engineering Department
University of Houston
Houston, Texas

Colombo Roberto
Remote Sensing of Environmental
    Dynamics Laboratory
DISAT
University of Milano Bicocca
Milano, Italy

Dar A. Roberts
Department of Geography
University of California
Santa Barbara, California

Keely L. Roth
Geospatial Research Department
The Climate Corporation
San Francisco, California

Alexei Solovchenko
Department of Bioengineering
Faculty of Biology
M.V. Lomonosov Moscow State University
Moscow, Russia

and

Michurin Federal Scientific Centre
Michurinsk, Russia

and

Eurasian Centre for Food Security
Moscow, Russia

Rei Sonobe
Faculty of Agriculture
Shizuoka University
Shizuoka, Japan

D. Stroppiana
Institute for Electromagnetic Sensing of
    the Environment
National Research Council
Milan, Italy

Jin Sun
School of Computer Science and Engineering
Nanjing University of Science and Technology
Nanjing, Jiangsu, China

Prasad S. Thenkabail
Research Geographer
Western Geographic Science Center
United States Geological Survey
Reston, Virginia

James C. Tilton
Computational and Information Sciences and
    Technology Office
NASA Goddard Space Flight Center
Greenbelt, Maryland

Quan Wang
Faculty of Agriculture
Shizuoka University
Shizuoka, Japan

Erin B. Wetherley
Department of Geography
University of California
Santa Barbara, California

Zebin Wu
School of Computer Science and Engineering
Nanjing University of Science and Technology
Nanjing, Jiangsu, China

Xia Yao
National Engineering and Technology Center
  for Information Agriculture (NETCIA)
Key Laboratory of Crop System Analysis and
  Decision Making
Ministry of Agriculture and Rural Affairs
Jiangsu Key Laboratory for Information
  Agriculture
Jiangsu Collaborative Innovation Center for
  Modern Crop Production
Nanjing Agricultural University
Nanjing, Jiangsu, China

Yi Zhang
School of Computer Science and Engineering
Nanjing University of Science and Technology
Nanjing, Jiangsu, China

Yongqin Zhang
Division of Mathematics and Sciences
Delta State University
Cleveland, Mississippi

Kai Zhou
National Engineering and Technology
  Center for Information Agriculture
  (NETCIA)
Key Laboratory of Crop System Analysis
  and Decision Making
Ministry of Agriculture and Rural Affairs
Jiangsu Key Laboratory for Information
  Agriculture
Jiangsu Collaborative Innovation Center for
  Modern Crop Production
Nanjing Agricultural University
Nanjing, Jiangsu, China

Yan Zhu
National Engineering and Technology
  Center for Information Agriculture
  (NETCIA)
Key Laboratory of Crop System Analysis and
  Decision Making
Ministry of Agriculture and Rural Affairs
Jiangsu Key Laboratory for Information
  Agriculture
Jiangsu Collaborative Innovation Center for
  Modern Crop Production
Nanjing Agricultural University
Nanjing, Jiangsu, China

Acronyms and Abbreviations

1DL_DGVI	First-order derivative green vegetation index derived using local baseline
1DZ_DGVI	First-order derivative green vegetation index derived using zero baseline
α	Absorbance
AA	Average accuracy
ACI	Anthocyanin content index
ACO	Ant colony optimization
AET	Actual evapotranspiration
AIC	Akaike information criterion
AISA	Airborne Imaging Spectrometer for Applications
AnC	Anthocyanins
ANN	Artificial neural network
AO	ArcGIS Online
APAR	Absorbed photosynthetically active radiation
API	Application programming interface
ARI	Anthocyanin reflectance index
ARVI	Atmospherically resistant vegetation index
AVIRIS	Airborne visible/infrared imaging spectrometer
BmND	Derivative-based modified normalized difference
BmSR	Derivative-based modified simple ratio
BRDI	*Bromus distachyon*
CAI	Cellulose absorption index
CAPY	*Carduus pychnocephalus*
Car	Carotenoid(s)
CARI	Chlorophyll absorption in reflectance index
CASI	Compact Airborne Spectrographic Imager
CCCI	Canopy chlorophyll content index
Chl	Chlorophyll(s)
CI$_{\text{red edge}}$	Chlorophyll red-edge index
CI$_{\text{Red}-\text{edge}}$	Red edge chlorophyll index
CK	Composite kernels
CNC	Canopy nitrogen content
CR	Continuum removal
CRI1 and 2	Carotenoid reflectance index
CSC	Canopy scattering coefficient
CWA	Continuous wavelet analysis
CWT	Continuous wavelet transform
DAG	Directed acyclic graph
DBFE	Decision boundary feature extraction
DD	Double difference
DGCI	Dark green color index
DNP	Double nearest proportion
DPMM	Dirichlet process mixture models
DT	Decision tree
EM	Expectation maximization
EMAPs	Extended morphological attribute profiles
EnMAP	Environmental Mapping and Analysis Programme
EO	Earth observation

ET	Evapotranspiration
ETM+	Enhanced Thematic Mapper plus
EVI	Enhanced vegetation index
FE	Feature extraction
Flv	Flavonoid
FPAR	Fraction of photosynthetically active radiation
FS	Feature selection
GA	Genetic algorithm
GEE	Google Earth Engine
GFS	Google file system
GMM	Gaussian mixture models
GO-RT	Geometrical optical and radiative transfer
GO	Geometrical optical
HDFS	Hadoop's distributed file system
HSeg	Hierarchical segmentation
HSWO	Hierarchical step-wise optimization
HVI	Hyperspectral vegetation index
IaaS	Infrastructure as a Service
ICA	Independent component analysis
IG	Inverted Gaussian
IGMM	Infinite Gaussian mixture model
IRRI	International Rice Research Institute
JD	Julian day
JSM	Joint storage matrix
κ	Kappa statistic
KPCA	Kernel principal component analysis
LAD	Leaf angle distribution
LAI	Leaf area index
LCC	Leaf chlorophyll content
LCC	Leaf color chart
LDA	Leaf angle distribution
LDA	Linear discriminant analysis
LFM	Live fuel moisture
LiDAR	Hyperspectral Light Detection and Ranging
LLE	Locally linear embedding
LMA	Leaf mass per area
LNC	Leaf nitrogen concentration
LNC_{area}	Area-based leaf nitrogen concentration
LNC_{mass}	Mass-based leaf nitrogen concentration
LUE	Light use efficiency
LUT	Lookup table
MAE	Mean absolute error
MARI	Modified anthocyanin reflectance index
MaxAE	Maximum absolute error
MB	Megabytes
MCARI	Modified chlorophyll absorption in reflectance index
MCARI	Modified chlorophyll absorption ratio index
MLR	Multinomial logistic regression
MLR	Multiple linear regression
mND	Modified normalized difference
MR	MapReduce

MSI	Moisture stress index
MSR	Modified simple ratio
MTCI	MERIS terrestrial chlorophyll index
MTVI2	Modified triangular vegetation index 2
N-H	Nitrogen-hydrogen
N	Nitrogen
ND	Normalized difference
NDA	Nonparametric discriminant analysis
NDI	Normalized differences index
NDII	Normalized difference infrared index
NDLI	Normalized difference lignin index
NDNI	Normalized difference nitrogen index
NDRE	Normalized difference red edge
NDVI	Normalized difference vegetation index
NDWI	Normalized difference water index
NIR	Near infrared
NIST	National Institute of Standards and Technology
NNI	N nutrition index
NPCI	Normalized pigments chlorophyll ratio index
NWFE	Nonparametric weighted feature extraction
OA	Overall accuracy
OC	Optimal combination
OLAP	Online analytical processing
OLTP	Online transaction processing
OMNBR	Optimized multiple narrow-band reflectance
OMP	Orthogonal matching pursuit
OSAVI	Optimized soil adjusted vegetation index
P	Phosphorus
PaaS	Platform as a Service
PAR	Photosynthetically active radiation
PC	Principal component
PCA	Principal component analysis
PCR	Principal component regression
PET	Potential evapotranspiration
PHI	Pushbroom Hyperspectral Imager
PLS	Partial least squares
PLSR	Partial least-squares regression
PNA	Plant nitrogen accumulation
PNC	Plant nitrogen concentration
PP	Projection pursuit
PPR	Plant pigment ratio
PRI	Photochemical reflectance index
PRISMA	PRecursore IperSpettrale della Missione Applicativa
PROSPECT	Leaf optical properties model
PSND	Pigment-sensitive normalized difference
PSO	Particle swarm optimization
PSRI	Plant senescence reflectance index
PSSR	Pigment-specific spectral ratio
QEA	Quantum-inspired evolutionary algorithm
QoS	Quality of service
ρ	Peflectance

RAM	Random access memory
RARS	Ratio analysis of reflectance spectra
RBF	Radial basis function
RDD	Resilient distributed dataset
REP	Red edge position
RF	Random forest
RGB	Red green blue
RGRI	Red/green ratio index
RHSeg	Recursive hierarchical segmentation
RMS	Root mean square
RMSE	Root mean square error
ROI	Return on investment
RT	Radiative transfer
RVSI	Red-edge vegetation stress index
RWC	Relative water content
SaaS	Software as a Service
SAM	Spectral angle mapper
SAVI	Soil adjusted vegetation index
SCSRC	Spatial correlation regularized sparse representation classification
SIPI	Structurally insensitive pigment index
SLA	Service-level agreement
SML	Stepwise multiple linear
SMLR	stepwise multiple linear regression
SNR	Signal-to-noise ratio
SPAD	Soil plant analysis development
SR	Simple ratio
SRC	Sparse representation-based classification
SVM	Support vector machine
SWIR	Short wave infrared
SZA	Solar zenith angle
TCARI/OSVAI	Transformed chlorophyll absorption in reflectance index/optimized soil-adjusted vegetation index
TCO	Total cost of ownership
TM	Thematic Mapper
TSAVI	Transformed soil adjusted vegetation index
UAV	Unmanned aerial vehicle
VARI	Vegetation atmospherically resistant index
VI	Vegetation index
VI$_{green}$	Vegetation index green
VIS	Visible
VM	Virtual machine
VMC	Volumetric moisture content
VNIR	Visible near infrared
VPD	Vapor pressure deficit
VR	Variable rate
VZA	View zenith angle
WBI	Water band index
WDVI	Weighted difference vegetation index
[AnC]	Anthocyanin content
[Car]	Carotenoid content
[Chl]	Chlorophyll content
[Flv]	Flavonoid content

Section I

Hyperspectral Vegetation Indices

1 Hyperspectral Vegetation Indices

Dar A. Roberts, Keely L. Roth, Erin B. Wetherley, Susan K. Meerdink, and Ryan L. Perroy

CONTENTS

1.1 INTRODUCTION

Vegetation properties are often measured by converting a reflectance spectrum into a single number value, known as a vegetation index (VI). Hyperspectral, or narrow-band (Elvidge and Chen, 1995) vegetation indices (HVIs) make use of the narrow-wavelength regions or bands captured by hyperspectral instruments (e.g., Galvao et al., 1999). Vegetation properties measured with HVIs can generally be divided into three main categories: (1) structure, (2) biochemistry, and (3) plant physiology. Frequently measured structural properties include fractional cover, green biomass, leaf area index (LAI), green LAI, senesced biomass, and fraction absorbed photosynthetically active radiation (FAPAR) (Jordan, 1969; Tucker, 1979; Sellers, 1985). A majority of the indices developed for structural analysis were formulated for broad-band systems and have narrow-band, hyperspectral equivalents. Biochemical properties include water, pigments (chlorophyll, carotenoids, anthocyanins), other nitrogen-rich compounds (e.g., proteins), and plant structural materials (lignin and cellulose) (Curran, 1989; Gamon and Surfus, 1999; Ustin et al., 2009; Thenkabail et al., 2014). Physiological and stress indices measure stress-induced changes in xanthophyll cycle pigments (Gamon et al., 1997), chlorophyll content (Horler et al., 1983), fluorescence (Zarco-Tejada et al., 2000), or leaf moisture (Hunt and Rock, 1989). In general, biochemical and physiological/stress indices were formulated using laboratory or field instruments ($\leq$10 nm spectral sampling) and are targeted at very fine spectral features. As a result, they are strictly hyperspectral. Indices developed for water are an exception.

Many structurally oriented VIs use some combination of near-infrared (NIR) and red reflectance, such as the NIR-to-red ratio, or simple ratio (SR; Jordan, 1969). These VIs rely on the fact that increases in LAI correspond with increases in chlorophyll absorption and NIR scattering and decreases in

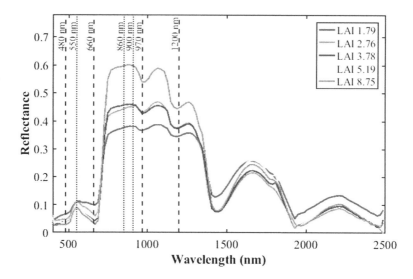

FIGURE 1.1   Reflectance spectra of *Populus trichocarpa* hybrids over a range in leaf area index (LAI). Vertical lines designate wavelengths related to absorption features (480, 660, 970, 1240 nm) or NIR scattering regions (860, 900 nm) typically used in combination to quantify structure. For details on field sampling, see Section 1.3.1.

exposed substrate, resulting in decreasing red and increasing NIR reflectance (Figure 1.1). Thus, equations for these VIs often compare reflectance at an absorbing wavelength to a nonabsorbing wavelength. However, more subtle changes also occur with an increase in LAI, including greater green reflectance (Gitelson et al., 2002a) and increasing absorption due to liquid water (Roberts et al., 2004).

Biochemical and stress-related indices rely on a similar comparison between absorbing and nonabsorbing wavelengths, varying the absorbing wavelength by the chemical of interest (Figure 1.2). For example, canopy moisture and moisture stress indices include wavelengths associated with liquid water absorption (e.g., 970, 1200 nm), while ligno-cellulose content indices utilize short-wave-infrared

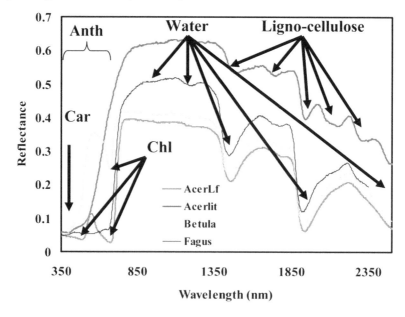

FIGURE 1.2   Reflectance spectra of leaves from a senesced birch (Betula), ornamental beech (Fagus), and healthy and fully senesced maple (AcerLf, Acerlit) illustrating carotenoid (Car), anthocyanin (Anth), chlorophyll (Chl), water, and ligno-cellulose absorptions.

(SWIR) wavelengths from 1500–1800 nm and 2000–2350 nm (Figure 1.2). By contrast, pigments (carotenoids, anthocyanins, xanthophylls, and chlorophylls) absorb in the visible and ultraviolet, with distinct but overlapping absorption features. Chlorophyll primarily absorbs blue and red light (*AcerLf*, Figure 1.2). Anthocyanins absorb all but red light (*Fagus*, Figure 1.2), and many carotenoids are yellow due to strong blue light absorption (*Betula*, Figure 1.2). Thus, pigment-sensitive VIs frequently include a combination of visible bands and sometimes a band located on the red edge (680–750 nm) (Horler et al., 1983).

In this chapter, we provide an overview of common HVIs associated with canopy structure (Section 1.2.1), canopy biochemistry (Section 1.2.2), and canopy physiology (Section 1.2.3). Canopy biochemistry is further divided into pigments (Section 1.2.2.1), moisture (Section 1.2.2.2), plant residues (Section 1.2.2.3), and nitrogen (Section 1.2.2.4). We also include a generalized index category (Section 1.2.4) using the optimized multiple narrow band reflectance (OMNBR) index (Thenkabail et al., 2000) as an example, in which optimal band combinations are determined through correlation. The OMNBR can be tuned for any structural, biochemical, or physiological value of interest. Many of the most commonly used HVIs, with their equations and key citations, are found in Table 1.1. We conclude with two applied examples: one study examining the relationship between LAI and HVIs for *Populus* (Section 1.3.1) and another evaluating the relationship between HVIs and seasonal environmental changes for two plant species (Section 1.3.2).

1.2 APPLICATIONS OF HYPERSPECTRAL VEGETATION INDICES

Most structural indices were developed for broad-band systems but have narrow-band ($\leq$10 nm) equivalents (Table 1.1). Exceptions include indices based on first derivatives and the water band index (WBI) (Penuelas et al., 1997), which includes wavelengths that are not sampled by broad-band systems (Table 1.1). In contrast, most biochemical and physiological indices are strictly hyperspectral, requiring narrow bands ($\leq$10 nm) and specific band centers that are not sampled by broad-band systems. Band centers and spectral sampling are typically defined by field or laboratory instrumentation used in the original research, but can be modified for alternative systems.

1.2.1 VEGETATION STRUCTURE

Many VIs have been designed to measure vegetation structure (e.g., LAI and FAPAR), including 1DL_DGVI, 1DZ_DGVI, ARVI, EVI, NDVI, NDWI, SAVI, SR, TSAVI, VARI, VI_{green}, and WDVI (Table 1.1). One of the first VIs for vegetation analysis dates back to Jordan (1969), who proposed the SR to estimate over-story LAI. Soon after, normalized forms of the SR, such as the NDVI (Rouse et al., 1973), were developed to reduce the impact of atmospheric scattering by using a normalized difference between two bands. Both the SR and NDVI are good predictors of wet and dry green biomass (Tucker, 1979), LAI (Jordan, 1969), FAPAR (Sellers, 1985), and fractional cover (Carlson and Ripley, 1997; Daughtry et al., 2004). However, NDVI saturates at high LAI values (Roberts et al., 2004) and varies with viewing geometry (Deering et al., 1994) and substrate reflectance (Huete et al., 1984). To compensate, narrow-band versions of the NDVI were proposed by Galvao et al. (1999) and Thenkabail et al. (2000) in which wavelength selection was optimized to reduce sensitivity to nonphotosynthetic vegetation (Galvao et al., 1999) or to improve crop-specific estimates of LAI, wet biomass, and canopy height (Thenkabail et al., 2000).

Other indices based on NDVI have been proposed to improve canopy structure estimates and minimize the impact of the atmosphere and substrate. For example, the soil adjusted vegetation index (SAVI) (Huete, 1988) includes an offset in the denominator designed to force the NDVI to radiate from the origin of a NIR and red scatterplot, making it independent of changes in substrate reflectance. The TSAVI (Baret et al., 1989) is an extension of SAVI, but includes offsets to adjust for soil influence. The WDVI (Clevers, 1991) also includes a weighted term to adjust for soil influence while measuring LAI. The atmospherically resistant vegetation index (ARVI)

TABLE 1.1

List of the Major Hyperspectral Vegetation Indices, Including Relevant Formulas and Key Citations

Index Name	Acronym	Equation	C
First-order derivative green vegetation index with local baseline (Elvidge and Chen, 1995)	1DL_DGVI	$\int_{\lambda_{626}}^{\lambda_{795}} R'(\lambda_i) - R'(\lambda_{626}) \mid \Delta\lambda_i$	S
First-order derivative green vegetation index with zero baseline (Elvidge and Chen, 1995)	1DZ_DGVI	$\int_{\lambda_{626}}^{\lambda_{795}} R'(\lambda_i) \mid \Delta\lambda_i$	S
Anthocyanin content index (Van den Berg et al., 2005)	ACI	R_{green}/R_{NIR}	B
Anthocyanin reflectance index (Gitelson et al., 2001)	ARI	$(1/R_{550}) - (1/R_{700})$	B
Atmospherically resistant vegetation index (Kaufman and Tanier, 1992)	ARVI	$(R_{NIR} - [R_{red} - \gamma^*(R_{blue} - R_{red})])/ (R_{NIR} + [R_{red} - \gamma^*(R_{blue} - R_{red})])$	S
Cellulose absorption index (Daughtry, 2001)	CAI	$0.5^*(R_{2019} + R_{2206}) - R_{2109}$	B
Chlorophyll absorption in reflectance index (Daughtry et al., 2000)	CARI	$[(R_{700} - R_{670}) - 0.2^*(R_{700} - R_{550})]$	B
Chlorophyll index red edge (Gitelson et al., 2003a)	$CI_{red\,edge}$	$R_{NIR}/R_{red\,edge} - 1$	B
Carotenoid reflectance index (Gitelson et al., 2002b)	CRI	$[(1/R_{510}) - (1/R_{550})]; [(1/R_{510}) - (1/R_{700})]$	B
Enhanced vegetation index (Huete et al., 1997)	EVI	$2.5^*(R_{NIR} - R_{red})/(R_{NIR} + 6^*R_{red} - 7.5^* R_{blue} + 1)$	S
Modified anthocyanin reflectance index (Gitelson et al., 2001)	MARI	$[(1/R_{550}) - (1/R_{700})]^*R_{NIR}$	B
Modified chlorophyll absorption in reflectance index (Daughtry et al. 2000)	MCARI	$[(R_{700} - R_{670}) - 0.2^*(R_{700} - R_{550})]^*(R_{700}/R_{670})$	B
Moisture stress index (Hunt and Rock, 1989)	MSI	R_{SWIR}/R_{NIR}	B;P
MERIS terrestrial chlorophyll index (Dash and Curran, 2004)	MTCI	$(R_{753.75} - R_{708.75})/(R_{708.75} - R_{681.25})$	B
Normalized difference infrared index (Hardisky et al., 1983)	NDII	$(R_{NIR} - R_{SWIR})/(R_{NIR} + R_{SWIR})$	B
Normalized difference lignin index (Serrano et al., 2002)	NDLI	$[\log(1/R_{1754}) - \log(1/R_{1680})]/[\log(1/R_{1754}) + \log(1/R_{1680})]$	B
Normalized difference nitrogen index (Serrano et al., 2002)	NDNI	$[\log(1/R_{1510}) - \log(1/R_{1680})]/[\log(1/R_{1510}) + \log(1/R_{1680})]$	B
Normalized difference red edge (Barnes et al., 2000)	NDRE	$(R_{790} - R_{720})/(R_{790} + R_{720})$	B
Normalized difference vegetation index (Rouse et al., 1973)	NDVI	$(R_{NIR} - R_{red})/(R_{NIR} + R_{red})$	S
Normalized difference water index (Gao, 1996)	NDWI	$(R_{860} - R_{1240})/(R_{860} + R_{1240})$	S;B

(Continued)

TABLE 1.1 (*Continued*)

List of the Major Hyperspectral Vegetation Indices, Including Relevant Formulas and Key Citations

Index Name	Acronym	Equation	C
Photochemical reflectance index (Gamon et al., 1997)	PRI	$(R_{531} - R_{570})/(R_{531} + R_{570})$	P
Pigment sensitive normalized difference (Blackburn, 1998)	PSND	$chl_a = [(R_{800} - R_{675})/(R_{800} + R_{675})];$ $chl_b = [(R_{800} - R_{650})/(R_{800} + R_{650})];$ $car = [(R_{800} - R_{500})/(R_{800} + R_{500})]$	B
Plant senescence reflectance index (Merzlyak et al., 1999)	PSRI	$(R_{678} - R_{500})/R_{750}$	B
Pigment-specific spectral ratio (Blackburn, 1998)	PSSR	$(R_{800}/R_{675}); (R_{800}/R_{650});$ (R_{800}/R_{500})	B
Red edge position (Horler et al., 1983)	REP	$\lambda(\text{max slope from } R_{680} - R_{750})$	P
Red/green ratio index (Gamon and Surfus, 1999)	RGRI	R_{red}/R_{green}	B;P
Red-edge vegetation stress index (Merton and Huntington, 1999)	RVSI	$[(R_{714} + R_{752})/2 - R_{733}]$	P
Soil-adjusted vegetation index (Huete, 1988)	SAVI	$[(R_{NIR} - R_{red})/(R_{NIR} + R_{red} + L)]* (1 + L)$	S
Structure-insensitive pigment index (Penuelas et al., 1995)	SIPI	$(R_{800} - R_{445})/(R_{800} - R_{680})$	B;P
Simple ratio (Jordan, 1969)	SR	R_{800}/R_{675}	S
Transformed soil-adjusted vegetation index (Baret et al., 1989)	TSAVI	$a*(R_{NIR} - a*R_{red} - b)/[a* R_{NIR} + R_{red} - a*b + 0.08*(1 + a^2)]$	S
Visible atmospherically resistant index (Gitelson et al., 2002a)	VARI	$(R_{green} - R_{red})/(R_{green} + R_{red} - R_{blue})$	S
Vegetation index using green band (Gitelson et al., 2002a)	VI$_{green}$	$(R_{green} - R_{red})/(R_{green} + R_{red})$	S
Weighted difference vegetation index (Clevers , 1991)	WDVI	$R_{NIR} - a*R_{red}$	S
Water band index (Peñuelas et al., 1997)	WBI	R_{900}/R_{970}	S;B

Note: The category of index—structural (S), biochemical (B), or physiological (P)—is indicated in the final column, labeled C.

(Kaufman and Tanier, 1992) includes a blue band in the numerator and denominator with a weighting factor to compensate for enhanced atmospheric scattering in red wavelengths. The enhanced vegetation index (EVI) (Huete et al., 1997) incorporates both a substrate reflectance correction and a blue band to compensate for the atmosphere. EVI has largely replaced NDVI as a primary global product because of improved resistance to the atmosphere and less evidence of saturation at high LAI (Huete et al., 2002), although it has been shown to be sensitive to albedo (Galvao et al., 2011).

Variants of the NDVI and ARVI that use visible and SWIR reflectance have also been proposed to measure structure. For example, the VI$_{green}$ (or VIG) and the vegetation atmospherically resistant index (VARI) (Gitelson et al., 2002a) replace the NIR band with a green band. VI$_{green}$ and VARI respond more linearly to changes in vegetation cover fraction than does NDVI (Gitelson et al., 2002a). VARI has also proven to be highly effective for estimating LAI and moisture stress in maize (Gitelson et al., 2003b; Perry and Roberts, 2008) and live fuel moisture (LFM) in shrub lands (Roberts et al., 2006). Indices such as the water band index (Peñuelas et al., 1997) and the normalized difference water index (NDWI) (Gao, 1996) are also sensitive to changes in LAI by contrasting liquid water absorption features in the SWIR to NIR reflectance. Perry and Roberts (2008) found WBI and NDWI to be the most sensitive of 15 narrow-band indices to a change in maize biomass.

More complex indices have been designed to detect changes in LAI that modify the shape and position of the red edge as well as the expression of liquid water in canopy spectra (Elvidge and Chen, 1995; Roberts et al., 2004). Elvidge and Chen (1995) proposed several indices based on the red edge, using first derivative spectra between 626 and 796 nm (1DZ_DGVI and 1DL_DGVI). The relationship between these indices and LAI is significantly improved (compared to other normalized indices) because the slope of the red edge is more sensitive to changes in LAI and the first derivative spectrum is less sensitive to changes in albedo. Although Elvidge and Chen (1995) proposed a range between 626 and 796 nm, hyperspectral systems offer numerous possibilities for alternative, narrower, or broader spectral ranges for integration that could be explored.

1.2.2 CANOPY BIOCHEMISTRY

1.2.2.1 Plant Pigments

Three types of plant pigments contribute significantly to visible reflectance in leaves and canopies: chlorophylls (a and b), carotenoids, and anthocyanin (Figure 1.2). A number of indices are designed to quantify specific pigments in plant leaves and canopies. Generalized indices that can be tuned to estimate concentrations of chlorophylls a or b, carotenoids, or carotenoid/chlorophyll ratios include PSSR, PSND, SIPI, and PSRI. These indices are formulated either as a simple ratio such as the pigment-specific spectral ratio (PSSR) (Blackburn, 1998) or as a normalized ratio such as the pigment-sensitive normalized difference (PSND) (Blackburn, 1998), and can target specific pigments by changing the absorbing wavelength. The structurally insensitive pigment index (SIPI) (Penuelas et al., 1995) calculates the ratio of carotenoids to chlorophyll a. Similarly, the plant senescence reflectance index (PSRI) (Merzlyak et al., 1999) changes in response to a change in the ratio of carotenoids to chlorophyll as plants senesce.

Chlorophyll is, arguably, the most critical plant pigment due to its fundamental role in photosynthesis and primary production. Indices developed to estimate chlorophyll content include the chlorophyll absorption in reflectance index (CARI) (Kim, 1994), modified CARI (MCARI) (Daughtry et al., 2000), chlorophyll red-edge index ($CI_{red\ edge}$) (Gitelson et al., 2003a), and MERIS terrestrial chlorophyll index (MTCI) (Dash and Curran, 2004). CARI quantifies the 670-nm chlorophyll absorption feature as the mathematical difference between 700 and 670 nm reflectance, adjusted by a weighted difference between 700 and 550 nm to compensate for nonphotosynthesizing materials. MCARI further adjusts the soil-compensating component by the ratio of NIR to red reflectance. It has proven effective in identifying nutrient stress in maize (Perry and Roberts, 2008) and drought stress in Amazonian forests (Asner et al., 2004). $CI_{red\ edge}$ is based on a three-band generalized model for quantifying pigments (Gitelson et al., 2006). In this model, the concentration of an absorber is quantified as the mathematical difference in reciprocal reflectance, $R_{\lambda 1}^{-1}$, within an absorption region and reciprocal reflectance at a second wavelength, $R_{\lambda 2}^{-1}$, outside of the main absorption region but with similar backscattering. This quantity is multiplied by R_{NIR} to compensate for backscatter-dependent variation in brightness, simplified in Table 1.1 because $R_{\lambda 2}^{-1}$ and R_{NIR} are the same. $CI_{red\ edge}$ has shown a near-linear relationship to chlorophyll content over a diversity of broadleaf tree species (Gitelson et al., 2009). The MTCI uses one red and two red-edge bands, taking advantage of the strong relationship between the position of the red edge and concentration of chlorophyll. MTCI has also been used to estimate green LAI (Vina et al., 2011) and N uptake in maize (Li et al., 2014). In addition to chlorophyll-specific indices, a number of the structural indices have also been used to effectively estimate chlorophyll at leaf scales, such as the SR (Sims and Gamon, 2002).

Anthocyanins are plant pigments that increase in response to environmental stress and may play a role in minimizing photoinhibition (Gitelson et al., 2001). Anthocyanin-specific indices include ACI, ARI, MARI, and RGRI. Two of these indices were proposed by Gitelson et al. (2001) and are based on a similar concept of reciprocal reflectance as developed for chlorophyll. These include the anthocynanin reflectance index (ARI), calculated as the difference between reciprocal green reflectance (540–560 nm) and reciprocal red-edge reflectance (690–710 nm), and the modified ARI, which weights ARI by NIR reflectance (760–800 nm). The red/green ratio index (RGRI) (Gamon and Surfus, 1999) and the anthocyanin content index (ACI) (Van den Berg and Perkins, 2005) are built upon the SR model and calculated as the ratio of green to NIR reflectance. RGRI is based on the concept that high anthocyanin content, which results in red leaves, will increase the red-to-green ratio, while ACI increases in response to larger concentrations of anthocyanins as green leaf reflectance drops.

Carotenoids aid in the process of light harvesting for photosynthesis and protect chlorophyll from photooxidation via the reversible conversion of the xanthophyll violaxanthin to zeaxanthin (Gamon et al., 1997; Gitelson et al., 2002b). They are most readily apparent in leaves during senescence, as they are retained while chlorophyll breaks down. Carotenoid indices include PRI, SIPI (Penuelas et al., 1995), and CRI, which consists of two reciprocal reflectance models proposed by Gitelson et al. (2002b). In the first, CRI uses the difference between reciprocal reflectance at 510 and 550 nm, while the second replaces the 550-nm band with a band at 700 nm (Gitelson et al., 2002b). The most widely used carotenoid index, however, is the photochemical reflectance index (PRI) (Gamon et al., 1997), which is designed to capture the shift from violaxanthin to zeaxanthin. This transition results in a subtle ($<1\%$) decrease in reflectance at 531 nm that can be quantified using a normalized difference index with 570 nm as the reference band. In this form, increasingly negative PRIs will occur with increasing plant stress. The PRI is discussed in more detail in Section 1.2.3.

1.2.2.2 Canopy Moisture

Plant canopy moisture varies as a function of the number of leaves within a crown (Figure 1.1) and the water content of individual leaves (Figure 1.2). Simple ratio and normalized indices have been proposed to compare the expression of subtle or strong liquid water bands relative to a reference nonabsorbing wavelength. These indices include NDII, NDWI, WBI, and MSI. The moisture stress index (MSI) (Hunt and Rock, 1989) and the WBI (Peñuelas et al., 1997) represent examples of the simple ratio indices. The MSI is calculated as the ratio of a SWIR band (1650 nm) to a NIR band (830 nm), while the WBI is calculated as the reflectance ratio of 900 to 970 nm. Increases in water content correspond to a decreasing MSI value but an increasing WBI value. Normalized versions of the MSI include the normalized difference infrared index (NDII) (Hardisky et al., 1983), in which the SWIR band can be at either a short (1650 nm) or long wavelength (2200 nm). The NDWI (Gao, 1996) is roughly equivalent to the normalized version of the WBI, although the 1240-nm water band takes the place of the 970-nm band.

Several studies have evaluated the relationship between moisture indices and relative water content (RWC) (Serrano et al., 2000), plant water content (Gao, 1996), LFM (Roberts et al., 2006), and moisture stress (Hardisky et al., 1983; Hunt and Rock, 1989; Perry and Roberts, 2008). For plant water content in shrublands, Serrano et al. (2000) found WBI and NDWI performed better than NDVI and NDII. In Mediterranean ecosystems, Peñuelas et al. (1997) found a strong correlation between the WBI and plant water content. For estimating LFM, Roberts et al. (2006) found WBI and NDWI superior to NDVI, EVI, and NDII. An additional study tested several forms of the WBI using water absorption features at 960, 1180, and 1450 nm to determine their ability to estimate water content for leaves, thin stems, and leaves and stems combined (Sims and Gamon, 2003). The 960- and 1180-nm WBI were the most strongly correlated with thin stem water ($R^2 > 0.75$) over a wide range of plant functional types, with slightly higher correlations found for 1180 nm. It should be noted that two greenness measures, the VI_{green} and VARI, have consistently shown stronger relationships to moisture stress than other greenness or water-based indices (Roberts et al., 2006;

Perry and Roberts, 2008), potentially because water stress also manifests as an increase in green reflectance (Zygielbaum et al., 2009), likely in response to chloroplast avoidance (Zygielbaum et al., 2012).

1.2.2.3   Lignin and Cellulose/Plant Residues

Specific absorption bands associated with proteins, starch, sugars, lignin, and cellulose make HVIs especially well suited for measuring the biochemistry of branches or senesced plant materials in the absence of water and pigments (Curran, 1989; Figure 1.2). Of these biochemicals, two of the most evident in reflectance spectra are lignin and cellulose, which produce many prominent absorption features in the SWIR. Several HVIs have been developed specifically to estimate the ligno-cellulose content or mass of senesced plant materials, including the cellulose absorption index (CAI) (Daughtry, 2001) and the normalized difference lignin index (NDLI) (Serrano et al., 2002). The CAI is a band-depth measure, calculated as the difference in reflectance within a strong cellulose absorption band at 2101 nm and average reflectance for two bands (2031 and 2211 nm) outside of this absorption feature. The NDLI targets a prominent lignin absorption band at 1754 nm and uses a combination of the normalized difference index formula with the natural logarithm of reciprocal reflectance, a common transform used in biochemical spectroscopy (e.g., Card et al., 1988). The nonabsorbing 1680-nm band is used as a reference band.

Daughtry et al. (2004) evaluated the performance of CAI and NDVI for estimating fractional cover of crop residues and green cover for corn, soybean, and wheat over several soils for cases of dry and wet residual biomass. They noted that CAI had a strong linear correlation with changes in the fraction of crop residues, and NDVI was strongly correlated to changes in green cover for all crops and conditions, while the two indices themselves were essentially uncorrelated. When applied to AVIRIS data, the combination of CAI and NDVI enabled them to accurately discriminate among conservation, reduced-tillage, and intensively tilled soils (Daughtry et al., 2005), which is critical for evaluating the potential for soil erosion or carbon uptake. Serrano et al. (2002) analyzed AVIRIS data acquired from drought deciduous and evergreen chaparral species, finding a strong linear relationship between NDLI and bulk lignin.

1.2.2.4   Nitrogen

Although plant nitrogen has no direct absorption features in leaves, it is an important component of many light-absorbing compounds in the visible to SWIR range, including chlorophyll and numerous proteins (Curran, 1989). Similar to the NDLI, Serrano et al. (2002) proposed an index targeted at the 1510-nm N absorption feature, called the normalized difference nitrogen index (NDNI), also referenced to the nonabsorbing 1680-nm band. Barnes et al. (2000), proposed the normalized difference red edge (NDRE), an index similar to the NDVI that replaces the red band with a band centered on the inflection point of the red edge, at 720 nm. The NDRE has been found to be highly correlated with nitrogen uptake in maize (Li et al., 2014). The NDRE is often reformulated as a canopy content chlorophyll index (CCCI; Fitzgerald et al., 2010), an index that has been used to estimate canopy N content in wheat and maize. The CCCI is calculated based on an NDRE (y), NDVI (x) plot in which data plot roughly as a triangle emanating from the origin, with the lower vector defined by low-NDRE, low-N plants and the upper limb of the triangle defined by the highest NDRE and highest NDVI. CCCI is calculated as the difference between the minimum NDRE for a given NDVI subtracted from the actual NDRE for that NDVI, divided by the possible range in NDRE for a given NDVI (NDRE maximum-NDRE minimum). A recent development in the estimation of plant N concentration and uptake is the use of OMNBR indices (see Section 1.2.4).

1.2.3   PLANT PHYSIOLOGY

Changes in leaf physiology and stress impact the position and shape of the red edge, shifting it toward either shorter wavelengths (blue shift) or longer wavelengths (red shift) (Horler et al., 1983). Blue shifts have been observed in response to heavy metal stress in plants (Horler et al., 1983; Rock et al., 1986), while red shifts typically occur during chlorophyll development and nutrient stress (Milton et al., 1991). NDRE, REP, and RVSI are physiology indices designed to detect these shifts. The red edge position (REP) index is unique because it reports the wavelength that is the maximum of the first derivative of reflectance between 650 and 750 nm, better known as the position of the red edge (Horler et al., 1983). The red-edge vegetation stress index (RVSI) (Merton and Huntington, 1999) captures variation in the shape of the red edge associated with plant stress by calculating the average canopy reflectance at 714 and 752 nm, minus reflectance at 733 nm. A concave upward red edge and slightly negative or positive RVSI occur in stressed plants, while a concave downward red edge and strongly negative RVSI indicate unstressed plants (Merton and Huntington, 1999). The NDRE index uses a normalized difference ratio using the red and red-edge bands to measure stress, primarily due to a change in chlorophyll content (Barnes et al., 2000)

Several other indices have been used to quantify different types of stress. The MSI has been used to measure plant stress in addition to plant water content. The PRI, another physiology index designed specifically to measure light use efficiency (LUE), has proven to be one of the most effective stress/physiology-oriented HVIs. As plants become progressively more stressed and are unable to utilize light absorbed by chlorophyll, reflectance at 531 nm drops, producing an increasingly negative PRI. RGRI and SIPI are also used as physiology indices to measure LUE.

Several studies have evaluated the relationship between indices and plant physiology. Naidu et al. (2009) studied leaf roll-infected grape vines and found stressed leaves had slightly less negative RVSI values than healthy leaves. In contrast, Perry and Roberts (2008) found RVSI to become less negative with higher leaf nitrogen. Gamon et al. (2005) evaluated carbon uptake by dry tropical forest species using canopy reflectance and found NDVI and SR responded primarily to changes in FAPAR, while PRI responded to photosynthetic down-regulation, becoming increasingly negative at midday. Rahman et al. (2001) combined the strengths of NDVI for FAPAR and PRI for LUE to develop a multiplicative model for estimating carbon uptake by boreal forests. This model showed a near-linear relationship between NDVI*PRI-predicted carbon uptake and CO_2 flux measured by several flux towers (Rahman et al., 2001). Zarco-Tejada et al. (2013), using UAV-mounted sensors, found a strong correlation between PRI and crown temperatures and used this relationship to estimate stomatal conductance in vineyards.

1.2.4 GENERALIZED INDICES

Given potentially hundreds of narrow-band measurements and tens of thousands of band combinations, the potential always exists for identifying a superior combination of ratios, normalized bands, three-band indices, and so on for a specific vegetation attribute of interest. Thenkabail et al. (2000) originally proposed the OMNBR as a means of identifying optimal band correlations between a measured variable of interest and spectra. While the OMNBR was first proposed using a normalized difference framework, the same basic approach applies to any formulation, such as simple ratios, three-band indices, or log-based indices. The principle of OMBNR is simple: all possible band combinations are calculated for a set of measured spectra that also include measurements for a physical variable of interest. Next, a measure of performance, such as R^2, t-test, or p-value, is calculated for each combination and stored in a diagonal matrix, typically with one wavelength plotted as x and the other y (for a two-band index). Using R^2 as an example, R^2 values are used to create an image that is scaled to highlight band combinations showing the highest R^2. This type of analysis would be expected to show high correlation between LAI and a combination of a NIR and red band (NDVI), but would also be expected to show similar high correlations for a red-green combination (VI_{green}).

OMNBR was first proposed to evaluate optimal band combinations for retrieval of canopy wet biomass and LAI from a selection of crops (Thenkabail et al., 2000), and was subsequently expanded to a greater diversity of crops and more structural and biochemical measures over a larger geographic region (Thenkabail et al., 2013). Recently it has been used to explore variation in N uptake for wheat

(Yao et al., 2010) and N content for rice (Tian et al., 2011) as they vary with crop stage. In general, an OMNBR index can be expected to outperform most standard HVIs, or at least match them, since all normalized indices are a subset of the possibilities presented by OMNBR. Furthermore, a more generalized form of an OMNBR index can be developed by analyzing multiple crops or plant species (Thenkabail et al., 2000). Thenkabail et al. (2014) provide a general overview hyperspectral remote sensing, identifying 28 wavelengths between 400 and 2500 nm associated with a diversity of biochemical, biophysical, and physiological plant properties, noting optimal wavelengths often vary between crop types. In Section 1.3.2, we provide an example using OMNBR to explore optimal band combinations for soil moisture and measures of evapotranspiration.

1.3 APPLICATIONS

1.3.1 Estimating Leaf Area Index Using Hyperspectral Vegetation Indices

We evaluated several structurally oriented HVIs for estimating the LAI of hybrid poplars (*Populus trichocarpa*). The original study, described in Roberts et al. (1998), explored the relationship between LAI and the spectral expression of liquid water in canopies of this hybrid. Here we focus strictly on HVIs, adding several that were not evaluated in the original study. Given that HVI-LAI relationships can vary by crop (Thenkabail et al., 2000), some caution should be used in generalizing these results to other types of vegetation.

The study site is located near Wallula, Washington (46° 4′N, 118° 54′W), at the Boise-Cascade Wallula Fiber Farm. Field work was conducted between 20 and 25 July 1997. Seventy-six young stump-sprouting plants ranging between 10 and 60 cm in height were sampled in a six-year old stand that had recently been harvested. Reflectance spectra were measured above each plant using an analytical spectral device (ASD) full-range instrument (Analytical Spectral Devices, Boulder, CO) and standardized to reflectance with a spectralon panel (Labsphere Inc. North Sutton, NH) measured at approximately 10-minute intervals. At least three replicates were measured for each set of observations. One to four sets of spectra were measured per plant depending on the size of the resprout at a height of 0.5 m above the canopies, translating to as few as 3 spectra to more than 12 per plant. In order to determine LAI, plants were destructively harvested, with five plants randomly sampled in each of five height classes, designed to ensure a range in LAI. Plant height and diameter along the major and minor axes were measured to determine ground area cover and plant volume for each plant. Leaf area was calculated for each plant using a linear equation relating stem diameter to leaf area. Stem diameter was measured for every stem on the sampled plants using calipers. In order to develop the linear equation relating leaf area to stem diameter, 1 out of every 10 stems was stored in a plastic bag, cooled, and then transported to the laboratory for analysis. Leaves from each stem were harvested, measured for leaf area, then regressed against stem diameter. This relationship was then combined with the resprout stem data to calculate total leaf area for each resprout, and LAI was calculated as leaf area divided by the areal projection of each resprout. The analysis included 3 soil spectra and 24 plant spectra, with each plant spectrum calculated as the average of all observations for that plant. For more details, see Roberts et al. (1998).

LAI increases led to dramatic changes in canopy reflectance, including decreasing reflectance for visible light, increasing NIR reflectance, and stronger expression of absorption features (Figure 1.1). Eight indices were calculated: two that respond to canopy water (NDWI, WBI), five to greenness (EVI, NDVI, SR, VARI, VIG), and one to the red edge (1DZ_DGVI). All of the indices were highly correlated with LAI (Table 1.2, Figure 1.3), and all but NDVI showed a linear relationship, producing R^2 values between 0.69 (WBI) and 0.78 (VARI: Figure 1.3). The NDVI-LAI scatter plot (Figure 1.3b) illustrates NDVI saturation at high LAI, beginning around an LAI of 4. The water-based indices had the lowest R^2 values, the NIR to red combinations (SR, NDVI, and EVI) intermediate, and VIG and VARI the highest. The red-edge–based index (1DZ_DGVI) outperformed the SR and linear NDVI models, but had a slightly lower R^2 than a nonlinear fit for NDVI or the linear relationship for the EVI.

TABLE 1.2
Linear and Nonlinear Relationship between LAI and 8 HVIs

Fit Metric	NDWI	WBI	EVI	NDVI	NDVI(P)	SR	VARI	VIG	1DZDG
Slope	34.31	44.51	7.28	7.23	8.12	0.38	8.11	11.32	10.15
Intercept	1.67	−43.19	−0.75	−1.11	0	0.27	1.24	1.18	−0.03
Exponent					2.87				
R²	0.73	0.69	0.77	0.72	0.77	0.74	0.78	0.78	0.76
RMS	0.20	0.21	0.18	0.20	0.18	0.19	0.18	0.18	0.19
MAE	0.82	0.90	0.69	0.77	0.67	0.80	0.64	0.65	0.71
MaxAE	2.57	2.29	3.57	3.47	3.03	2.17	3.24	3.13	3.65
MinAE	0.01	0.09	0.05	0.07	0.00	0.03	0.02	0.03	0.06
Soil Error	0.18	0.23	0.25	0.42	0.00	0.72	0.14	0.07	0.21

Note: Fit metrics include R^2, root mean squared (RMS) error, mean absolute error (MAE), and variants on the MAE to quantify LAI prediction errors at high LAI (MaxAE) and over bare soils (soil error). Only the NDVI-LAI relationship was improved by a nonlinear model (designated with the letter P, for power function).

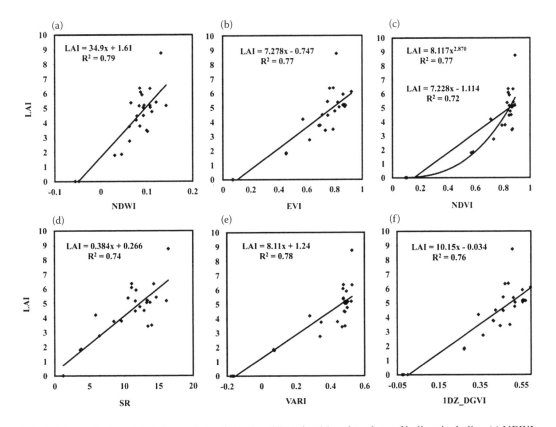

FIGURE 1.3   Scatter plots between LAI of the clonal Populus (y) and a subset of indices including (a) NDWI, (b) EVI, (c) NDVI, (d) SR, (e) VARI, and (f) 1DZ_DGVI. Linear and nonlinear models are shown for NDVI. Other indices, including VARI and EVI, were also tested for nonlinearity, but improvements were minor compared to NDVI.

Error metrics were calculated to examine model differences. The maximum absolute error (MaxAE) captures a model's over- or underprediction across the range of LAI values. Based on this error metric, the NDWI and WBI produced considerably lower errors than the greenness measures except the SR, demonstrating that although the greenness measures fit the total population better, they underpredicted high LAI values. This suggests that water-based indices and SR are likely to perform better in high-LAI forests. Soil error was calculated as the average mean absolute error for predictions of LAI over bare soils. A high value indicates that an index will produce an error in LAI when cover is sparse or absent. Based on this metric, the lowest error occurred for a power function fit of NDVI, followed closely by VIG and VARI. The highest soil error observed was for SR, equal to nearly one LAI, suggesting that SR is not effective for sparsely covered areas. The second highest error was for the NDVI linear model, equal to 0.5 LAI. EVI had a lower soil error than NDVI, as would be expected given the formulation of EVI.

1.3.2 Hyperspectral Vegetation Indices, Soil Moisture, and Evapotranspiration

In the second example, we focus on how seasonal changes in soil moisture affect the reflectance spectra and HVI values for two invasive species. We extend the analysis to evaluate the relationship between several HVIs and measures of potential (PET) and actual (AET) evapotranspiration, calculated using the Penman Monteith equation (Monteith and Unsworth, 1990) and the Bowen ratio (Bowen, 1926), respectively. Additional data shown include calculations from webcam imagery used to track changes in canopy greenness for annual and perennial plants. Canopy greenness was calculated using solar noon images analyzed using software described in Bradley et al. (2010) to generate estimates of percent greenness for uniform regions of annuals and perennials.

This study was conducted at Coal Oil Point Reserve (COPR), California, one of four areas sampled by the Innovative Datasets for Environmental Analysis by Students (IDEAS) network (www.geog. ucsb.edu/ideas: Roberts et al., 2010). COPR is located at 34.41386°N and119.8802°W at an elevation of 6 m and is dominated by a mixture of native and invasive annual grasses, forbs, and perennial shrubs. The micrometeorological tower at COPR measures all variables needed to calculate PET and AET, including wind speed and direction, air temperature and relative humidity (at 0.75 and 2.85 m), and net radiation using a four-channel net radiometer. Additional instrumentation includes a tipping bucket rain gauge, fog collector, and leaf wetness sensor. Three below-ground sensors measure soil temperature and volumetric water content (VMC) at 10, 20, and 50 cm. Soil heat flux, G, was estimated from a combination of soil temperature at 10 cm; thermal conductivity, K (calculated using soil VMC and texture); and surface temperature (calculated from the four-channel radiometer using an assumed emissivity of 0.98 and corrected for reflected downwelling radiation). Soils at COPR are clay-loams with clay content increasing from 29.6% to 35.7% from 10 to 50 cm. For more details on instrumentation and site properties, or to access data, see www.geog.ucsb.edu/ideas.

Plotted over two years starting in November 2007, seasonal environmental data illustrate a typical Mediterranean climate defined by winter precipitation and summer drought (Figure 1.4a). Rainfall was highly variable between years, with almost all of the rainfall in the 2007–2008 hydrological year falling in a single month. In 2008–2009, the rains started earlier and persisted longer, but totaled less. Soil moisture responded rapidly to precipitation, with the greatest seasonal fluctuations at the shallowest depths and a rapid increase followed by a more gradual dry down (Figure 1.4c). Overall, soil moisture remained relatively high due to the high clay content of soils in the study region.

A webcam deployed in the late spring of 2008 shows the pronounced but distinct seasonal cycle of annual and perennial vegetation (Figure 1.4b). PET, AET, and the ratio of AET to PET also show pronounced seasonal cycles, as would be expected (Figure 1.4d). PET appears to be primarily driven by a combination of net radiation and vapor pressure deficit (VPD) and thus peaks in the summer when the skies are clear and the air is dry. AET, by contrast, is controlled by available moisture as well as net radiation and VPD, and thus peaks in the spring, when soil water is plentiful and can either be directly evaporated from the surface or transpired by green plants. AET can also be high following fog events,

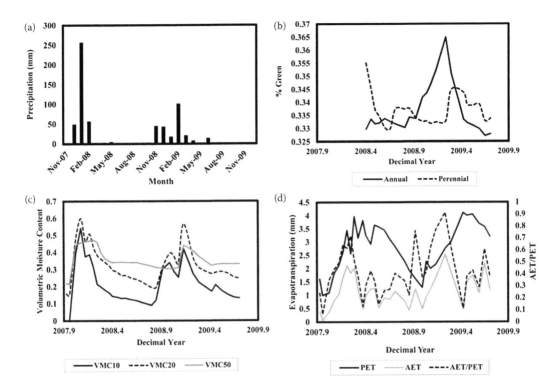

FIGURE 1.4 Showing (a) precipitation, (b) webcam percent green, (c) volumetric soil moisture (VMC) at 10, 20, and 50 cm and (d) PET, AET, and AET/PET.

which are common for this coastal site. Based on the ratio of AET to PET, between 70% and 90% of the atmospheric moisture demand was met by ET in the spring, yet typically less than 30% over the summer.

Reflectance spectra were collected of two invasive plants species over two hydrological years starting in October 2007. Species measured included *Brachypodium distachyon* (BRDY: Purple false brome) and *Carduus pycnocephalus* (CAPY: Italian thistle). Spectra were collected over five individuals of each species by measuring 25 spectra for each individual plant, 5 at the plant center and 5 in each of the four cardinal directions from approximately 0.5–1 m above the canopy. Spectra were acquired within 2 hrs of solar noon at two- to four-week intervals using an ASD spectrometer and a spectralon standard.

Biweekly reflectance spectra demonstrated significant variation in plant responses to changes in moisture, temperature, and net radiation (Figure 1.5). BRDY started the year highly senesced, beginning to green up by Julian day (JD) 77 (Figure 1.5a). Peak greenness was reached by JD137 and followed by a rapid senescence. While chlorophyll and water absorption features are evident, they were never expressed as clearly as might be expected for an individual leaf. Furthermore, ligno-cellulose absorptions were evident throughout the year, becoming the dominant absorption feature by JD167. After senescence, BRDY spectra continued to evolve, showing a gradual decrease in reflectance and change in convexity of the visible-NIR region as stems decomposed.

Unlike BRDY, CAPY showed a well-defined, steeply sloping red edge with very pronounced chlorophyll and liquid water absorptions (Figure 1.5b). Furthermore, CAPY greened up earlier than BRDY and remained active far longer into the growing season. For example, on JD167, when BRDY was fully senesced, CAPY still showed a pronounced red-edge and chlorophyll absorptions, although ligno-cellulose bands also suggest it was senescing. By JD197, CAPY was senesced, but had lower reflectance than BRDY because of its taller canopy.

A subset of HVIs were calculated from reflectance spectra to explore the relationship between HVIs and soil moisture and ET (Figure 1.6). In this figure, BRDY is shown on the left and CAPY on the right. Temporal plots for three indices, $CI_{red\,edge}$ (a chlorophyll index), WBI (a water index), and RVSI (a stress

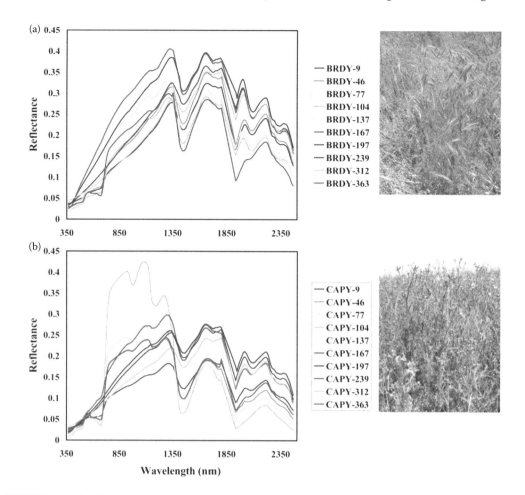

FIGURE 1.5 Reflectance spectra of *Brachypodium distachyon* (BRDY) and *Carduus pychnocephalus* (CAPY) for 2009. The number to the right on the legend reports Julian day. Photographs to the right show what the two species look like, taken on Julian day 138.

index) are plotted with VMC20 and the ratio of AET to PET. All indices show a clear seasonal pattern that corresponds well to soil moisture, illustrating the strong dependence of plant growth on soil moisture in a Mediterranean climate. However, VMC20 and the HVIs are not perfectly aligned. For example, in 2008, the WBI significantly lags behind the peak in soil moisture, while $CI_{red\ edge}$ lacks VMC20 asymmetry. RVSI, interestingly, appears to capture some of the asymmetry in soil moisture in 2008.

In 2009, all three indices were better aligned with VMC20. Furthermore, indices calculated for CAPY tended to better track seasonal changes in VMC20, particularly in 2008.

The HVIs align far better with the ratio of AET to PET than with VMC20. The peak in $CI_{red\ edge}$ and WBI, and the trough in RVSI, align very well with peak AET/PET for CAPY. For BRDY, $CI_{red\ edge}$ and RVSI match AET/PET, but WBI is clearly lagged in 2008, in which peak AET occurred well before peak WBI. Better alignment for CAPY suggests that AET at the site may be largely controlled by the seasonal response in this species, which was already photosynthetically active by January and thus was able to take advantage of early winter rains and remain green as long as enough moisture was available to support high AET. Better alignment for AET than VMC20 is not surprising, given that VMC can be high when plants are completely senesced, yet AET will only be high when plants are green if transpiration is a major contributor to ET.

Statistical relationships between HVIs and environmental measures were evaluated by regressing a selection of HVIs against five measures for each species (Figure 1.7; Tables 1.3 and 1.4). Many

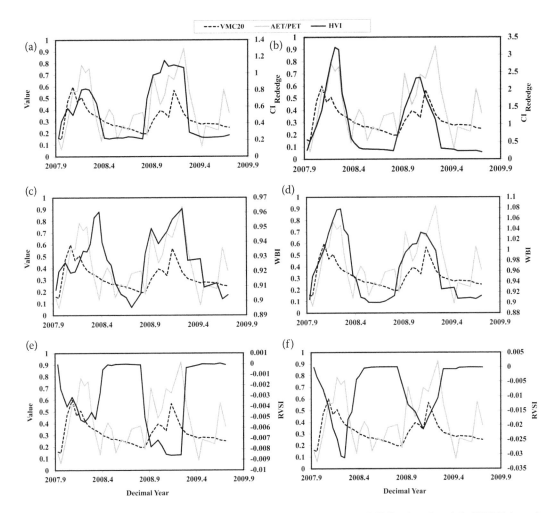

FIGURE 1.6 Time series plots of VMC20 (dashed), AET/PET (solid), and HVI values (gray) for BRDY (a, c, e) and CAPY (b, d, f) for CI$_{red\,edge}$ (a, b), WBI (c, d), and RVSI (e, f), as labeled on the right axis of each plot.

HVIs showed statistically significant relationships to VMC at 10 or 20 cm depth, PET, or AET/PET. For example, MTCI had the highest R^2 value with soil moisture, followed by CRI1 and NDVI for BRDY (Table 1.3).

For this shallow-rooted species, higher correlations were observed for 10 than 20 cm, with MTCI showing the highest correlation of all indices. Other indices that showed a strong correlation with soil moisture included all of the pigment and greenness measures: NDNI, NDLI, and RVSI. The water-based indices showed a weaker relationship, and no significant relationship was observed for PRI. Most indices were positively correlated to VMC and AET, although RVSI was negatively correlated, consistent with increasingly positive RVSI with increasing plant stress (Merton and Huntington, 1999).

Several indices were strongly correlated to PET, most likely driven by a strong relationship to net radiation. These included VARI, SIPI, NDNI, and CAI. The weakest relationships were observed for AET. By contrast, the highest R^2 values were observed for AET/PET, with most greenness measures producing R^2 values greater than 0.32 and NDRE producing the highest R^2 value of 0.421, followed closely by ARI2.

Similar correlations were observed for CAPY (Table 1.4), although R^2 values tended to be higher, as might be expected based on the better temporal match between VMC, AET, and reflectance changes in CAPY. The highest R^2 value for this species was 0.518, between the NDLI and VMC20.

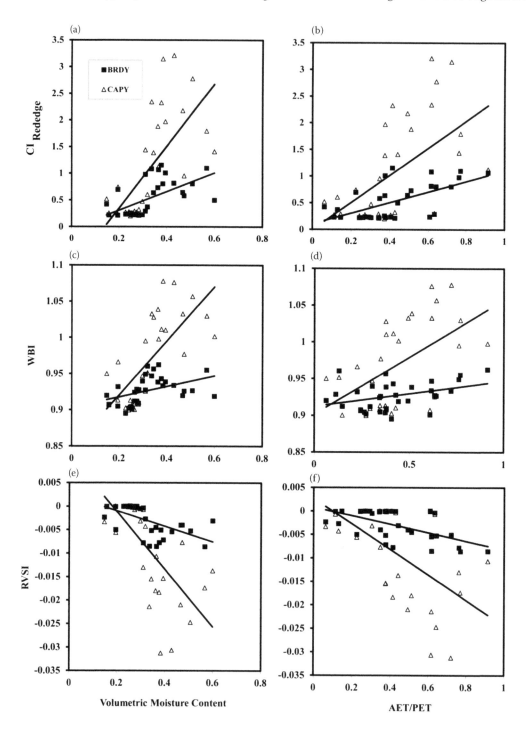

FIGURE 1.7   Scatterplots between volumetric moisture content (a, c, e) and AET/PET (b, d, f) for three HVIs, $CI_{red\ edge}$, WBI, and RVSI for BRDY (solid squares) and CAPY (open triangles).

High R^2 values were also observed between greenness measures and VMC, with VMC at 20 cm showing a slightly stronger relationship (in contrast to BRDY). Pigment measures were similarly highly correlated with VMC, although CRI1 had the highest correlation ($R^2 = 0.51$) with VMC10. Contrary to BRDY, HVIs for CAPY were also highly correlated with AET, and water-based indices

TABLE 1.3

Statistical Relationships Using Linear Regression for BRDY for 21 HVSIs

	R^2					F-Stat				
	VMC10	VMC20	PET	AET	AET/PET	VMC10	VMC20	PET	AET	AET/PET
Greenness										
DL1_DGVI	0.291	0.284	0.019	0.121	0.321	11.9**	11.5**	0.6	4.0	13.7***
DZ1_DGVI	0.352	0.306	0.218	0.055	0.337	15.8***	12.8**	8.1**	1.7	14.7***
EVI	0.362	0.323	0.116	0.087	0.362	16.4***	13.9***	3.8	2.8	16.4***
NDVI	0.404	0.358	0.204	0.069	0.354	19.7***	16.1***	7.4*	2.1	15.9***
SR	0.360	0.303	0.194	0.071	0.369	16.3***	12.6**	7.0*	2.2	17.0***
VARI	0.285	0.199	0.463	0.000	0.169	13.7**	11.3*	15.6***	0.5	8.4*
Pigments										
ARI2	0.298	0.312	0.006	0.220	0.403	12.3**	13.2**	0.2	8.2**	19.6***
$CI_{red\ edge}$	0.367	0.326	0.192	0.082	0.375	16.8***	14.0***	6.9*	2.6	17.4***
CRI	0.421	0.388	0.164	0.087	0.364	21.0***	18.4***	5.7*	2.7	16.6***
MCARI	0.342	0.274	0.204	0.034	0.303	15.1***	11.0**	7.4*	1.0	12.6**
SIPI	0.321	0.281	0.350	0.017	0.225	13.7***	11.3**	15.6***	0.5	8.4**
MTCI	0.614	0.638	0.041	0.040	0.206	46.2***	51.1***	1.2	1.2	7.5*
Nitrogen										
NDNI	0.381	0.308	0.516	0.005	0.129	17.8***	12.9**	31.0***	0.1	4.3*
NDRE	0.239	0.270	0.007	0.257	0.421	9.1**	10.7**	0.2	10.0**	21.1***
Water										
MSI	0.161	0.184	0.020	0.156	0.223	5.6*	6.5*	0.6	5.3*	8.3**
NDII	0.125	0.144	0.016	0.162	0.239	4.2	4.9*	0.5	5.6*	9.1**
NDWI	0.065	0.077	0.033	0.090	0.128	2.0	2.4	1.0	2.9	4.3*
WBI	0.142	0.173	0.009	0.039	0.134	4.8*	6.1*	0.3	1.2	4.5*
Cellulose/Residues										
CAI	0.288	0.184	0.512	0.002	0.162	11.7**	6.5*	30.5***	0.1	5.6*
NDLI	0.317	0.332	0.004	0.174	0.331	13.4***	14.4***	0.1	6.1*	14.4***
Plant Stress/Physiology										
PRI	0.052	0.093	0.245	0.150	0.062	1.6	3.0	9.4**	5.1*	1.9
RVSI	0.361	0.322	0.208	0.059	0.341	16.4***	13.8***	7.6*	1.8	15.0***
Generalized										
OMNBR	0.596	0.553	0.806	0.396	0.565	42.8***	36***	120.2***	19***	37.6***
Band 1 (nm)	2405	2405	1691	1721	1462					
Band 2 (nm)	2036	2036	1672	1701	1452					

Note: Statistical significance is reported as (* = 0.05, ** = 0.01, *** = 0.001). Values in bold represent the highest correlation for an index type.

were significantly correlated with soil moisture. Finally, R^2 values between HVIs and AET/PET were similar for CAPY and BRDY.

Recognizing the limitations of existing narrow-band indices, we calculated all possible OMNBR indices between PET and VMC20 for both species (Figure 1.8). In this plot, all potential normalized

TABLE 1.4

Statistical Relationships Using Linear Regression for CAPY for 21 HVSIs

	R²					F-Stat				
	VMC10	VMC20	PET	AET	AET/PET	VMC10	VMC20	PET	AET	AET/PET
Greenness										
DL1_DGVI	0.346	0.452	0.014	0.221	0.367	15.4**	24.0**	0.4	8.2	16.8***
DZ1_DGVI	0.369	0.447	0.061	0.144	0.328	16.9***	23.5**	1.9**	4.9	14.1***
EVI	0.388	0.476	0.038	0.174	0.352	18.4***	26.4***	1.2	6.1	15.8***
NDVI	0.480	0.530	0.113	0.091	0.308	26.7***	32.7***	3.7*	2.9	12.9***
SR	0.326	0.401	0.037	0.153	0.302	14.1***	19.4**	1.1*	5.3	12.5***
VARI	0.432	0.453	0.180	0.049	0.244	18.1**	19.3*	9.5***	0.3	5.9*
Pigments										
ARI2	0.218	0.334	0.013	0.327	0.377	8.1**	14.5**	0.4	14.1**	17.5***
$CI_{red\ edge}$	0.387	0.450	0.068	0.123	0.303	18.3***	23.7***	2.1*	4.1	12.6***
CRI	0.448	0.510	0.079	0.114	0.309	23.6***	30.2***	2.5*	3.7	13***
MCARI	0.327	0.423	0.027	0.193	0.345	14.1***	21.3**	0.8*	6.9	15.3**
SIPI	0.385	0.400	0.246	0.011	0.169	18.1***	19.3**	9.5***	0.3	5.9**
MTCI	0.185	0.169	0.012	0.001	0.019	6.6***	5.9***	0.4	0.0	0.6*
Nitrogen										
NDNI	0.492	0.512	0.114	0.099	0.330	28.1***	30.5**	3.7***	3.2	14.3*
NDRE	0.185	0.303	0.000	0.265	0.331	6.6**	12.6**	0.0	10.4**	14.3***
Water										
MSI	0.443	0.517	0.083	0.083	0.272	23.1*	31.1*	2.6	2.6*	10.8**
NDII	0.404	0.504	0.044	0.149	0.327	19.7	29.5*	1.3	5.1*	14.1**
NDWI	0.420	0.508	0.057	0.130	0.316	21.0	30.0	1.7	4.3	13.4*
WBI	0.419	0.504	0.081	0.122	0.312	20.9*	29.4*	2.5	4.0	13.1*
Cellulose/Residues										
CAI	0.333	0.380	0.174	0.040	0.225	14.5**	17.8*	6.1***	1.2	8.4*
NDLI	0.402	0.518	0.027	0.175	0.343	19.5***	31.2***	0.8	6.2*	15.1***
Plant Stress/Physiology										
PRI	0.109	0.056	0.601	0.098	0.004	3.6	1.7	43.7**	3.1*	0.1
RVSI	0.391	0.470	0.060	0.141	0.324	18.6***	25.7***	1.9*	4.8	13.9***
Generalized										
OMNBR	0.65	0.649	0.693	0.519	0.532	53.9***	53.5***	65.5***	31.2***	33.0***
Band 1 (nm)	889	782	501	501	1652					
Band 2 (nm)	733	753	433	482	1642					

Note: Statistical significance is reported as (* = 0.05, ** = 0.01, *** = 0.001).

ratio band combinations are shown, with the y axis corresponding to the first wavelength and the x axis corresponding to second wavelength. Using NDVI for an example, the x axis would be the red band and the y axis the NIR band. Plots are symmetrical around the diagonal because swapping the bands does not change the correlation coefficient, but would reverse the sign of the correlation. Visually, the best combination of bands is readily apparent for each species. For example, the highest

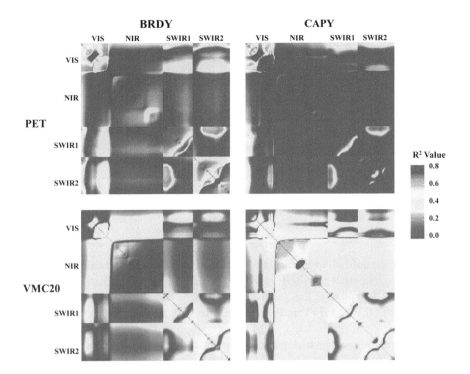

FIGURE 1.8 Correlation matrices for each band pair with band 1 plotted on the y axis and band 2 plotted on the x axis. Wavelength regions are labeled as visible (400–700 nm), NIR (700–1350 nm), SWIR1 (1422–1801 nm), and SWIR2 (1965–2450 nm). Strong water vapor absorption regions have been discarded from the analysis and used to define the wavelength regions.

correlations for BRDY and PET are found in much of the visible parts of the spectrum and select regions of the SWIR1 and SWIR2. Overall, CAPY shows weak correlations, but with the highest correlations found for very narrow-band combinations in the visible.

The OMNBR index results for each of the five parameters (shown in Tables 1.3 and 1.4) confirm what is visually apparent in Figure 1.8. For CAPY, the OMNBR index produced the highest correlation for all variables, ranging from 0.520 (Band 1 = 501 nm, Band 2 = 482 nm) for AET, to 0.693 (Band 1 = 501 nm, Band 2 = 433 nm) for PET. For BRDY, the OMNBR indices had highest correlations for PET, AET, and AET/PET, ranging from a low of 0.396 for AET (Band 1 = 1721 nm, Band 2 = 1701 nm), to a high of 0.806 for PET (Band 1 = 1691 nm, Band 2 = 1672 nm). For VMC20, the three-band combination of MTCI actually showed a higher correlation than any two-band combination, although correlations were still high, exceeding 0.55 for a SWIR2 band combination consisting of 2405 and 2036 nm.

Significant differences in band selection by species are not uncommon (Thenkabail et al., 2000), illustrating the challenge of developing generalized models that do not vary by species. However, it is also possible to identify band pairs that, in combination, provide the highest average correlation across species (Table 1.5). In this analysis, the best band combination for VMC for both species included two SWIR bands (2395 and 2045 nm), likely focused on significant ligno-cellulose absorptions in the SWIR2. For PET, the best composite band combination involved a green (501 nm) and blue (443 nm) band, while AET had the weakest correlations involving a NIR (1343 nm) and blue (433 nm) combination. The AET/PET ratio selected two SWIR1 bands (1482 and 1442 nm), with the shorter wavelength located on the wings of a strong liquid water absorption. In general, CAPY showed better correlations than BRDY for the combined selection (Table 1.5), but poorer correlations when analyzed individually (Tables 1.3 and 1.4).

TABLE 1.5

Optimal Band Pairs for Analysis for Both Plant Species

	Band1	Band2	BRDY R^2	CAPY R^2
VMC10	2395	2045	0.54	0.57
VMC20	2395	2045	0.53	0.63
PET	501	443	0.6	0.66
AET	1343	433	0.34	0.45
AET/PET	1482	1442	0.49	0.44

Note: Band wavelengths are expressed in nm.

In this example, we used all available dates to explore correlations and did not apply a lag. There are alternative approaches that can also be used and could be explored further. For example, Liu et al. (2012) noted that HVIs tended to lag soil moisture, finding the highest correlations when the HVI was lagged by 20 days (i.e., the HVI tends to change after soil moisture has increased). Peterson et al. (2008) found that relationships between LFM and VIs improved if the analysis was restricted to the dry-down period, the period in which plants declined from peak LFM in late winter to a seasonal low in midsummer. In this analysis, we did not lag HVIs and did not limit the analysis to a subset of dates. To evaluate whether a more restrictive analysis, similar to Peterson et al. (2008), would improve results, we experimented with restricting the analysis to include only observations from March to July, when plants are most dynamic, finding significantly higher correlations for VMC, but slightly lower correlations for PET (not shown). We suggest remote sensing analysis targeted at estimating soil moisture will likely be most successful if it focuses on time periods where both soil moisture and vegetation are changing, while those focused on PET may be better suited using the full year, since PET changes throughout the year and vegetation changes spectrally even following senescence (Figure 1.5).

Overall, we conclude that: (1) HVIs are better predictors of the ratio of AET to PET than of PET or AET alone. Direct correlation with VMC tended to be higher than correlations with measures of ET, although this may also be a product of additional variables used in calculating ET, which may contribute error (e.g., G, K, surface temperature); (2) relationships varied significantly between plant species with different phenologies; (3) pigment- and greenness-based indices were effective across species, while the performance of water-based indices was more species dependent; and (4) optimal band pairs that outperform existing HVIs can be identified, except in the case where three bands were used to generate an index (e.g., MTCI). These results show that HVIs can be of great use in applications quantifying a correlated environmental variable where field data are sparse or absent.

1.4 CONCLUSIONS

Most hyperspectral systems collect a large volume of data in wavelengths that are highly correlated (Thenkabail et al., 2014). Hyperspectral vegetation indices are one way to distill this wealth of information into a few physically meaningful variables. Hyperspectral systems also present the opportunity to create new indices that incorporate wavelengths not sampled by any broad-band system and synthesize broad-band systems or identify narrow wavelengths better suited for the trait of interest (Thenkabail et al., 2014). While narrow-band equivalents of broad-band indices have not necessarily improved index performance (e.g., Elvidge and Chen, 1995), a change in the wavelength position of one or more narrow bands within the index, better tuned for a specific absorption, has been shown to significantly improve performance and account for variation between plant species

(Galvao et al., 1999; Thenkabail et al., 2000). In many cases, these HVIs have also proven to be less sensitive to atmospheric contamination or changes in lighting/viewing geometry. For example, NIR- and SWIR-based indices should be less sensitive to atmospheric contamination because they sample spectral regions with reduced atmospheric scattering, while indices that include wavelengths restricted to a strong absorption region (e.g., VARI) or a strong scattering region (e.g., NDWI) may be less sensitive to viewing geometry (Perry and Roberts, 2008). Red-edge–based indices and indices that include three bands appear particularly promising for some traits as well. Finally, as shown in the second case study, OMNBR indices offer the potential of capturing the strength of multiple narrow bands, tuned to maximize our ability to explain variability in the data.

A potentially greater contribution of hyperspectral systems is their ability to measure absorptions specific to important vegetation traits. Some of the most promising HVIs are intended to quantify key plant physiological responses, such as the PRI, which is used as a proxy for LUE, and RVSI, which is highly sensitive to seasonal changes in environmental stress. There is also the possibility of combining several indices through multivariate regression to improve a biophysical or biochemical retrieval or use HVIs as inputs into classification (Fuentes et al., 2001) and species discrimination (Clark and Roberts, 2012). Such an approach takes advantage of the greater diversity of spectral features available through hyperspectral sensors to improve a retrieval, and offers the potential of classifying vegetation based on inferred biochemistry and structure as in Fuentes et al. (2001).

In this chapter, we provided a review of the most commonly used and best-adopted HVIs reported in the literature. We strongly linked the formulation of HVIs to their corresponding physical basis, which typically includes at least one strong absorption band and a nonabsorbing reference band, targeted over a narrow wavelength range to maximize absorption and reflectivity while minimizing averaging over broad spectral regions. To illustrate the potential of a few indices, we supplied two case studies. As the number and volume of hyperspectral sensors and data increase, HVIs will become ever more prevalent and critical for vegetation research.

REFERENCES

Asner, G.P., Nepstad, D., Cardinot, G. and Ray, D. 2004. Drought stress and carbon uptake in an Amazon forest measured with spaceborne imaging spectroscopy, *Proceedings National Academy of Sciences*, 101(16), 6039–6044.

Baret, F., Guyot, G. and Major, D.J. 1989. TSAVI: A vegetation index which minimizes soil brightness effects on LAI And APAR estimation. *12th Canadian Symposium on Remote Sensing Geoscience and Remote Sensing Symposium*, 3(2), 1355–1358.

Barnes, E., Clarke, T., Richards, S., Colaizzi, P.D., Haberland, J., Kostrzewski, M. and Moran, M.S. 2000. Coincident detection of crop water stress, nitrogen status and canopy density using ground-based multispectral data. *Proceedings of the Fifth International Conference on Precision Agriculture*, ASA-CSSA-SSSA, Madison, Wisconsin, 16 pp.

Blackburn, G.A. 1998. Spectral indices for estimating photosynthetic pigment concentrations: A test using senescent tree leaves, *International Journal of Remote Sensing*, 19, 657–675.

Bowen, I.S. 1926. The ratio of heat losses by conduction and by evaporation from any water surface, *Physical Review*, 27, 779.

Bradley, E., Still, C. and Roberts, D. 2010. Design of an image analysis website for phenological and meteorological monitoring. *Environmental Modelling & Software*, 25, 107–116.

Card, D.H., Peterson, D.L., Matson, P.A. and Aber, J.D. 1988. Prediction of leaf chemistry by the use of visible and near infrared reflectance spectroscopy, *Remote Sensing of Environment*, 26, 123–147.

Carlson, T.N. and Ripley, D.A. 1997. On the relation between NDVI, fractional vegetation cover and leaf area index, *Remote Sensing of Environment*, 62, 241–252.

Clark, M.L., and Roberts, D.A. 2012. Species-level differences in hyperspectral metrics among tropical rainforest trees as determined by a tree-based classifier, *Remote Sensing*, 4, 1820–1855, doi:10.3390/rs4061820.

Clevers, J.G.P.W. 1991. Application of the WDVI in estimating LAI at the generative stage of barley, *ISPRS Journal of Photogrammetry and Remote Sensing*, 46(1), 37–47.

Curran, P.J. 1989. Remote sensing of foliar chemistry, *Remote Sensing of Environment*, 30, 271–278.

Dash, J. and Curran, P.J. 2004. The MERIS terrestrial chlorophyll index, *International Journal of Remote Sensing*, 25(23), 5403–5413.

Daughtry, C.S.T. 2001. Discriminating crop residues from soil by shortwave infrared reflectance, *Agronomy Journal*, 93, 125–131.

Daughtry, C.S.T., Hunt Jr., E.R., Doraiswamy, P.C. and McMurtrey III, J.E. 2005. Remote sensing the spatial distribution of crop residues, *Agronomy Journal*, 97(3), 864–871.

Daughtry, C.S.T., Hunt Jr., E.R. and McMurtrey III, J.E. 2004. Assessing crop residue cover using shortwave infrared reflectance. *Remote Sensing of Environment*, 90, 126–134.

Daughtry, C.S.T., Walthall, C.L. Kim, M.S., de Colstoun, E.B. and McMurtrey, J.E. 2000. Estimating corn leaf chlorophyll concentration from leaf and canopy reflectance. *Remote Sensing of Environment*, 74, 229–239.

Deering, D.W., Middleton, E.M. and Eck, T.F. 1994. Reflectance anisotropy for a spruce-hemlock forest canopy, *Remote Sensing of Environment*, 47, 242–260.

Elvidge, C.D. and Chen, Z. 1995. Comparison of broad-band and narrow-band red and near-infrared vegetation indices, *Remote Sensing of Environment*, 54, 38–48.

Fitzgerald, G., Rodriguez, D. and O'Leary, G. 2010. Measuring and predicting canopy nitrogen nutrition in wheat using a spectral index, The canopy chlorophyll content index (CCCI), *Field Crops Research*, 116, 318–324.

Fuentes, D., Gamon, J.A., Qiu, H.-L., Sims, D. and Roberts, D. 2001. Mapping Canadian Boreal forest vegetation using pigment and water absorption features derived from the AVIRIS sensor, *Journal Geophysical Research Atmospheres*, 106(D24), 33,565–33, 577.

Galvao, L.S., dos Santos, J.R., Roberts, D.A., Breunig, F.M., Toomey, M., and de Moura, Y.M. 2011. On intra-annual EVI variability in the dry season of tropical forest: A case study with MODIS and hyperspectral data, *Remote Sensing of Environment*, 115, 2350–2359.

Galvao, L.S., Viterello, I. and Almeida Filho, R. 1999. Effects of band positioning and bandwidth on NDVI Measurements of Tropical Savannas, *Remote Sensing of Environment*, 67, 181–193.

Gamon, J.A., Kitajima, K., Mulkey, S.S., Serrano, L. and Wright, S.J. 2005. Diverse optical and photosynthetic properties in a neotropical dry forest during the dry season: Implications for remote estimation of photosynthesis. *Biotropica*, 37(4), 547–560.

Gamon J.A., Serrano, L. and Surfus, J.S. 1997. The photochemical reflectance index: An optical indicator of photosynthetic radiation-use efficiency across species, functional types, and nutrient levels, *Oecologia*, 112, 492–501.

Gamon J.A. and Surfus J.S. 1999. Assessing leaf pigment content and activity with a reflectometer, *New Phytologist*, 143, 105–117.

Gao, B. 1996. NDWI: A normalized difference water index for remote sensing of vegetation liquid water from space, *Remote Sensing of Environment*, 58, 257–266.

Gitelson, A.A., Chivkunova, O.B. and Merzlyak, M.N. 2009. Nondestructive estimation of anthocyanins and chlorophylls in anthocyanic leaves, *American Journal of Botany*, 96(10), 1861–1868.

Gitelson, A.A., Gritz, Y. and Merzlyak, M.N. 2003a. Relationship between leaf chlorophyll content and spectral reflectance and algorithms for non-destructive chlorophyll assessment, in higher plants, *Journal of Plant Physiology*, 160, 271–282.

Gitelson A.A., Kaufman Y.J., Stark R. and Rundquist, D. 2002a. Novel algorithms for remote estimation of vegetation fraction, *Remote Sensing of Environment*, 80, 76–87.

Gitelson, A.A., Keydan, G.P. and Merzlyak, M.N. 2006. Three band model for noninvasive estimation of chlorophyll, carotenoids, and anthocyanin contents in higher plant leaves, *Geophysical Research Letters*, 33, L11402.

Gitelson, A.A., Merzlyak, M.N. and Chivkunova, O.B. 2001. Optical properties and non-destructive estimation of anthocyanin content in plant leaves, *Photochemistry and Photobiology*, 74(1), 38–45.

Gitelson, A.A., Viña, A., Arkebauer, T.J., Rundquist, D.C., Keydan, G. and Leavitt, B. 2003b. Remote estimation of leaf area index and green leaf biomass in maize canopies, *Geophysical Research Letters*, 30(5), 1248, doi:10.1029/2002GL016450.

Gitelson, A.A., Zur, Y., Chivkunova, O.B. and Merzlyak, M.N. 2002b. Assessing Carotenoid Content in Plant Leaves with Reflectance Spectroscopy, *Photochemistry and Photobiology*, 75(3), 272–281.

Hardisky, M.A., Klemas, V. and Smart, R.M. 1983. The influence of soil salinity, growth form and leaf moisture on spectral radiance of *Spartina alterniflora* canopies, *Photogrammetric Engineering and Remote Sensing*, 49, 77–83.

Horler, D.N.H., Dockray, M. and Barber, J. 1983. The red-edge of plant leaf reflectance, *International Journal of Remote Sensing*, 4, 273–288.

Huete, A.R. 1988. A soil adjusted vegetation index (SAVI), *Remote Sensing of Environment*, 25, 295–309.

Huete, A.R., Didan, K., Miura, T., Rodriguez, E.P., Gao, X. and Ferreira, L. 2002. Overview of the radiometric and biophysical performance of the MODIS vegetation indices, *Remote Sensing of Environment*, 83, 195–213.

Huete, A.R., Jackson, R.D. and Post, D.F. 1984. Spectral response of a plant canopy with different soft backgrounds, *Remote Sensing of Environment*, 17, 37–53.

Huete, A.R., Liu, H.Q., Batchily, K. and van Leeuwen, W. 1997. A comparison of vegetation indices over a global set of TM images for EOS-MODIS, *Remote Sensing of Environment*, 59, 440–451.

Hunt Jr., E.R., and Rock, B.N. 1989. Detection of changes in leaf water content using near- and middle-infrared reflectances, *Remote Sensing of Environment*, 30, 43–54.

Jordan, C.F. 1969. Leaf-area index from quality of light on the forest floor, *Ecology*, 50(4), 663–666.

Kaufman, Y.J. and Tanier, D. 1992. Atmospherically resistant vegetation index (ARVI) for EOS-MODIS, *IEEE Transactions on Geoscience and Remote Sensing*, 30(2), 261–270.

Kim, M.S. 1994. The use of narrow spectral bands for improving remote sensing estimation of fractionally absorbed photosynthetically active radiation (fAPAR). *Masters thesis*. Department of Geography, University of Maryland, College Park, MD.

Li, F., Miao, Y., Feng, G., Yuan, F., Yue, S., Gao, X., Liu, Y., Liu, B., Ustin, S.L. and Chen, X. 2014. Improving estimation of summer maize nitrogen status with red edge-based spectral vegetation indices, *Field Crops Research*, 157.

Liu, S., Roberts, D.A., Chadwick, O.A., and Still, C.J. 2012. Spectral response to plant available soil moisture in a Californian grassland, *International Journal of Applied Earth Observation and Geoinformation*, 19, 31–44.

Merton, R. and Huntington, J. 1999. Early simulation results of the ARIES-1 satellite sensor for multi-temporal vegetation research derived from AVIRIS. Available at ftp://popo.jpl.nasa.gov/pub/docs/workshops/99_docs/41.pdf, NASA Jet Propulsion Lab., Pasadena, CA.

Merzlyak, M.N., Gitelson, A.A., Chivkunova, O.B. and Rakitin, Y. 1999. Non-destructive optical detection of pigment changes during leaf senescence and fruit ripening, *Physiologia Plantarum*, 105, 135–141.

Milton, N.M., Eiswerth, B.A. and Ager, C.M. 1991. Effect of phosphorus deficiency on spectral reflectance and morphology of soybean plants, *Remote Sensing of Environment*, 36, 121–127.

Monteith, J.L. and Unsworth, M.H. 1990. *Principles of Environmental Physics*, 2nd ed.: London: Edward Arnold.

Naidu, R.A., Perry, E.M., Pierce, F.J. and Mekuria, T. 2009. The potential of spectral reflectance technique for the detection of grapevine leafroll-associated virus-3 in two red-berried wine grape cultivars, *Computers and Electronics in Agriculture*, 66, 38–45.

Peñuelas, J., Baret, F. and Filella, I. 1995. Semi-empirical indices to assess carotenoids/chlorophyll a ratio from leaf spectral reflectance. *Photosynthetica*, 31, 221–230.

Peñuelas, J., Pinol, J., Ogaya, R. and Lilella, I. 1997. Estimation of plant water content by the reflectance water index WI (R900/R970), *International Journal of Remote Sensing*, 18, 2869–2875.

Perry, E.M. and Roberts, D.A. 2008. Sensitivity of narrow-band and broad-band indices for assessing nitrogen availability and water stress in annual crop, *Agronomy Journal*, 100(4), 1211–1219.

Peterson, S.H., Roberts, D.A., and Dennison, P.E. 2008. Mapping live fuel moisture with MODIS data: A multiple regression approach, *Remote Sensing of Environment*, 112(12), 4272–4284.

Rahman, A.F., Gamon, J.A., Fuentes, D.A., Roberts, D.A. and Prentiss, D. 2001. Modeling spatial distributed ecosystem flux of boreal forests using hyperspectral indices from AVIRIS imagery, *Journal of Geophysical Research Atmospheres*, 106(d24), 33579–33591.

Roberts, D.A., Bradley, E.S., Roth, K., Eckmann, T. and Still, C. 2010. Linking physical geography education and research through the development of an environmental sensing network and project-based learning, *Journal of Geoscience Education*, 58, 262–274.

Roberts, D.A., Brown, K.J., Green, R., Ustin, S. and Hinckley, T. 1998. Investigating the relationship between liquid water and leaf area in clonal Populus. *Proc. 7th AVIRIS Earth Science Workshop JPL 97-21*, Pasadena, CA, pp. 335–344.

Roberts, D.A., Dennison, P.E., Peterson, S., Sweeney, S. and Rechel, J. 2006. Evaluation of AVIRIS and MODIS measures of live fuel moisture and fuel condition in a shrubland ecosystem in Southern California, *Journal of Geophysical Research Biogeosciences*, 111, G04S02. doi:10.1029/2005JG000113, 16 pp.

Roberts, D.A., Ustin, S.L., Ogunjemiyo, S., Greenberg, J., Dobrowski, S.Z., Chen, J. and Hinckley, T.M. 2004. Spectral and structural measures of Northwest forest vegetation at leaf to landscape scales, *Ecosystems*, 7, 545–562.

Rock, B.N., Vogelmann, J.E., Williams, D.L., Vogelmann, A.F., and Hoshizaki, T. 1986. Detection of forest damage, *BioScience*, 36(7), 439,445.

Rouse, J.W., Haas, R.H., Schell, J.A. and Deering, D.W. 1973. Monitoring vegetation systems in the great plains with ERTS. *Third ERTS Symposium, NASA SP-351, NASA*, Washington, DC, Vol. 1, pp. 309–317.

Sellers, P.J. 1985. Canopy reflectance, photosynthesis and transpiration, *International Journal of Remote Sensin*, 6, 1335–1372.

Serrano, L., Penuelas, J. and Ustin, S.L. 2002. Remote sensing of nitrogen and lignin in Mediterranean vegetation from AVIRIS data: Decomposing biochemical from structural signals, *Remote Sensing of Environment*, 81, 355–364.

Serrano, L., Ustin, S.L., Roberts, D.A., Gamon, J.A. and Penuelas, J. 2000. Deriving water content of chaparral vegetation from AVIRIS data, *Remote Sensing of Environment*, 74, 570–581.

Sims, D.A. and Gamon, J.A. 2002. Relationships between leaf pigment content and spectral reflectance across a wide range of species, leaf structures and developmental stages, *Remote Sensing of Environment*, 81, 337–354.

Sims, D.A. and Gamon, J.A. 2003. Estimation of vegetation water content and photosynthetic tissue area from spectral reflectance: A comparison of indices based on liquid water and chlorophyll absorption, *Remote Sensing of Environment*, 84, 526–537.

Thenkabail, P.S., Gumma, M.K., Teluguntla, P. and Mohammed, I.A. 2014. Hyperspectral remote sensing of vegetation and agricultural crops. Highlight article, *Photogrammetric Engineering and Remote Sensing*, 80(4), 697–709. IP-052042.

Thenkabail, P.S., Mariotto, I., Gumma, M.K., Middleton, E.M., Landis, D.R. and Huemmrich, K.F. 2013. Selection of hyperspectral narrowbands (HNBs), and composition of hyperspectral two-band vegetation indices (HVIs) for biophysical characterization and discrimination of crop types using field reflectance and Hyperion/EO-1 data, *IEEE Journal of Selected Topics in Applied Earth Observations and Remote Sensing*, 6(2), 427–439.

Thenkabail P.S., Smith, R.B. and De-Pauw, E. 2000. Hyperspectral vegetation indices for determining agricultural crop characteristics, *Remote Sensing of Environment*, 71, 158–182.

Tian, Y.C., Yao, X., Yang, J., Cao, W.C. Hannaway, D. B. and Zhu, Y. 2011. Assessing newly developed and published vegetation indices for estimating rice leaf nitrogen concentration with ground- and space-based hyperspectral reflectance, *Field Crops Research*, 120, 299–310.

Tucker, C.J. 1979. Red and photographic infrared linear combinations for monitoring vegetation, *Remote Sensing of Environment*, 8, 127–150.

Ustin, S.L., Gitelson, A.A., Jacquemoud S., Schaepman, M., Asner, G.P., Gamon J.A. and Zarco-Tejada, P. 2009. Retrieval of foliar information about plant pigment systems from high resolution spectroscopy, *Remote Sensing of Environment*, 113, S67–S77.

Van den Berg, A.K. and Perkins, T.D. 2005. Non-destructive estimation of anthocyanin content in autumn sugar maple leaves, *Horticultural Science*, 40(3), 685–685.

Vina, A., Gitelson, A.A., Nguy-Robertson, A.L. and Peng, Y. 2011. Comparison of different vegetation indices for remote assessment of green leaf area index of crops, *Remote Sensing of Environment*, 115, 3468–3478.

Yao, X., Zhu, Y., Tian, Y., Feng, W. and Cao, W. 2010. Exploring hyperspectral bands and estimation indices for leaf nitrogen accumulation in wheat, *International Journal of Applied Earth Observations and Geoinformation*, 12, 89–100.

Zarco-Tejada, P.J., Gonzalez-Dugo, V., Suarez, L., Berni, J.A.J., Goldhamer, D. and Fereres, E. 2013. A PRI-based water stress index combining structure and chlorophyll effects: Assessment using diurnal narrow-band airborne imagery and the CWSI thermal index. *Remote Sensing of Environment*, 138, 38–50.

Zarco-Tejada, P.J., Miller, J.R., Mohammed, G.H. and Noland, T.L. 2000. Chlorophyll fluorescence effects on vegetation: Apparent reflectance: Leaf-level measurements and model simulation, *Remote Sensing of Environment*, 74, 582–595.

Zygielbaum, A.I., Arkebauer, T.J., Walter-Shea, E.A. and Scoby, D.L. 2012. Detection and measurement of vegetation photoprotection stress response using PAR reflectance, *Israel Journal of Plant Sciences*, 60, 37–47.

Zygielbaum, A.I., Gitelson, A.A., Arkebauer, T.J. and Rundquist, D.C. 2009. Non-destructive detection of water stress and estimation of relative water content in maize. *Geophysical Research Letters*, 36, L12403, doi: 10.1029/2009GL038906.

2 Derivative Hyperspectral Vegetation Indices in Characterizing Forest Biophysical and Biochemical Quantities

Quan Wang, Jia Jin, Rei Sonobe, and Jing Ming Chen

CONTENTS

2.1   INTRODUCTION

Hyperspectral remote sensing, which provides information on targets with high spectral and spatial resolutions, is a useful tool to identify conditions of targets (Ben-Dor et al. 2013), particularly in ecosystem monitoring (Goetz 2009). In the last 50 years, various methods have been developed to derive biophysical and biochemical parameters from remote sensing data (Dorigo et al. 2007), among which hyperspectral remote sensing data are receiving special attention due to their prominent ability to provide spectral details of target objects. Specifically, hyperspectral remote sensing has been successfully applied to quantifying biophysical parameters such as leaf area index (Haboudane et al. 2004, Gonsamo and Pellikka 2012, Delegido et al. 2013, Li and Wang 2013), biomass (Hansen and Schjoerring 2003, Cho et al. 2007, Fu et al. 2014), biochemical parameters like pigment contents (Blackburn 1998a, Li and Wang 2012, Cheng et al. 2013, Yi et al. 2014), plant nitrogen content (Hansen and Schjoerring 2003, Huang et al. 2004, Nguyen and Lee 2006, Ryu et al. 2009, Ryu et al. 2011, Inoue et al. 2012), and nitrogen and phosphorus concentrations (Ramoelo et al. 2013, Zhang et al. 2013), as well as ecophysiological status such as plant water status and transpiration (Marino et al. 2014, Sun et al. 2014, Wang and Jin 2015, Jin and Wang 2016, Marshall et al. 2016).

Traditionally, empirically/statistically based methods and physically based radiative transfer model (RTM) methods were developed to interpret reflected spectral information. Empirically/statistically based methods typically involve establishing relationships between spectral information and biophysical or biochemical parameters, which in general are obtained from synchronous measurements on the ground (Dorigo et al. 2007). These methods are simple and effective for a site-specific dataset but hardly as much so when extrapolated to other datasets (Liang 2005, Li and Wang 2012). As a comparison, physically based methods typically estimate targets' parameters from obtained reflectance by inverting RTMs (Jacquemoud 1993, Li and Wang 2011). RTMs simulate the interactions between light and plants based on physical laws and thus can account for cause-effect relationships. However, RTMs are usually complex, with many input variables, leading to "ill-posed" problems in their inversion for the retrieval of biophysical and biochemical parameters (Combal et al. 2003, Wang et al. 2007, Yebra and Chuvieco 2009, Li and Wang 2011).

Robustness of empirical/statistical methods and operational difficulty of RTM inversions are two barriers hindering wide applications of remote sensing data. A hybrid approach is emerging, which is able to overcome these barriers by making full use of data mining and physical modeling for developing statistical relationships between targets' parameters and spectral data generated from synthetic simulations using RTMs (le Maire et al. 2008, Feret et al. 2011, Wang and Li 2012a). The approach will largely improve the representativeness of data, which is critical to developing robust empirical linkages of targeted parameters with spectral reflectance and has thus had great impacts on traditional empirical methods like vegetation indices.

Traditional empirical approaches, such as those based on vegetation indices (VIs), stepwise multiple linear regression, partial least-squares regression, and so on, have been widely employed for their easy manipulation (Dorigo et al. 2007). Among them, vegetation indices, consisting of several reflected or transformed spectral bands to enhance spectral signals or to remove perturbations caused by background factors, are the most common ones used for relationship analysis (Baret and Guyot 1991, Dorigo et al. 2007). Broadband VIs, which are mainly based on multispectral remotely sensed data, can be traced back to the 1970s. Popular broadband VIs, including ratio vegetation index (Pearson and Miller 1972), normalized difference vegetation index (NDVI) (Rouse et al. 1974), and soil-adjusted vegetation index (SAVI) (Huete 1988), were largely developed to remove the effects of atmosphere and soil background. Similarly, with the advent of hyperspectral remote sensing, many hyperspectral VIs (HVIs) based on narrow bands have been proposed. They have been widely used for retrieving biophysical and biochemical quantities (Rodriguez-Perez et al. 2007, Wang and Li 2012a,b, Zhang et al. 2012, Li and Wang 2013, Marshall et al. 2016). Even though some of them are directly borrowed or mimicked from their broadband counterparts, it has been suggested that narrow bands can provide additional information with significant advantages over broad bands in quantifying biophysical parameters (Thenkabail et al. 2000).

Although simple to understand and implement, VIs, however, have several inherent limitations (Baret and Guyot 1991, Liang 2005, Li and Wang 2011). The efficiency of vegetation indices depends preliminarily upon the amount and quality of the measurement database (le Maire et al. 2008, Feret et al. 2011), and, for example, indices calibrated on a certain species may fail on other species (le Maire et al. 2008, Wang and Li 2012a,b).

Such shortcomings may only be alleviated by either increasing the amount or improving the quality of data contained in the dataset used for developing VIs. While the newly emerging hybrid approach greatly increased the data volume and thus the possibility of developing a more generally applicable VI, a number of transformed formats of spectra, for example, transmittance and derivative spectra, have also proved to be helpful in deriving biophysical and biochemical parameters (Rady et al. 2014). Among them, derivative techniques have the advantages of minimizing additive constants and linear functions (Imanishi et al. 2004) and have been shown to be feasible for estimating plant biophysical and biochemical parameters. For instance, the red-edge position (REP), which is the wavelength of the maximum first derivative in the range 690–750 nm, has been successfully used in plant condition detection (Curran et al. 1990, Imanishi et al. 2004), and the peak of the first derivative of red-edge reflectance could be two or more sources of information (Smith et al. 2004). As a result, various derivative hyperspectral indices (dHVIs) have been developed and used for obtaining biophysical and biochemical quantities (Demetriades-Shah et al. 1990, Imanishi et al. 2004, Yao et al. 2014, Jin and Wang 2016). Indices based on derivative spectra have proven to be more effective than reflectance-based indices (Demetriades-Shah et al. 1990, Zarco-Tejada et al. 2003a,b), due primarily to their reduction of the background signal as well as to the separation of overlapping spectra through the use of various differentiation techniques (Demetriades-Shah et al. 1990).

Nonetheless, the advantages of dHVIs over reflectance-based VIs and the differences among derivatives of different orders have not been fully investigated. Here we explain the pros and cons of dHVIs and provide state-of-the-art methods of developing and using dHVIs to estimate forest biophysical and biochemical quantities both at the leaf and canopy scales. Pros and cons of dHVIs are revealed not only by comparing VIs based on original and derivative spectra for typical parameters using *in situ* measured datasets, but also via a series of virtual experiments based on RTMs at different scales.

2.2 DERIVATIVE HYPERSPECTRAL VEGETATION INDICES: HISTORY, TYPES, AND NEW EMERGING DEVELOPING APPROACHES

2.2.1 Vegetation Index History, from Multiple to Hyperspectral, and Main Vegetation Index Types (Models)

Vegetation indices usually combine information from different spectral channels (Bannari et al. 1995) and can be classified into broadband VIs based on multispectral bands and hyperspectral VIs based on narrow bands (Dorigo et al. 2007). Using vegetation indices to evaluate vegetation status can be traced back to as early as the launch of Earth Resources Technology Satellite 1 (Landsat-1) in 1972. The best-known indices in the early days were the ratio vegetation index (RVI) (Pearson and Miller 1972) and the normalized difference vegetation index (Rouse et al. 1974), which are based on the reflectance in the broad red and near-infrared (NIR) bands. Since then, remarkable efforts have been put into developing new vegetation indices in different formulas, such as the soil-adjusted vegetation index (Huete 1988, Qi et al. 1994) and the enhanced vegetation index (EVI) (Huete et al. 2002), for enhancing vegetation response and minimizing background effects (Wiegand et al. 1991, Bannari et al. 1995). Such broadband VIs have been proved to be correlated with vegetation cover parameters such as leaf area index (LAI), leaf pigment contents, and above-ground biomass (Bioucas-Dias et al. 2013). However, they also have significant limitations; for example, they may saturate at high vegetation levels, most are only on red and NIR spectral information, and only a few VIs have ever used information from other bands (Thenkabail et al. 2012).

The concept of hyperspectral remote sensing began in the mid-1980s (Liang 2005). Compared with multiple-band remote sensing, hyperspectral remote sensing collects spectra in many relatively narrow spectral wavelengths (Im and Jensen 2008), offering unprecedented abundant information. A similar philosophy is then applied naturally to hyperspectral remote sensing data, producing a greater number of narrowband spectral indices, which can better compensate for background effects, particularly where vegetation cover is low (Kalacska and Sanchez-Azofeifa 2008). During the last 40 years, numerous HVIs under different formulae and band combinations have been successfully used for tracing biophysical and biochemical parameters (Haboudane et al. 2004, le Maire et al. 2004, Zarco-Tejada et al. 2005, Zhao et al. 2005, Main et al. 2011, Wang and Li 2012a,b).

As hyperspectral remote sensing can acquire continuous spectra of the target with high spectral resolutions, apart from reflectance-based narrowband index development, new approaches based on various transformations including derivative spectra, spectral shape, and the spectral absorption features have been proposed in parallel (Demetriades-Shah et al. 1990, Tsai and Philpot 1998, Kokaly and Clark 1999, Mutanga and Skidmore 2004, Mutanga et al. 2005). The idea of using derivative spectra was proposed several decades ago, and derivative analysis has been an established technique in analytical chemistry (Demetriades-Shah et al. 1990, Kochubey and Kazantsev 2012). But derivative analysis was not widely used in monitoring ecosystems before hyperspectral remote sensing was developed (Kochubey and Kazantsev 2012). Recently, studies demonstrate that dHVIs are superior to simple reflectance-based HVIs in the field of monitoring plant biophysical and biochemical parameters (Demetriades-Shah et al. 1990, Diaz and Blackburn 2003, Zarco-Tejada et al. 2004, Kochubey and Kazantsev 2012).

Although there are various HVIs, most of them can be categorized into several general types (le Maire et al. 2008, Li and Wang 2013, Sonobe and Wang 2017). Most published indices are expressed as reflectance or a first-order derivative at a given wavelength (R) (Boochs et al. 1990, Gamon and Surfus 1999, Carter and Knapp 2001), wavelength difference (D) (Tanaka et al. 2015, Sonobe and Wang 2016), simple ratio (SR) (Jordan 1969, Chappelle et al. 1992, Penuelas et al. 1993, Vogelmann et al. 1993, Carter 1994, McMurtrey et al. 1994, Smith et al. 1995, Gitelson and Merzlyak 1996, Lichtenthaler et al. 1996, Blackburn 1998a,b, Datt 1998, 1999a,b, Carter and Knapp 2001, Zarco-Tejada et al. 2003a,b), normalized difference (ND) (Gitelson and Merzlyak 1994, Blackburn 1998a,b, Gandia et al. 2005, le Maire et al. 2008) or double differences (DDn) (le Maire et al. 2008).

For original reflectance-based hyperspectral indices, their formulae can be shown as:

$$R(\lambda_1) = R_{\lambda_1}$$

$$D(\lambda_1, \lambda_2) = R_{\lambda_1} - R_{\lambda_2}$$

$$SR(\lambda_1, \lambda_2) = R_{\lambda_1} / R_{\lambda_2}$$

$$ND(\lambda_1, \lambda_2) = \left(R_{\lambda_1} - R_{\lambda_2}\right) / \left(R_{\lambda_1} + R_{\lambda_2}\right)$$

$$DDn(\lambda_1, \Delta\lambda) = 2R_{\lambda_1} - R_{\lambda_1 - \Delta\lambda} - R_{\lambda_1 + \Delta\lambda} \qquad (2.1)$$

For derivative spectra-based indices, another type of inverse reflectance differences (IDs) such as the anthocyanin reflectance index (ARI) or carotenoid reflectance index (CRI) (Gitelson et al. 2001, 2002) have been proposed as well.

$$dR(\lambda_1) = dR_{\lambda_1}$$

$$dD(\lambda_1, \lambda_2) = dR_{\lambda_1} - dR_{\lambda_2}$$

$$dSR(\lambda_1, \lambda_2) = dR_{\lambda_1}/dR_{\lambda_2}$$

$$dND(\lambda_1, \lambda_2) = \left(dR_{\lambda_1} - dR_{\lambda_2}\right)/\left(dR_{\lambda_1} + dR_{\lambda_2}\right)$$

$$dDDn(\lambda_1, \Delta\lambda) = 2dR_{\lambda_1} - dR_{\lambda_1 - \Delta\lambda} - dR_{\lambda_1 + \Delta\lambda}$$

$$dID(\lambda_1, \lambda_2) = 1/dR_{\lambda_1} - 1/dR_{\lambda_2} \tag{2.2}$$

where dR is the derivative spectra and the suffixes (λ_1 or λ_2) are wavelength (nm).

All derivative spectra-based hyperspectral vegetation indices used in this study are listed in Table 2A.1.

2.2.2 Spectral Transformations and Derivative Spectra at Different Orders

Reflectance is the most commonly used spectral mode in hyperspectral remote sensing, but reflectance is not always effective in reducing noise (Tsai and Philpot 1998). In addition to original reflectance, various transformations such as derivative spectrum, shape features, and pseudo absorbance (Log 1/R) from reflectance spectra have been demonstrated as powerful options for quantifying plant pigments and/or physical parameters (Yoder and Pettigrew-Crosby 1995, Blackburn 1998a, Madeira et al. 2000, Imanishi et al. 2004, Kochubey and Kazantsev 2012, Yao et al. 2014). Among various transformations, derivative spectrum analysis has the advantage of reducing background effects and has been widely used for detecting plant biophysical and biochemical parameters from hyperspectral data (Imanishi et al. 2004).

The process of finding derivatives is known as differentiation (Hoffman and Frankel 2001). One simple method to calculate derivatives is finite divided difference approximation (Chapra and Canale 1988, Tsai and Philpot 1998). The first-order derivative is calculated as:

$$d = \left.\frac{ds}{d\lambda}\right|_i \approx \frac{s(\lambda_i) - s(\lambda_j)}{\Delta\lambda} \tag{2.3}$$

where $s(\lambda_i)$ and $s(\lambda_j)$ are the values of the spectrum at wavelengths λ_i and λ_j, respectively, and $\Delta\lambda$ is the wavelength increment between λ_i and λ_j.

Higher-order derivatives can then be calculated from lower-order derivatives iteratively and can be expressed as:

$$\left.\frac{d^n s}{d\lambda^n}\right|_j = \left.\frac{d}{d\lambda}\left(\frac{d^{(n-1)}s}{d\lambda^{(n-1)}}\right)\right|_j \tag{2.4}$$

As derivatives are sensitive to noise, various smoothing methods, including Savitzky-Golay (Savitzky and Golay 1964) and Kawata-Minami smoothing (Kawata and Minami 1984), have been introduced into hyperspectral data analysis to minimize random noise (Tsai and Philpot 1998). On the other hand, smoothing involves fitting functions to the discrete data, and if the fitting is inaccurate, the following differentiation turns to be an inaccurate process (Hoffman and Frankel 2001).

2.2.3 DATA-MINING APPROACH (FRAMEWORK FOR DEVELOPING VEGETATION INDICES BASED ON DERIVATIVE SPECTRA)

The aforementioned five common types of indices based on the original reflected spectra, as well as Log(1/R) and the first- to sixth-order derivative spectra from the original reflectance data, are implemented in this study: (1) a simple reflectance (R) or derivative spectra (dR) index, (2) a simple ratio index, (3) a simple difference (SD) index, (4) a normalized difference index, and (5) a double-difference index (le Maire et al. 2004, 2008, Wang and Li 2012a,b, Jin and Wang 2016).

The ratio of performance to deviation (RPD, Equation 2.5), expressed as a ratio of the standard error in prediction to the standard deviation of the samples (Williams 1987), is an indicator for assessing the goodness of fit and has been applied to judge if the indices are applicable. The indices are then assigned to categories A (RPD > 2.0), B ($1.4 \leq$ RPD ≤ 2.0), and C (RPD < 1.4) (Chang et al. 2001). It has been claimed that category B can be improved by using different calibration strategies, but properties in category C may not be reliably predicted (Chang et al. 2001).

$$RPD = Sd/SEP \tag{2.5}$$

where SEP is standard error of prediction and Sd is standard deviation of the samples.

In the meantime, the widely applicable information criterion (WAIC) (Watanabe 2010), asymptotically equivalent to the leave-one-out cross-validation in the Bayes estimation, has been calculated in addition to the root mean square errors (RMSEs) and the coefficient of determination (R^2) for assessment. The selection is made by the criterion of WAIC.

2.3 PROS AND CONS OF DERIVATIVE HYPERSPECTRAL VEGETATION INDICES FOR QUANTIFYING FOREST BIOPHYSICAL AND BIOCHEMICAL PROPERTIES: AN ANALYSIS THROUGH VIRTUAL EXPERIMENTS BASED ON LEAF- AND CANOPY-SCALE RADIATIVE TRANSFER MODELS

In order to reveal the inherent advantage of using dHVIs, we designed four virtual experiments based on RTMs at different scales. Among them, three are based on a widely applied leaf-scale RTM model (PROSPECT-4) by adding a dummy input variable (C_{dummy}) with convertible specific absorption coefficients, while the remaining one is based on the canopy-scale multiple-layer canopy radiative transfer model (MRTM). At the leaf scale, we examined different performances of dHVIs for retrieving this dummy variable with either weak or strong absorption coefficients under the background of chlorophyll, as well as the robustness of dHVIs to estimate leaf chlorophyll content under the effects of unknown biochemistry with different absorption intensities and central wavelengths, while the canopy-scale experiment focused on examining the performance of indices to estimate LAI under various conditions.

2.3.1 LEAF-SCALE EXPERIMENTS BASED ON THE MODIFIED PROSPECT-4

2.3.1.1 Modified PROSPECT-4 and Experimental Designs

PROSPECT is a radiative transfer model for simulating leaf directional-hemispherical reflectance and transmittance from 400 to 2500 nm (Jacquemoud and Baret 1990, Feret et al. 2008). In its PROSPECT-4 version, the reflectance is calculated as a function of leaf structure index (N), leaf chlorophyll content (C_{ab}, μg/cm^2), leaf water content (C_w, g/cm^2), leaf mass area (C_m, g/cm^2), and the specific absorption coefficients K of each component (le Maire et al. 2004). The total absorption coefficient at wavelength λ [k(λ)] for one layer is estimated from:

$$k(\lambda) = \frac{C_{ab}}{N} \cdot K_{ab}(\lambda) + \frac{C_w}{N} \cdot K_w(\lambda) + \frac{C_m}{N} \cdot K_m(\lambda) \tag{2.6}$$

where $K_{ab}(\lambda)$, $K_w(\lambda)$, and $K_m(\lambda)$ are the specific absorption coefficients at wavelength λ of chlorophyll a + b, water, and dry matter, respectively.

In this study, we add a dummy variable (C_{dummy}) with convertible specific absorption coefficients into the original PROSPECT-4 so that we can compare the performances of VIs based on various artificially generated cases. The specific absorption coefficients of dummy variables (K_{dummy}) were generated using the Gaussian function:

$$K_{dummy}(\lambda) = a \cdot e^{-\frac{(\lambda-b)^2}{2c^2}} \qquad (2.7)$$

where a is the height of the absorption peak, b is the wavelength of the absorption peak, and c is the standard deviation, which controls the width of absorption bands. Thus, the total absorption coefficient at wavelength λ [$k(\lambda)$] for one layer in our modified PROSPECT-4 is expressed as:

$$k(\lambda) = \frac{C_{ab}}{N} \cdot K_{ab}(\lambda) + \frac{C_w}{N} \cdot K_w(\lambda) + \frac{C_m}{N} \cdot K_m(\lambda) + \frac{C_{dummy}}{N} \cdot K_{dummy}(\lambda) \qquad (2.8)$$

We aimed to reveal the differential performance of dHVIs from three aspects: (1) their abilities in tracing a new biochemical with weak absorption, where weak absorption is somewhat compared with that of chlorophyll, a main spectral shape determiner in the visible domain; (2) their abilities in tracing a new biochemical with comparable absorption with that of chlorophyll; and (3) their robustness in deriving chlorophyll under various contents of unknown (dummy) biochemical with gentle or comparable absorptions. Different sets of experiments were designed accordingly:

Experiment #1

In this set of experiments, an artificially added dummy variable with different specific absorption coefficients was introduced into the modified PROSPECT-4. Two respective series of specific absorption coefficients with different central absorption wavelengths (moving from 400 to 800 nm with a step of 10 nm) and with different absorption peak values of 0.03 cm²/μg or 0.04 cm²/μg (determined by a in the Gaussian function previously presented) were examined. The central wavelength of the absorption peak (b in the Gaussian function) shifted from 400 to 800 nm at a step of 10 nm, and Figure 2.1 illustrates only two cases when the central wavelength was located at 660 nm and 690 nm, respectively. The width of the "bells," which are controlled by c in the Gaussian function, was set to10 consistently. For each case, 1000 simulations were conducted.

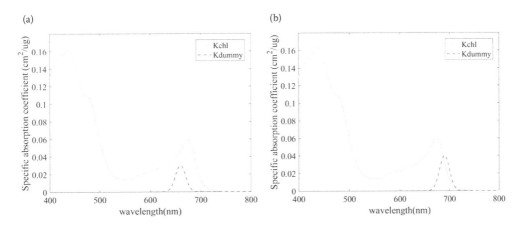

FIGURE 2.1 Specific absorption coefficients of chlorophylls (green solid lines) in PROSPECT-4 and artificially added dummy variable (blue dashed lines).

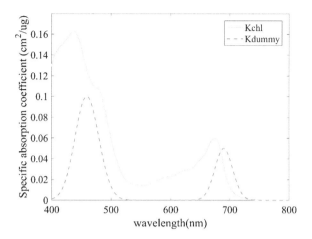

FIGURE 2.2   Specific absorption coefficients of chlorophylls (green solid line) in PROSPECT-4 and artificially added dummy variable (blue dashed line) in Experiment #2.

Experiment #2

This set of experiments focused on examining the performances of different spectra indices to identify a new biochemical with similar specific absorption coefficients (in the context of both intensity and pattern) of chlorophyll. In the experiments, we defined the absorption coefficients of the dummy variable having similar peak features as chlorophylls, as shown in Figure 2.2. The absorption peaks of the dummy variable were set to 460 nm and 690 nm, near the absorption peaks of chlorophyll, with the peak values of 0.10 $cm^2/\mu g$ and 0.05 $cm^2/\mu g$, respectively.

Experiment #3

Derivative analysis has the advantage of reducing additive constants and minimizing background effects (Curran et al. 1990, Imanishi et al. 2004). To address the point, we designed the third set of experiments to reveal the robustness of dHVIs in retrieving chlorophyll under various interfering conditions with different varying absorption dummy biochemicals.

The specific absorption coefficients of chlorophyll and the artificially added dummy variable involved in this set of experiments are shown in Figure 2.3.

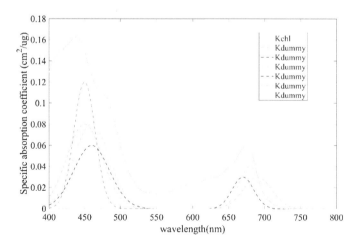

FIGURE 2.3   Specific absorption coefficients of chlorophylls (green solid line) in PROSPECT-4 and artificially added dummy variables (dashed lines) in Experiment #3.

As the refractive index and the absorption coefficient were smoothed in PROSPECT (le Maire et al. 2004, Feret et al. 2008), the simulated reflectance spectra were inherently different from field-measured ones. To simulate the radiometric observation noise, we also added a random Gaussian noise to each spectrum at each wavelength with a standard deviation of 1% of simulated reflectance values in this experiment (le Maire et al. 2008). The Savitzky-Golay smoothing method (Savitzky and Golay 1964) was applied to simulated reflectance values to calculate accurate derivatives. For each spectrum with 1 nm spectral resolution, the size of the moving window was set to 17, and the fit polynomial order was 4 (le Maire et al. 2004).

2.3.1.2 Simulation Datasets

For all sets of experiments, we generated simulation datasets using the modified PROSPECT-4. There are two common sampling strategies for parameterization. The first sampling strategy is to define a range of variation and the number of sampling levels for each input parameter (le Maire et al. 2008) and is helpful to study the responses of reflectance or derivative spectra to input parameters. The second sampling strategy, however, takes the actual correlations between leaf constituents into account and was first proposed by Feret et al. (2011). In this study, the second parameterization strategy was adopted, in which the covariations between leaf constituents were calculated based on the ANGERS leaf optical properties database (June 2003) (Feret et al. 2008). For each combination, the artificially added dummy input variable (C_{dummy}) was randomly sampled between 0 to 0.8 times chlorophyll a + b content ($0.8 \times C_{ab}$). The statistical distribution of inputting parameters based on the second sampling strategy is shown in Figure 2.4. This method could avoid unrealistic combinations of leaf constituents.

The reflectance spectrum was then computed for each combination with the modified PROSPECT-4. A typical set of generated spectra are illustrated in Figure 2.5, in which 100 reflectance spectra are presented as examples. These 100 reflectance curves were generated when N was set to 1 with 100 uniformly distributed C_{ab} varying from 20 to 100 µg/cm². C_w and C_m were fixed to be 0.0116 and 0.0052 g/cm², respectively, which are the average values of all samples in the ANGERS leaf optical properties database.

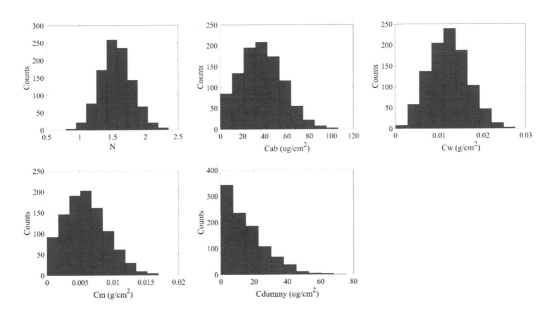

FIGURE 2.4 Distributions of model input parameters (N: leaf structure parameter, C_{ab}: chlorophyll a + b content, C_w: equivalent water thickness, C_m: dry matter content, C_{dummy}: artificially added dummy input variable content) based on the second sampling strategy for parameterizing.

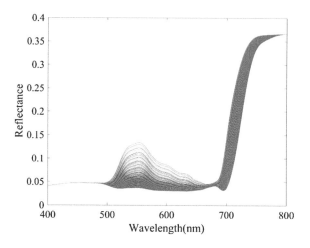

FIGURE 2.5 Reflectance simulated by modified PROSPECT-4 with the N = 1 and 100 uniformly sampled C_{ab} between 20 and 100 µg/cm².

Different orders of derivative spectra were then calculated from the reflectance values, as shown in Figure 2.6. It clearly indicates that the first- to third-order derivative spectra are very sensitive to the change of chlorophyll around the wavelengths of 550 and 700 nm, since the derivative spectra of different samples changed significantly.

2.3.2 CANOPY SCALE EXPERIMENT WITH MULTIPLE-LAYER CANOPY RADIATIVE TRANSFER MODEL

Experiment #4

Although le Maire et al. (2004) suggested that derivative-based indices are not necessary for leaf-level study, they claimed that derivative-based indices may perform significantly better in a canopy-level study. To verify the performance of dHVIs at the canopy scale, we designed and conducted a simulation experiment using a multiple-layer canopy radiative transfer model (Wang and Li 2013).

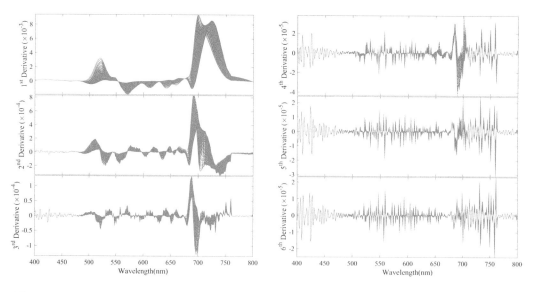

FIGURE 2.6 Derivative spectra calculated from reflectance simulated by modified PROSPECT-4 with the N = 1 and 100 uniformly sampled C_{ab} between 20 and 100 µg/cm².

In this experiment, we focused on the performances of indices in estimating LAI, and five input parameters, including LAI and soil parameters (s1, s2, s3, s4), were sampled using the Latin hypercube sampling scheme (Benoist et al. 1994, Feret et al. 2011) for generating a simulation database. This method can maintain the statistical properties of these five parameters and decrease the number of simulations (Feret et al. 2011). The range of LAI was from 0.01 to 6.00. The minimum values of soil parameters s1, s2, s3, and s4 were set to 0.05, -0.10, -0.05, and -0.04, respectively, while the maximum values of s1, s2, s3, and s4 were 0.40, 0.10, 0.05, and 0.04, respectively. Each parameter was equally distributed within the range, and a total of 1000 combinations were generated in this experiment.

2.3.3 Pros and Cons of Derivative Hyperspectral Vegetation Indices

The best indices were then identified following the framework as illustrated in Section 2.3. The performances of dHVIs based on spectra of different derivative orders, as well as those based on reflectance, Log(1/R), or the first- to sixth-order derivatives of reflectance, are examined in detail.

2.3.3.1 Derivative Hyperspectral Vegetation Indices to Trace a New Biochemical with Gentle Absorptions: Analysis of Experiment #1 Results

This set of experiments contains the cases of a dummy biochemical with different central absorption peaks and intensities. Two cases are illustrated in Figure 2.1. The central absorption peaks in Cases 1 (Figure 2.1a) and 2 (Figure 2.1b) were 660 and 690 nm, with different peak values of 0.03 and 0.04 cm²/µg, respectively. For each case, 1000 simulations were conducted using the modified PROSPECT-4.

The derivatives of these two cases are shown in Figure 2.7. Results clearly indicate that the first derivatives successfully identified the absorption peaks if we use derivative analysis only.

As a further step, we examined the relationships of the content of the dummy variable with different transformation-based HVIs. Again, the above two cases were used as examples here. Table 2.1 presents the performances of different types of HVIs developed from different transformations of spectra. In general, there were significant relationships (<0.01) between the content of the dummy variable with different types of HVIs, ignorant of transformations. However, it clearly shows that dHVIs based on higher orders do not show any superiority over the first-order dHVIs, the Log (1/R)-transformed, or the original reflectance-based HVIs. Also, the performances of all HVIs are apparently different when the central absorption peak of the dummy variable is different.

2.3.3.2 Derivative Hyperspectral Vegetation Indices to Trace a New Biochemical with Strong Absorptions: Analysis of Experiment #2 Results

In this experiment, 1000 combinations of input parameters were generated using a Latin hypercube sampling method and led to 1000 simulations using the modified PROSPECT-4. Results suggest that the content of an artificially added dummy variable (C_{dummy}) can be captured using hyperspectral indices. For each type of index, Log(1/R) and derivative spectra-based indices were more effective than the original reflectance-based indices. Moreover, the best type of index for tracing C_{dummy} among the five types was found to be the double-difference index, which was linked to the "peak jump" feature in the red-edge region when chlorophyll varies (le Maire et al. 2004). As the specific absorption coefficients defined in this experiment were quite similar to those of chlorophyll, the DDn type of index also has higher correlations with C_{dummy} (Table 2.2).

In Figure 2.8, we further present the differences of R^2 between the first-order derivative spectra-based dR indices and the original reflectance-based R indices at each wavelength. Results show that the first-order derivative spectra-based dR indices showed great improvement around 715 nm. The absorption coefficients of chlorophyll and the artificially added dummy variable were all below 0.02 cm²/µg at 715 nm, but with different slopes. The results indicate that the first-order derivative spectra can be a potential tool to differentiate parameters with low absorption coefficients but different structural features.

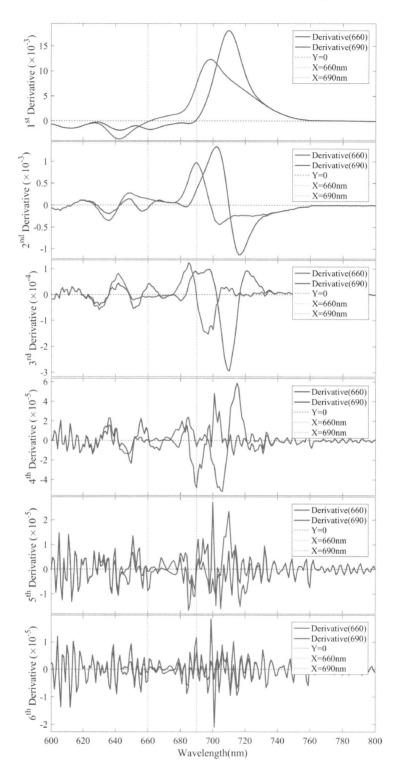

FIGURE 2.7 The first- to sixth-order derivatives of synthetic spectra. The blue solid lines are results for specific absorption coefficients of artificially added dummy variable in Figure 2.1a (left panel). The red solid lines are results for specific absorption coefficients of artificially added dummy variable in Figure 2.1b (right panel).

TABLE 2.1
Results of Different Types of Spectra-Based Indices for Dummy Variable (C_{dummy}) from Experiment #1

Indices		$\lambda 1$	$\lambda 2$ (or Δ)	R^2	RMSE	P	AICc	$\lambda 1$	$\lambda 2$ (or Δ)	R^2	RMSE	P	AICc
			Case #1 (Central Absorption Peak = 660 nm)						Case #2 (Central Absorption Peak = 690 nm)				
Reflectance	R	660		0.02	3.97	<0.01	3.76	700		0.13	3.74	<0.01	3.64
	D	655	665	0.05	3.91	<0.01	3.73	710	720	0.56	2.65	<0.01	2.95
	SR	515	640	0.41	3.07	<0.01	3.25	705	710	0.56	2.67	<0.01	2.97
	ND	515	640	0.42	3.04	<0.01	3.23	705	710	0.57	2.64	<0.01	2.95
	DDn	645	15	0.18	3.62	<0.01	3.58	705	5	0.85	1.56	<0.01	1.90
Log(1/R)	R	660		0.02	3.97	<0.01	3.76	700		0.13	3.73	<0.01	3.64
	D	515	640	0.42	3.05	<0.01	3.23	705	710	0.57	2.63	<0.01	2.94
	SR	660	665	0.10	3.81	<0.01	3.68	710	715	0.83	1.67	<0.01	2.03
	ND	660	665	0.10	3.80	<0.01	3.68	710	715	0.83	1.67	<0.01	2.03
	DDn	645	35	0.30	3.36	<0.01	3.43	720	15	0.79	1.85	<0.01	2.24
1st derivative	dR	660		0.05	3.91	<0.01	3.73	715		0.56	2.65	<0.01	2.95
	dD	620	660	0.43	3.03	<0.01	3.22	700	710	0.81	1.75	<0.01	2.13
	dSR	445	670	0.22	3.54	<0.01	3.54	700	705	0.55	2.67	<0.01	2.97
	dND	445	670	0.25	3.47	<0.01	3.49	700	705	0.55	2.69	<0.01	2.99
	dDDn	660	20	0.28	3.40	<0.01	3.45	715	15	0.83	1.67	<0.01	2.03
2nd derivative	dR	645		0.15	3.70	<0.01	3.62	705		0.81	1.75	<0.01	2.12
	dD	450	645	0.18	3.64	<0.01	3.59	640	705	0.85	1.57	<0.01	1.90
	dSR	650	665	0.38	3.17	<0.01	3.31	695	700	0.61	2.51	<0.01	2.84
	dND	655	665	0.34	3.25	<0.01	3.36	695	700	0.64	2.41	<0.01	2.77
	dDDn	645	100	0.22	3.54	<0.01	3.53	630	75	0.84	1.62	<0.01	1.98
3rd derivative	dR	660		0.13	3.74	<0.01	3.64	715		0.79	1.82	<0.01	2.20
	dD	635	660	0.34	3.25	<0.01	3.36	475	715	0.85	1.53	<0.01	1.86
	dSR	450	620	0.15	3.70	<0.01	3.62	685	710	0.58	2.58	<0.01	2.90
	dND	600	660	0.32	3.30	<0.01	3.39	705	710	0.53	2.75	<0.01	3.03
	dDDn	635	30	0.33	3.29	<0.01	3.39	585	130	0.85	1.54	<0.01	1.87
4th derivative	dR	645		0.11	3.78	<0.01	3.66	705		0.79	1.85	<0.01	2.24
	dD	645	740	0.16	3.68	<0.01	3.61	705	740	0.84	1.61	<0.01	1.96
	dSR	615	630	0.29	3.38	<0.01	3.44	695	700	0.79	1.84	<0.01	2.22
	dND	585	630	0.38	3.17	<0.01	3.31	695	700	0.81	1.73	<0.01	2.11
	dDDn	645	95	0.16	3.68	<0.01	3.61	705	30	0.83	1.67	<0.01	2.04
5th derivative	dR	655		0.04	3.93	<0.01	3.74	700		0.82	1.71	<0.01	2.07
	dD	625	635	0.13	3.73	<0.01	3.64	600	700	0.84	1.60	<0.01	1.94
	dSR	615	625	0.16	3.66	<0.01	3.60	705	710	0.67	2.29	<0.01	2.66
	dND	615	625	0.24	3.50	<0.01	3.51	705	710	0.71	2.14	<0.01	2.53
	dDDn	620	20	0.08	3.85	<0.01	3.70	635	65	0.84	1.62	<0.01	1.97
6th derivative	dR	630		0.13	3.73	<0.01	3.63	720		0.75	1.99	<0.01	2.38
	dD	565	630	0.18	3.62	<0.01	3.58	620	695	0.84	1.58	<0.01	1.92
	dSR	595	635	0.08	3.83	<0.01	3.69	695	700	0.77	1.91	<0.01	2.30
	dND	560	665	0.15	3.68	<0.01	3.61	695	705	0.83	1.63	<0.01	1.98
	dDDn	630	105	0.18	3.64	<0.01	3.59	720	10	0.82	1.69	<0.01	2.06

TABLE 2.2

Results of Different Types of Spectra-Based Indices for Dummy Variable (C_{dummy}) from Experiment #2

	Indices	$\lambda 1$	$\lambda 2$(or Δ)	R^2	RMSE ($\mu g/cm^2$)	P	AICc
Reflectance	R	710		0.39	6.76	<0.01	4.83
	D	515	710	0.64	5.17	<0.01	4.29
	SR	715	720	0.70	4.77	<0.01	4.13
	ND	715	720	0.71	4.67	<0.01	4.09
	DDn	715	5	0.84	3.48	<0.01	3.50
Log(1/R)	R	705		0.43	6.55	<0.01	4.76
	D	715	720	0.71	4.66	<0.01	4.08
	SR	720	725	0.78	4.04	<0.01	3.80
	ND	720	725	0.77	4.13	<0.01	3.84
	DDn	725	5	0.87	3.18	<0.01	3.32
1st derivative	dR	725		0.61	5.37	<0.01	4.37
	dD	710	715	0.83	3.55	<0.01	3.54
	dSR	555	705	0.75	4.37	<0.01	3.95
	dND	535	700	0.76	4.22	<0.01	3.89
	dDDn	705	5	0.92	2.38	<0.01	2.74
2nd derivative	dR	715		0.81	3.74	<0.01	3.64
	dD	530	715	0.91	2.56	<0.01	2.88
	dSR	695	735	0.80	3.83	<0.01	3.69
	dND	580	690	0.86	3.27	<0.01	3.37
	dDDn	715	10	0.94	2.14	<0.01	2.53
3rd derivative	dR	705		0.90	2.68	<0.01	2.97
	dD	710	720	0.94	2.14	<0.01	2.52
	dSR	705	745	0.77	4.20	<0.01	3.88
	dND	685	745	0.86	3.23	<0.01	3.35
	dDDn	555	150	0.94	2.18	<0.01	2.57
4th derivative	dR	715		0.94	2.14	<0.01	2.52
	dD	470	715	0.94	2.04	<0.01	2.44
	dSR	570	730	0.55	5.78	<0.01	4.51
	dND	695	710	0.72	4.60	<0.01	4.06
	dDDn	715	20	0.95	1.95	<0.01	2.34
5th derivative	dR	725		0.92	2.45	<0.01	2.80
	dD	705	725	0.95	1.93	<0.01	2.32
	dSR	580	685	0.76	4.20	<0.01	3.88
	dND	685	740	0.84	3.46	<0.01	3.49
	dDDn	725	20	0.95	1.84	<0.01	2.23
6th derivative	dR	750		0.92	2.51	<0.01	2.84
	dD	715	735	0.93	2.29	<0.01	2.66
	dSR	585	730	0.58	5.61	<0.01	4.46
	dND	680	730	0.88	3.03	<0.01	3.22
	dDDn	715	20	0.94	2.04	<0.01	2.43

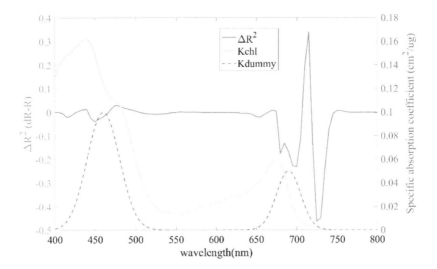

FIGURE 2.8 Improvement of the first-order derivative spectra-based dR index compared with the original reflectance-based R index for tracing an artificially added dummy variable (C_{dummy}). The blue solid line shows the difference of R^2 between dR index and R index.

2.3.3.3 Robustness of Derivative Hyperspectral Vegetation Indices to Estimate Chlorophyll: Analysis of Experiment #3 Results

Indices calibrated to the simulated database in Experiment #3 are shown in Table 2.3. Although indices with high accuracies to trace the content of chlorophyll based on different types of spectra have been identified differently, band combinations of these best indices are clustered near the wavelength domains of 570 and 750 nm. The artificially added dummy variable should have only marginal effects on chlorophyll estimation indices using these two bands, since the specific absorption coefficients of the dummy variable in this experiment are close to 0, as shown in Figure 2.3.

The best index among all identified best indices was found with the D type index based on Log (1/R)-transformed spectra, followed by the dND (520, 745) of the first-order derivative spectra. Different bands were used for the identified best indices of different types of spectra.

To show the strength of derivative analysis, the differences of R^2 were calculated among various dR-type indices based on the first- to sixth-order derivatives and reflectance-based R-type index at each wavelength (at a spectral resolution of 5 nm). The first-order derivative-based dR index was more effective ($\Delta R^2 > 0.30$) than the reflectance-based index around 720 nm. This result was also achieved in Experiment #1, indicating that first-order derivative spectra are more effective in differentiating parameters with low absorption coefficients but having different structural features than the original reflectance. Another noticeable improvement was around 675 nm ($\Delta R^2 \approx 0.15$), which is consistent with the absorption peak of chlorophyll.

The most significant improvement of the second-order derivative-based dR index was found near 645 nm ($\Delta R^2 = 0.54$ at 645 nm), followed by $\Delta R^2 = 0.43$ at 515 nm. It can be easily recognized that the specific absorption coefficients of chlorophyll are higher than those of artificially added dummy variables. The second-order derivative spectra could hence capture the signal of targeted parameters even under noisy background conditions.

The third- to sixth order-derivatives had similar levels of performance, as shown in Figure 2.9c through f. Significant improvements were recognized between 500–570 and 690–730 nm.

The results in Table 2.3 also show that the ND type of index was more effective among these five types of index. To identify the strength of derivative-based indices, we have calculated and illustrated the differences of R^2 between the derivative-based dND indices and the original reflectance-based index with each combination of wavebands (shown in Figure 2.10).

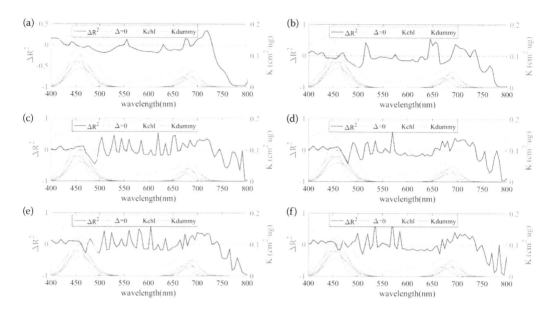

FIGURE 2.9 Improvements of dR indices based on the first- to sixth- (panels a to f) order derivative spectra over the original reflectance-based R indices for tracing chlorophyll content in synthetic database generated in Experiment #3. The blue solid lines show the differences of R^2 between the first- to sixth-order derivative-based dR type indices and reflectance-based R type indices.

Derivative-based dND indices were much more effective (those areas marked in red) than the reflectance-based ND index under band combinations of around 680 nm and between 520 and 650 nm. As a comparison, the reflectance-based ND index performed better when the two bands were both around 600 nm. Overall, about 31% of the first-order derivative-based indices were greatly improved ($\Delta R^2 > 0.2$). Similarly, over 44% of the second- to sixth-order derivative-based indices had higher R^2 values than reflectance-based ones, among which over 35% were greatly improved ($\Delta R^2 > 0.2$).

2.3.3.4   Derivative Hyperspectral Vegetation Indices to Trace Canopy Leaf Area Index: Analysis of Experiment #4 Results

The best indices based on different types of spectra for canopy LAI calibrated to the simulation dataset generated in Experiment #4 are shown in Table 2.4. The indices based on derivatives, especially R, D, SR, and ND types, had significant improvement in tracing canopy LAI over reflectance-based ones. Although the highest coefficient of determination (R^2) of the R-type indices based on reflectance reached 0.77 at 915 nm, its counterpart of the simple derivative spectra (dR) index had better correlation with LAI ($R^2 = 0.88$ vs. 0.77). The coefficients of determination of the second dR indices were even higher than 0.92. Similarly, the D, SR, and ND types of dHVIs all had significant improvement over their reflectance-based counterparts for estimating canopy LAI. Furthermore, higher-orders dHVIs had higher correlations. On the other hand, the DDn type of dHVIs only had marginal improvement (ΔR^2 between 0.02 and 0.04) for tracing LAI over reflectance-based ones.

2.3.3.5   Sensitivities of Derivative Hyperspectral Vegetation Indices to Noise Levels and Smoothing Methods

The best indices calibrated to the database generated in Experiment #3 with random Gaussian noise were examined for their sensitivities to different noise levels and smoothing methods. One sample reflectance with 1% random Gaussian noise and smoothed reflectance with the Savitzky-Golay method is shown in Figure 2.11 as an example. The Savitzky-Golay smoothing method with a window size of 17 and fourth-degree polynomial can significantly decrease the noise effect.

TABLE 2.3

Results of Different Types of Spectra-Based Indices for Chlorophyll (C$_{chl}$) for Simulation Dataset Generated in the Set of Experiment #3 (without Random Gaussian Noise)

	Indices	λ1	λ2(or Δ)	R^2	RMSE (μg /cm^2)	P	AICc
Reflectance	R	555		0.73	15.02	<0.01	6.42
	D	585	595	0.82	12.36	<0.01	6.03
	SR	400	610	0.85	11.17	<0.01	5.83
	ND	555	745	0.87	10.56	<0.01	5.72
	DDn	570	25	0.82	12.40	<0.01	6.04
Log(1/R)	R	565		0.83	11.97	<0.01	5.96
	D	555	745	0.92	8.08	<0.01	5.18
	SR	575	670	0.84	11.63	<0.01	5.91
	ND	510	520	0.84	11.68	<0.01	5.92
	DDn	570	170	0.88	10.09	<0.01	5.62
1st derivative	dR	590		0.82	12.36	<0.01	6.03
	dD	520	745	0.84	11.46	<0.01	5.88
	dSR	575	595	0.87	10.23	<0.01	5.65
	dND	520	745	0.88	10.20	<0.01	5.65
	dDDn	575	5	0.82	12.13	<0.01	5.99
2nd derivative	dR	590		0.81	12.47	<0.01	6.05
	dD	570	585	0.82	12.16	<0.01	6.00
	dSR	575	580	0.85	11.07	<0.01	5.81
	dND	580	595	0.86	10.66	<0.01	5.73
	dDDn	585	10	0.82	12.23	<0.01	6.01
3rd derivative	dR	575		0.82	12.29	<0.01	6.02
	dD	520	580	0.83	12.04	<0.01	5.98
	dSR	570	590	0.87	10.21	<0.01	5.65
	dND	560	580	0.86	10.67	<0.01	5.74
	dDDn	540	20	0.83	12.07	<0.01	5.98
4th derivative	dR	585		0.82	12.23	<0.01	6.01
	dD	530	555	0.83	11.82	<0.01	5.94
	dSR	570	575	0.85	11.34	<0.01	5.86
	dND	570	590	0.86	10.79	<0.01	5.76
	dDDn	580	5	0.84	11.50	<0.01	5.89
5th derivative	dR	575		0.82	12.10	<0.01	5.99
	dD	520	580	0.83	11.87	<0.01	5.95
	dSR	520	565	0.85	11.14	<0.01	5.82
	dND	565	590	0.87	10.51	<0.01	5.70
	dDDn	560	55	0.85	11.33	<0.01	5.86
6th derivative	dR	510		0.83	11.99	<0.01	5.97
	dD	515	520	0.83	11.78	<0.01	5.93
	dSR	515	570	0.86	10.76	<0.01	5.75
	dND	585	600	0.86	10.70	<0.01	5.74
	dDDn	570	55	0.84	11.67	<0.01	5.91

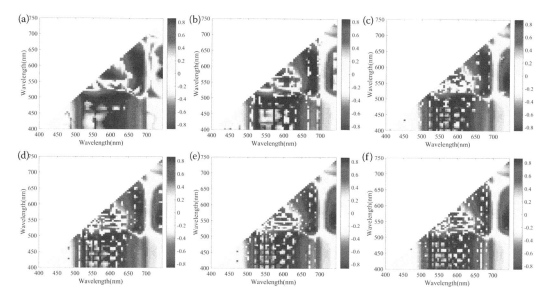

FIGURE 2.10 Improvements of the first- to sixth- (panel a to f) order derivative-based dND index compared with the original reflectance based ND index for tracing chlorophyll content in the synthetic database generated in Experiment #3.

In addition to the popular Savitzky-Golay method used in this study, we also examined another three smoothing methods. Methods include moving average, local regression using weighted linear least squares and a first-degree polynomial model (LOWESS), and local regression using weighted linear least squares and a second-degree polynomial model (LOESS), all for revealing the sensitivities of dHVIs to different smoothing methods.

Derivative analysis results showed that derivatives computed from the original reflectance with 1% Gaussian noise fluctuated sharply throughout the wavelength, and smoothing can greatly reduce the abrupt spectral variations. Smoothing results also showed that the moving average method (with a filter window size of 17) obtained the strongest smoothing effect among these four methods, but it also greatly decreased the accuracy on the other hand, especially around the peaks and troughs of the spectrum. The LOWESS method showed a higher accuracy than the moving average method, but remained unchanged with results around peaks and troughs. Additionally, the Savitzky-Golay and LOESS smoothing methods achieved quite similar results, and these two methods could reduce random noise and preserve details of the spectra.

To investigate the sensitivity of indices to different smooth methods, the performances of indices with all combinations generated from different smoothed spectra to predict chlorophyll were verified. Figure 2.12 illustrates the R^2 between chlorophyll and indices of the ND type based on the first-order derivatives of different spectra, including the original reflectance (Figure 2.12a) simulated by modified PROSPECT-4 in Experiment #3, reflectance with 1% Gaussian noise (Figure 2.12b), and smoothed ones with four different smoothing methods (Figure 2.12c–f). Compared with dND indices based on the first-order derivative of original reflectance (Figure 2.12a), indices from derivatives of reflectance with random noise (Figure 2.12b) showed lower accuracy in chlorophyll estimation. Smoothing can remove random noise; thus, indices based on the first-order derivatives of smoothed reflectance (Figure 2.12c–f) achieved better performance than those based on unsmoothed spectra (Figure 2.12b). But as all these four smoothing methods could not remove noise and resolve real spectral features, the accuracy of these indices generated from smoothed spectra was still lower than indices of original simulated spectra with modified PROSPECT-4.

Only the first-order dHVIs showed marginal improvement for estimates of chlorophyll content relative to those based on reflectance. Although smoothing was conducted, indices based on

TABLE 2.4

Results of Different Types of Spectra-Based Indices for Canopy LAI for Simulation Dataset Generated in the Set of Experiment #4

	Indices	$\lambda 1$	$\lambda 2$(or Δ)	R^2	RMSE	P	AICc
Reflectance	R	915		0.77	0.83	<0.01	0.63
	D	745	750	0.88	0.60	<0.01	0.00
	SR	1895	2355	0.78	0.81	<0.01	0.58
	ND	1895	2360	0.78	0.82	<0.01	0.61
	DDn	940	10	0.94	0.41	<0.01	−0.78
Log(1/R)	R	920		0.59	1.10	<0.01	1.20
	D	1895	2360	0.78	0.82	<0.01	0.61
	SR	675	920	0.92	0.47	<0.01	−0.49
	ND	745	750	0.89	0.57	<0.01	−0.13
	DDn	1070	10	0.84	0.70	<0.01	0.29
1st derivative	dR	745		0.88	0.61	<0.01	0.02
	dD	935	945	0.94	0.41	<0.01	−0.79
	dSR	805	1830	0.94	0.44	<0.01	−0.66
	dND	525	935	0.98	0.25	<0.01	−1.76
	dDDn	970	35	0.96	0.34	<0.01	−1.14
2nd derivative	dR	940		0.94	0.41	<0.01	−0.79
	dD	645	780	0.98	0.27	<0.01	−1.62
	dSR	805	830	0.94	0.43	<0.01	−0.70
	dND	435	1220	0.98	0.23	<0.01	−1.94
	dDDn	935	15	0.97	0.29	<0.01	−1.48
3rd derivative	dR	745		0.94	0.41	<0.01	−0.78
	dD	1085	2295	0.98	0.23	<0.01	−1.97
	dSR	740	1085	0.97	0.29	<0.01	−1.44
	dND	505	1500	0.99	0.17	<0.01	−2.50
	dDDn	1520	805	0.98	0.27	<0.01	−1.63
4th derivative	dR	935		0.94	0.41	<0.01	−0.79
	dD	770	1025	0.98	0.26	<0.01	−1.69
	dSR	960	1650	0.94	0.42	<0.01	−0.72
	dND	505	2055	0.99	0.16	<0.01	−2.64
	dDDn	735	240	0.97	0.29	<0.01	−1.46
5th derivative	dR	945		0.97	0.30	<0.01	−1.42
	dD	515	695	0.98	0.24	<0.01	−1.84
	dSR	945	1335	0.96	0.36	<0.01	−1.05
	dND	780	1060	0.99	0.19	<0.01	−2.36
	dDDn	935	55	0.97	0.30	<0.01	−1.40
6th derivative	dR	815		0.92	0.48	<0.01	−0.46
	dD	925	985	0.97	0.29	<0.01	−1.46
	dSR	815	1170	0.98	0.23	<0.01	−1.93
	dND	495	2375	0.99	0.19	<0.01	−2.36
	dDDn	960	50	0.97	0.28	<0.01	−1.56

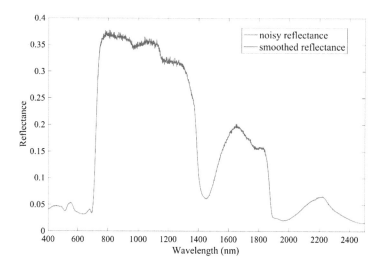

FIGURE 2.11   Noisy reflectance and smoothed reflectance with Savitzky-Golay method.

higher-order derivatives did not show improvements as great as those obtained in previous experiments. The analysis results on the second-order derivatives of different spectra clearly show that the indices based on the second-order derivatives are quite sensitive to random noise. Even if different smoothing methods have been applied to the spectra with 1% Gaussian noise, the generated indices are not very effective in tracing chlorophyll. Except for R-type indices, the other four index types based on third- to sixth-order derivative spectra have lower precision to

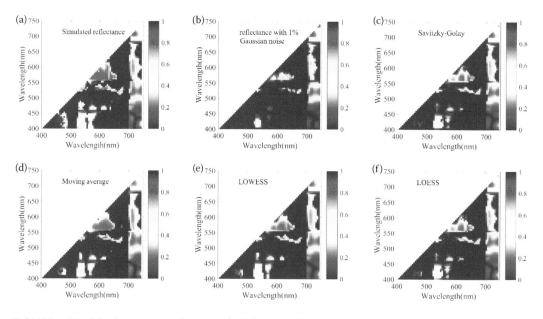

FIGURE 2.12   Matrices representing the R^2 of chlorophyll estimation with ND type of indices based on the first-order derivatives of different spectra. (a) dND indices based on the first-order derivative of original reflectance simulated in Experiment #3, (b) dND indices based on the first derivative of reflectance with 1% Gaussian noise, (c) dND indices based on the first derivative of smoothed spectra with Savitzky-Golay method, (d) dND indices based on the first derivative of smoothed spectra with moving average method, (e) dND indices based on the first derivative of smoothed spectra with LOWESS method, (f) dND indices based on the first derivative of smoothed spectra with LOESS method.

estimate chlorophyll content than reflectance-based ones, indicating higher-order derivatives are quite sensitive to noise and random variations in reflectance spectra (Singer and Geissler 1988, Cloutis 1996).

We also verified the impact of different noise levels on the performances of dHVIs. The 0.5% and 3% levels of Gaussian noise were introduced into the modeled original reflectance values generated in Experiment #3. Under 0.5% Gaussian noise, indices based on reflectance as well as the first- and second-order derivatives can trace the content of chlorophyll with good accuracy ($R^2 > 0.8$); the performances of the third and higher order derivatives were inferior for the same purpose. When 3% Gaussian noise was introduced into simulated reflectance values, only indices based on the first-order derivative showed commensurate accuracy with reflectance-based indices to trace chlorophyll ($R^2 \approx 0.8$). The R^2 values between chlorophyll and indices based on the second and higher order derivatives were less than 0.8.

2.4   VALIDATION OF DERIVATIVE HYPERSPECTRAL VEGETATION INDICES FOR QUANTIFYING BIOPHYSICAL AND BIOCHEMICAL PROPERTIES: APPLICATIONS IN BEECH FORESTS AND/OR OTHERS (LOPEX, ANGERS)

2.4.1   MEASURED DATASETS USED FOR VALIDATING AND DEVELOPING DERIVATIVE HYPERSPECTRAL VEGETATION INDICES

We collected four independently measured datasets in this study for evaluating different vegetation indices. They include two *in situ* measured datasets, one in Naeba consisting of measurements on beech forests (*Fagus crenata*) during the period from June 2007 to August 2013 in Mt. Naeba (Japan), and one in Nakagawane (Japan) comprising measurements from 29 deciduous species in 2014. In addition, two public online datasets, LOPEX (Hosgood et al. 1994) and ANGERS (Feret et al. 2008), were also used. Although there are more data from various vegetation types in LOPEX and ANGERS, we focused on deciduous species only.

The main characteristics of these datasets are summarized in Table 2.5. For LOPEX, five repetitions were made for each spectral measurement, corresponding to each physical and biological measurement (Hosgood et al. 1994); thus, the averaged reflectance was adopted for any further analysis in this study. For the measurements obtained in Naeba, chlorophyll a was estimated to vary from 10.36 to 45.49 μg/cm² with an average of 28.41 μg/cm², while chlorophyll b ranged from 3.53 to 17.18 μg/cm² with a mean of 9.38 μg/cm². In the same dataset, carotenoids varied from 2.44 to 12.37 μg/cm² with a mean of 7.13 μg/cm². The biophysical parameter of LMA covered a range of 0.00379–0.00905 g/cm². Except for a much wider range of LMA (from 0.00261 to 0.01152 g/cm²) in Nakagawane, other biochemical parameters (chlorophyll a, b and carotenoids) had similar statistical results. As a comparison, the two online datasets have nearly two times the standard deviations, with the contents ranging from 0.41 to 77.60 μg/cm², 0.30 to 29.88 μg/cm², 0 to 10.31 μg/cm², 0.0029 to 0.05249, and 0.00166 to 0.03310 for the three pigments, EWT, and LMA, respectively.

2.4.2   VALIDATION RESULTS OF REPORTED AND NEWLY DEVELOPED DERIVATIVE HYPERSPECTRAL VEGETATION INDICES

Here, both currently reported and newly developed dHVIs were validated for their potential in estimating forest biophysical and biochemical parameters. Both currently reported and developed dHVIs were evaluated using the aforementioned four independent datasets and, for newly developed dHVIs, that also included the dataset they were generated from. The results are categorized into two groups, one for biophysical and the other for biochemical, which are detailed below.

TABLE 2.5

Main Characteristics of Four Measured Datasets

		Naeba	Nakagawane	LOPEX	ANGERS
Number of samples		135	123	330	276
	Conifer	0	0	5	0
	Deciduous broadleaf	135	123	165	233
	Evergreen broadleaf	0	0	25	40
	Herbaceous	0	0	135	1
	Semi-evergreen broadleaf	0	0	0	2
Chlorophyll-a (μg/cm^2)	Mean value	28.41	28.02	35.84	25.44
	Maximum value	45.49	46.71	77.60	76.83
	Minimum value	10.36	13.79	0.90	0.41
	Standard deviation	7.86	6.18	13.18	16.10
	Median value	29.08	27.16	33.56	22.15
	Skewness	-0.15	0.66	0.50	0.95
	NA	0	0	10	0
Chlorophyll-b (μg/cm^2)	Mean value	9.38	11.27	11.44	8.44
	Maximum value	17.18	19.79	23.28	29.88
	Minimum value	3.53	4.79	0.47	0.30
	Standard deviation	2.53	2.63	4.34	5.68
	Median value	9.32	11.06	10.87	7.10
	Skewness	0.09	0.62	0.59	1.23
	NA	0	0	10	0
Carotenoids (μg/cm^2)	Mean value	7.13	7.28	10.31	8.66
	Maximum value	12.37	13.49	28.35	25.28
	Minimum value	2.44	3.63	3.45	0.00
	Standard deviation	1.79	1.70	4.22	5.07
	Median value	7.15	7.09	9.65	7.53
	Skewness	0.16	0.86	1.45	1.21
	NA	0	0	0	0
EWT (g/cm^2)	Mean value	0.00905	0.00989	0.01113	0.01162
	Maximum value	0.01629	0.01800	0.05249	0.03400
	Minimum value	0.00379	0.00400	0.00029	0.00439
	Standard deviation	0.00257	0.00218	0.00698	0.00486
	Median value	0.00917	0.01000	0.00949	0.01027
	Skewness	-0.16329	0.39955	2.34894	2.20636
	NA	22	0	0	0
LMA (g/cm^2)	Mean value	0.00723	0.00429	0.00530	0.00524
	Maximum value	0.01152	0.00863	0.01573	0.03310
	Minimum value	0.00261	0.00190	0.00171	0.00166
	Standard deviation	0.00262	0.00132	0.00247	0.00367
	Median value	0.00831	0.00429	0.00470	0.00441
	Skewness	-0.66528	0.74694	1.32289	3.73211
	NA	22	0	0	0

2.4.2.1 Biophysical Properties: Leaf Mass per Area

Leaf mass per area (LMA) is the ratio of a leaf dry mass to its surface area, and it is also known to be expressed as the reciprocal of specific leaf area (SLA). LMA is related to the life strategies of plants (Westoby et al. 2002) and is an important factor reflecting plant physiological processes ranging from light capture to growth rates (Poorter et al. 2006).

Despite its importance, we hardly found any reported dHVI suitable for it. Hence, we newly developed dHVIs following the methodology described in Section 2.2.3. Regression analysis was performed for all possible combinations of wavelengths for the index type based on the dataset in Naeba and then validated using all four datasets. Results revealed that in total, there were 28 indices applicable for three of the four datasets. In particular, the determination coefficient of 0.51 ($p < 0.001$, RMSE = 0.00156 g/cm^2, WAIC = −5080.25, RPD = 1.44) between dID (1408, 1869) and LMA, which had the best performance for all measured datasets, was confirmed (Figure 2.13).

For comparison, we also validated an HVI based on original reflected spectra, the NDLMA, which is expressed as = $(R_{2260} − R_{1490})/(R_{2260} + R_{1490})$ for assessing leaf-scale LMA (le Maire

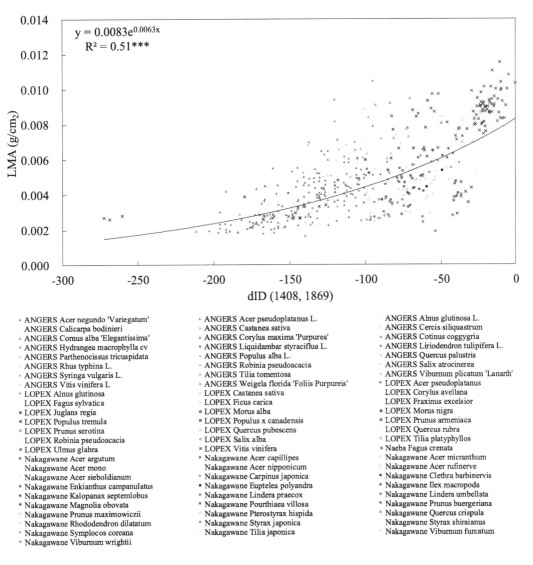

FIGURE 2.13   Relationship between dID (1408,1869) and LMA.

et al. 2008). Although significant correlations ($p < 0.001$) were also confirmed for all datasets, its performances were not well judged from RPD (all were categorized as "C"), except for Naeba dataset, for which it had an RPD of 2.81 and was categorized as "A."

2.4.2.2 Biochemical Properties: Total Chlorophyll and Carotenoid Contents

2.4.2.2.1 Total Chlorophyll

A total of 15 dHVIs were evaluated for their potential for tracing chlorophyll content (Table 2.6). Since the red-edge region of 680–750 nm and the trough at about 630 nm were caused by the absorption spectrum by chlorophyll pigments, variations of chlorophyll content thus induce changes in these features (Miller et al. 1990). Based on these features, indices calculated as the sum of the reflectance at the red and infrared regions were proposed, including Sum1 (Filella et al. 1995), Sum2 (Elvidge and Chen 1995), RE (Filella et al. 1996), REIP (Collins 1978, Horler et al. 1983), mREIP (Miller et al. 1990), and some single-derivative spectra at the red edge such as D703 and D720 (Boochs et al. 1990). Also, simple ratios (dSRs) of derivative spectra have been proposed based on linear regressions between chlorophyll contents and these ratios (Vogelmann et al. 1993, Main et al. 2011). Furthermore, EGFN and EGFR using the normalized difference/ratio between the maximum of the first derivative of reflectance in the red edge and that in the green were proposed as well (Penuelas et al. 1994). Indices expressed as normalized differences (e.g., the first derivative-based normalized difference vegetation index — FDNDVI) were also proposed (Zhao et al. 2014, Zheng et al. 2014). In a recent work, Sonobe and Wang (2017) utilized the normalized difference between the green peak and the end of the red edge [dND(522, 728)] and proved that it was the most effective index for quantifying chlorophyll concentrations not only for field-measured data (the four datasets), but also for a simulated dataset from PROSPECT 5, indicating its general applicability. However, since the dHVIs reported in Sonobe and Wang (2017) were developed and validated using the four datasets listed above, they are treated as newly developed.

TABLE 2.6
Current Derivative Indices for Chlorophyll Content Estimation

Index	Reference	Formula
REIP	Collins (1978), Horler et al. (1983)	The position of the red-edge inflection point
D703	Boochs et al. (1990)	D_{703}
D720		D_{720}
mREIP	Miller et al. (1990)	The index based on the Gaussian fit of the red-edge derivative
Vogelmann	Vogelmann et al. (1993)	D_{715}/D_{705}
EGFN	Penuelas et al. (1994)	$(dRE-dG)/(dRE+dG)$
EGFR		dRE/dG
Sum1	Filella et al. (1995)	The sum of the amplitudes between 680 and 780 nm in the first derivative of the reflectance spectra
Sum2	Elvidge and Chen (1995)	The sum of derivative values between 626 nm and 795 nm
RE	Filella et al. (1996)	Amplitude of the main peak in the first derivative of the reflectance spectra
Datt	Datt (1999b)	D_{754}/D_{704}
DPI	Main et al. (2011)	$(D_{688}*D_{710})/D_{697}{}^2$
dSR1		D_{730}/D_{706}
dSR2		D_{705}/D_{722}
FDNDVI	Zhao et al. (2014)	$(D_{630}-D_{723})/(D_{630}+D_{723})$

Note: dRE refers to the maximum of the first derivative of reflectance in the red edge, while dG denotes the maximum of the first derivative of reflectance in the green domain.

Evaluation results indicate that mREIP, Vogelmann, Datt, DPI, and dSR1 have good performance with the dataset of ANGERS and are all categorized as "A." In particular, EGFN and dSR1 are significantly correlated with the measurements of all datasets except Naeba and are categorized as "A" or "B."

According to the validation results from Sonobe and Wang (2017) for the newly developed dHVIs, the dD and dND index types are generally applicable for all datasets for those indices that use the combinations of reflectance near 740 nm and near 520 or 695 nm. The wavelength near 740 nm is located at the end of the red edge, and reflectance at this domain has often been used in a set of indices, as listed in Section 2.3.1.2. The reflectance at green peak has also been used for estimating chlorophyll content in previous studies (Gitelson et al. 1996, Datt 1998, Broge and Leblanc 2001, Carter and Knapp 2001). However, few studies have ever reported the effectiveness of reflectance near 520 nm. This may be primarily due to the red-edge positions of deciduous species considered in the study, which were generally shorter than 740 nm, and their variation among different species is larger than that of the green peak positions of different species. Thus, the combination of the first derivative at the green peak and that at the end of the red edge is proper, as proved by the performance of dND (522, 728), which had the best performance for all measured datasets, with a determination coefficient of 0.862 ($p < 0.001$, RMSE = 15.69 µg/cm^2, WAIC = 18507.6).

2.4.2.2.2 *Carotenoids*

Although several indices have been proposed for assessing carotenoid content, few have ever used derivative spectra. In a previous study using the same datasets, Sonobe and Wang (2017) identified a total of seven dHVIs to quantify carotenoid contents based on RPD, including five combinations of dD types and two combinations of dND types. They were based on the first-order derivatives at 516–517 and 744–750 nm. When all the datasets were combined, the best performance was confirmed for dND (516,744) (R^2 = 0.475, WAIC = 2430.1, and RPD = 1.45) using exponential regression.

In order to compare this newly proposed index, 13 earlier reported HVIs (Table 2.7) were evaluated for their correlations with carotenoid contents based on the four datasets. In these indices, Chappelle (Chappelle et al. 1992), PRI (Gamon et al. 1992), Blackburn1, Blackburn2 (Blackburn

---

TABLE 2.7

Current Hyperspectral Indices for Carotenoid Content Estimation

Index	Reference	Formula
Chappelle	Chappelle et al. (1992)	R_{760}/R_{500}
PRI	Gamon et al. (1992)	$(R_{530} - R_{570})/(R_{530} - R_{570})$
Blackburn1	Blackburn (1998a)	R_{800}/R_{470}
Blackburn2		$(R_{800} - R_{470})/(R_{800} + R_{470})$
Datt2	Datt (1998)	$0.0049[R_{672}/(R_{550}*R_{708})]$
Gitelson1	Gitelson et al. (2002)	R_{NIR}/R_{510}
CRI$_{550}$		$1/R_{510} - 1/R_{550}$
CRI$_{700}$		$1/R_{510} - 1/R_{700}$
CARrededge	Gitelson et al. (2006)	$(1/R_{510 - 520} - 1/R_{690-710})*R_{NIR}$
CARgreen		$(1/R_{510-520} - 1/R_{560-570})*R_{NIR}$
AVIcar	Fassnacht et al. (2015)	$AVI2(R_{410}, R_{530}, R_{550})$
Fassnacht1		scale(AVIcar) + scale(Chappelle)
Fassnacht2		scale(AVIcar) + scale(CARred-edge)

Note: The use of scaled index values from 0–1 and the angular vegetation index were proposed by Fassnacht et al. (2015). Gitelson et al. (2006) defined NIR as the reflectance at 760–800 nm.

1998a), and Datt2 (Datt 1998) were compared with Giltson1, CRI550, and CRI700 (Gitelson et al. 2002). Furthermore, Gitelson et al. (2006) proposed CARrededge and CARgreen. On the other hand, Chappelle, CARrededge, AVIcar, Fassnacht1, and Fassnacht2 were analyzed for their potential for assessing carotenoids by Fassnacht et al. (2015).

The evaluation of 13 reported indices revealed that Chappelle, Datt2, Gitelson1, AVIcar, Fassnacht1, and Fassnacht2 have good performance with ANGERS. Furthermore, Datt2, Fassnacht1, and Fassnacht2 were significantly correlated with the measurements of all distinctive datasets. However, no index was applicable for estimating carotenoid content of Naeba, Nakagawane, and LOPEX.

2.5 DISCUSSION AND FUTURE DEVELOPMENTS

2.5.1 DERIVATIVE-BASED INDICES VS. REFLECTANCE-BASED INDICES

Derivative analysis has been used in analytical chemistry for several decades as an established technique to eliminate background signals and resolve overlapping spectral features (Demetriades-Shah et al. 1990). In remote sensing, analysis based on derivative spectra has been suggested to be powerful for detecting biochemical parameter chlorophyll concentration and biophysical parameter LAI (Imanishi et al. 2004), as well as for vegetation stress monitoring, such as the most commonly used red edge of reflectance (Demetriades-Shah et al. 1990, Imanishi et al. 2004, le Maire et al. 2004).

Here we have investigated the utility of dHVIs for detecting contents of three biochemical parameters (chlorophyll, carotenoids, and an artificially added dummy variable) and two biophysical parameters (LMA at the leaf scale and LAI at the canopy scale).

For the artificially added dummy variable, the first derivatives can successfully identify the absorption peaks (results from Experiment #1 based on modified PROSPECT-4). Virtual experiments clearly indicated that when the specific absorption coefficients of a biochemical (the artificially added dummy variable in the experiments) are similar to those of chlorophyll (Experiment #2), both Log(1/R)- and derivative-based indices were more effective than those indices developed from the original reflected spectra.

For chlorophyll with various interfering conditions (Experiment #3), derivative-based indices were much more effective than original reflectance under the same band combinations, although band combinations of identified best indices have been clustered near the wavelength domains where the dummy variable should have only a marginal effect. Taking the ND type index as an example, we estimate that more than 31% of derivative-based indices performed better than reflectance-based indices ($\Delta R^2 > 0.2$). Analysis on four independent measured datasets also revealed that the dD and dND index types were generally applicable for all datasets, suggesting a certain robustness of derivative-based indices.

For the representative physical parameter LAI, we examined the utility of dHVIs based on a simulated database generated from MRTM (Experiment #4). The R, D, SR, and ND types of dHVIs showed significant improvement in tracing canopy LAI over those of reflectance-based indices. Similarly, for the biophysical parameter LMA at the leaf scale, high determination coefficients between the derivative-based index dID(1408,1869) and LMA were confirmed for all measured datasets. In addition, another study on 3-year-old potted *Quercus glauca* and *Q. serrata* revealed that the best single bands for detecting LAI were 676.0 nm in the second derivative (r = 0.828), and the initial part of the red-edge peak performed better than the top (i.e., REP or the first derivative at REP) for the independent detection of LAI (Imanishi et al. 2004) and had the further advantage of dHVIs on estimating biophysical parameters.

2.5.2 SENSITIVITIES OF DERIVATIVE HYPERSPECTRAL VEGETATION INDICES TO DIFFERENT NOISE LEVELS AND SMOOTHING METHODS

Although the superiority of dHVIs has been confirmed, we emphasize that derivatives, particularly high-order derivatives, are increasingly sensitive to noise (Singer and Geissler 1988). Noise in remote sensing data degrades the interpretability of the data (Corner et al. 2003). The sensitivity of dHVIs has been confirmed from the above results, in which different Gaussian noise levels (0.5%, 1%, and 3%) were added into each spectrum at each wavelength generated in Experiment #3. Under the relatively gentle noise level (with 0.5% Gaussian noise being added), indices based on reflectance as well as the first- and second-order derivatives can trace the content of chlorophyll with good accuracy ($R^2 > 0.8$), while the performances of the third and higher order derivatives are inferior for the same purpose. When under strong noise conditions (with 3% Gaussian noise being added), only indices based on the first-order derivative show commensurate accuracy with reflectance-based indices to trace chlorophyll ($R^2 \approx 0.8$), while indices based on the second and higher order derivatives are poorer. These results suggest that remote sensing of plant leaf and canopy parameters using indices based on lower-order derivatives is a possibility, since indices based on lower-order derivatives are more effective than the original reflectance and, in the meantime, are not very sensitive to noise.

Furthermore, the use of a proper smoothing method is a major issue for generating accurate derivatives (Tsai and Philpot 1998), as derivative analysis results indicated that derivatives computed from the original reflectance with Gaussian noise fluctuated sharply throughout the wavelength, while smoothing can greatly reduce the abrupt spectral variations. In this study, we examined the popular Savitzky-Golay method as well as three other smoothing methods and found that the moving average method (with a filter window size of 17) obtained the strongest smoothing effect among these four methods. On the other hand, we found less satisfactory results concerning the accuracy, especially around the peaks and troughs of the spectrum. The LOWESS method had a higher accuracy than the moving average method, but remained unchanged with results around peaks and troughs. As a comparison, the Savitzky-Golay and LOESS smoothing methods could reduce random noise and preserve details of the spectra, suggesting they might be good choices for smoothing.

2.5.3 Future Potential Large-Scale Applications with Satellite-Borne Hyperspectral Data

Although the idea of using reflectance derivatives instead of original reflectance was proposed several decades ago, it has not been widely used for regional-scale monitoring (Kochubey and Kazantsev 2012). One apparent limitation is data availability, since most satellites have never been equipped with hyperspectral sensors, while airborne hyperspectral data have only been available since the early 1980s (Kruse et al. 2003). For instance, the airborne visible/infrared imaging spectrometer (AVIRIS), the first imaging sensor to measure the solar reflected spectrum ranging from 400 to 2500 nm at 10-nm intervals (Vane et al. 1993, Green et al. 1998), was proposed to the National Aeronautics and Space Administration (NASA) in 1983. Back to space-borne hyperspectral data, the launch of NASA's EO-1 Hyperion sensor in 2000 started space-borne hyperspectral imaging activities (Kruse et al. 2003). The EO-1 Hyperion imaging spectrometer covers the 400–2500 nm spectral range with 242 spectral bands and was the first imaging spectrometer that could routinely acquire Earth observations (Pearlman et al. 2003).

Moreover, satellite images always contain noisy signals from various sources (Al-Amri et al. 2010). The quality of digital remote sensing data is directly related to the level of system noise relative to signal strength (Kruse et al. 2003). Furthermore, atmospheric effects, thermal effects, sensor saturation, and so on all may be introduced into the remote sensing data obtained (Corner et al. 2003, Schowengerdt 2006). In addition, modern remote sensors generally have optical or electrical cross-talk effects between bands, thus affecting spectral characteristics. So-called "spectral cross-talk" has been found in ASTER, MODIS, and Hyperion data (Schowengerdt 2006). To make things worse, image noise is often hard to diminish because of its wide variety of forms (Corner et al. 2003). Many proposed noise reduction techniques, such as smoothing methods or filters, are only designed for a specific problem or task (Schowengerdt 2006).

Even so, with the coming of unprecedented hyperspectral remote sensing data, it can be foreseen that dHVIs will be a dominant approach for deriving biophysical and biochemical quantities in the near future. However, to achieve this, further extensive work is needed to refine this technology in order to verify its feasibility at the regional scale with space-borne hyperspectral images, for example, proper methods to limit or reduce different types of identified image noise.

2.6 CONCLUSIONS

Derivative hyperspectral vegetation indices have been investigated for their potentialities in characterizing forest biophysical and biochemical parameters at different spatial scales, based on either simulation results from radiative transfer models or *in situ* measurements. Results clearly indicated that dHVIs are superior over HVIs for estimating leaf biochemical constituents with either weak or strong absorption capabilities. Results also showed dHVIs performed better than corresponding HVIs. However, results also suggested dHVIs are sensitive to noise, especially for higher-order dHVIs, but such sensitivities can be largely depressed or eliminated by applying proper smoothing methods for generating derivative spectra. In particular, first-order dHVIs show a certain balance between accuracy and robustness on retrieving both biophysical and biochemical parameters, suggesting they might be good choices for future applications.

Even though dHVIs are not popular yet in large spatial-scale applications, due primarily to data poverty of hyperspectral remote sensing, with the rapid development of airborne and satellite-borne hyperspectral sensors and associated hyperspectral remote sensing data being archived, dHVIs will no doubt gain prevalence in characterization of forest biophysical and biochemical quantities, even though more work is needed to perfect this technology.

APPENDIX

All derivative hyperspectral vegetation indices used in this study have been tabularized in Table 2A.1.

TABLE 2A.1
Derivative Hyperspectral Vegetation Indices (dHVIs) Used in This Study

Index	Reference	Formula
D660	This study	D_{660}
dD(620, 660)	This study	$D_{620} - D_{660}$
dSR(445, 670)	This study	D_{445}/D_{670}
dND(445, 670)	This study	$(D_{445} - D_{670})/(D_{445} + D_{670})$
dDDn(660, 20)	This study	$2D_{660} - D_{680} - D_{640}$
dIID(1408, 1869)	This study	$1/D_{1408} - 1/D_{1869}$
D645*	This study	D_{645}
dD(450, 645)*	This study	$D_{450} - D_{645}$
dSR(650, 665)*	This study	D_{650}/D_{665}
dND(655, 665)*	This study	$(D_{655} - D_{665})/(D_{655} + D_{665})$
dDDn(645, 100)*	This study	$2D_{645} - D_{545} - D_{745}$
D660**	This study	D_{660}
dD(635, 660)**	This study	$D_{635} - D_{660}$
dSR(450, 620)**	This study	$D_{450} - D_{620}$
dND(600, 660)**	This study	$D_{600} - D_{660}$
dDDn(635, 30)**	This study	$2D_{635} - D_{605} - D_{665}$
D645***	This study	D_{645}
dD(645, 740)***	This study	$D_{645} - D_{740}$

(Continued)

TABLE 2A.1 (*Continued*)
Derivative Hyperspectral Vegetation Indices (dHVIs) Used in This Study

Index	Reference	Formula
dSR(615, 630)***	This study	D_{615}/D_{630}
dND(585, 630)***	This study	$(D_{585} - D_{630})/(D_{585} + D_{630})$
dDDn(645, 95)***	This study	$2D_{645} - D_{740} - D_{550}$
D655****	This study	D_{655}
dD(625, 635)****	This study	$D_{625} - D_{635}$
dSR(615, 625)****	This study	D_{615}/D_{625}
dND(615, 625)****	This study	$(D_{615} - D_{625})/(D_{615} + D_{625})$
dDDn(620,20)****	This study	$2D_{620} - D_{600} - D_{640}$
D630*****	This study	D_{630}
dD(565, 630)*****	This study	$D_{565} - D_{630}$
dSR(595, 635) *****	This study	D_{595}/D_{635}
dND(560, 665) *****	This study	$(D_{560} - D_{665})/(D_{560} + D_{665})$
dDDn(630, 105) *****	This study	$2D_{630} - D_{735} - D_{525}$
D725	This study	D_{725}
dD(710, 715)	This study	$D_{710} - D_{715}$
dSR(555, 705)	This study	D_{555}/D_{705}
dND(535, 700)	This study	$(D_{535} - D_{700})/(D_{535} + D_{700})$
dDDn(705, 5)	This study	$2D_{705} - D_{710} - D_{700}$
D715*	This study	D_{715}
dD(530, 715)*	This study	$D_{530} - D_{715}$
dSR(695, 735)*	This study	D_{695}/D_{735}
dND(580, 690)*	This study	$(D_{580} - D_{690})/(D_{580} + D_{690})$
dDDn(715, 10)*	This study	$2D_{715} - D_{705} - D_{725}$
D705**	This study	D_{705}
dD(710, 720) **	This study	$D_{710} - D_{720}$
dSR(705, 745) **	This study	D_{705}/D_{745}
dND(685, 745) **	This study	$(D_{685} - D_{745})/(D_{685} + D_{745})$
dDDn(555, 150) **	This study	$2D_{555} - D_{405} - D_{705}$
D715***	This study	D_{715}
dD(470, 715) ***	This study	$D_{470} - D_{715}$
dSR(570, 730) ***	This study	D_{570}/D_{730}
dND(695, 710) ***	This study	$(D_{695} - D_{710})/(D_{695} + D_{710})$
dDDn(715, 20) ***	This study	$2D_{715} - D_{735} - D_{695}$
D725****	This study	D_{725}
dD(705, 725) ****	This study	$D_{705} - D_{725}$
dSR(580, 685) ****	This study	D_{580}/D_{685}
dND(685, 740) ****	This study	$(D_{685} - D_{740})/(D_{685} + D_{740})$
dDDn(725, 20) ****	This study	$2D_{725} - D_{745} - D_{705}$
D750*****	This study	D_{750}
dD(715, 735) *****	This study	$D_{715} - D_{735}$
dSR(585, 730) *****	This study	D_{585}/D_{730}
dND(680, 730) *****	This study	$(D_{680} - D_{730})/(D_{680} + D_{730})$
dDDn(715, 20) *****	This study	$2D_{715} - D_{695} - D_{735}$
D590	This study	D_{590}
dD(520, 745)	This study	$D_{520} - D_{745}$
dSR(575, 595)	This study	D_{575}/D_{595}
dND(520, 745)	This study	$(D_{520} - D_{745})/(D_{520} + D_{745})$
dDDn(575, 5)	This study	$2D_{575} - D_{580} - D_{570}$
D590*	This study	D_{590}

(*Continued*)

TABLE 2A.1 (*Continued*)

Derivative Hyperspectral Vegetation Indices (dHVIs) Used in This Study

Index	Reference	Formula
dD(570, 585)*	This study	$D_{570} - D_{585}$
dSR(575, 580)*	This study	D_{575}/D_{580}
dND(580, 595)*	This study	$(D_{580} - D_{590})/(D_{580} + D_{590})$
dDDn(585, 10)*	This study	$2D_{585} - D_{595} - D_{575}$
D575**	This study	D_{575}
dD(520, 580) **	This study	$D_{520} - D_{580}$
dSR(570, 590) **	This study	D_{570}/D_{590}
dND(560, 580) **	This study	$(D_{560} - D_{580})/(D_{560} + D_{580})$
dDDn(540, 20) **	This study	$2D_{540} - D_{560} - D_{520}$
D585***	This study	D_{585}
dD(530, 555) ***	This study	$D_{530} - D_{555}$
dSR(570, 575) ***	This study	D_{570}/D_{575}
dND(570, 590) ***	This study	$(D_{570} - D_{590})/(D_{570} + D_{590})$
dDDn(580, 5) ***	This study	$2D_{580} - D_{585} - D_{575}$
D575****	This study	D_{575}
dD(520, 580) ****	This study	$D_{520} - D_{580}$
dSR(520, 565) ****	This study	D_{520}/D_{565}
dND(565, 590) ****	This study	$(D_{565} - D_{590})/(D_{565} + D_{590})$
dDDn(560, 55) ****	This study	$2D_{560} - D_{615} - D_{505}$
D510*****	This study	D_{510}
dD(515, 520) *****	This study	$D_{515} - D_{520}$
dSR(515, 570) *****	This study	D_{515}/D_{570}
dND(585, 600) *****	This study	$(D_{585} - D_{600})/(D_{585} + D_{600})$
dDDn(570, 55) *****	This study	$2D_{570} - D_{625} - D_{515}$
D745	This study	D_{745}
dD(935, 945)	This study	$D_{935} - D_{945}$
dSR(805, 1830)	This study	D_{805}/D_{1830}
dND(525, 935)	This study	$(D_{525} - D_{935})/(D_{525} + D_{935})$
dDDn(970, 35)	This study	$2D_{970} - D_{1005} - D_{935}$
D940*	This study	D_{940}
dD(645, 780)*	This study	$D_{645} - D_{780}$
dSR(805, 830)*	This study	D_{805}/D_{830}
dND(435, 1220)*	This study	$(D_{435} - D_{1220})/(D_{435} + D_{1220})$
dDDn(935, 15)*	This study	$2D_{935} - D_{950} - D_{920}$
D745**	This study	D_{745}
dD(1085, 2295) **	This study	$D_{1085} - D_{2295}$
dSR(740, 1085) **	This study	D_{740}/D_{1085}
dND(505, 1500) **	This study	$(D_{505} - D_{1500})/(D_{505} + D_{1500})$
dDDn(1520, 805) **	This study	$2D_{1520} - D_{2325} - D_{715}$
D935***	This study	D_{935}
dD(770, 1025) ***	This study	$D_{770} - D_{1025}$
dSR(960, 1650) ***	This study	D_{960}/D_{1650}
dND(505, 2055) ***	This study	$(D_{505} - D_{2055})/(D_{505} + D_{2055})$
dDDn(735, 240) ***	This study	$2D_{735} - D_{975} - D_{499}$
D945****	This study	D_{945}
dD(515, 695) ****	This study	$D_{515} - - D_{695}$
dSR(945, 1335) ****	This study	D_{945}/D_{1335}
dND(780, 1060) ****	This study	$(D_{780} - D_{1060})/(D_{780} + D_{1060})$
dDDn(935, 55) ****	This study	$2D_{935} - D_{990} - D_{880}$

(*Continued*)

TABLE 2A.1 (*Continued*)
Derivative Hyperspectral Vegetation Indices (dHVIs) Used in This Study

Index	Reference	Formula
D815*****	This study	D_{815}
dD(925, 985) *****	This study	$D_{925} - - D_{985}$
dSR(815, 1170) *****	This study	D_{815}/D_{1170}
dND(495, 2375) *****	This study	$(D_{495} - D_{2375})/(D_{495} + D_{2375})$
dDDn(960, 50) *****	This study	$2D_{960} - D_{1010} - D_{910}$
REIP	Collins (1978), Horler et al. (1983)	The position of the red-edge inflection point
D703	Boochs et al. (1990)	D_{703}
D720		D_{720}
mREIP	Miller et al. (1990)	The index based on the Gaussian fit of the red-edge derivative
Vogelmann	Vogelmann et al. (1993)	D_{715}/D_{705}
EGFN	Penuelas et al. (1994)	(dRE-dG)/(dRE + dG)
EGFR		dRE/dG
Sum1	Filella et al. (1995)	The sum of the amplitudes between 680 and 780 nm in the first derivative of the reflectance spectra
Sum2	Elvidge and Chen (1995)	The sum of derivative values between 626 nm and 795 nm
RE	Filella et al. (1996)	Amplitude of the main peak in the first derivative of the reflectance spectra
Datt	Datt (1999b)	D_{754}/D_{704}
DPI	Zarco-Tejada et al. (2003a,b)	$(D_{688}*D_{710})/D_{697}^2$
dSR1		D_{730}/D_{706}
dSR2		D_{705}/D_{722}
FDNDVI	Zhao et al. (2014)	$(D_{630} - D_{723})/(D_{630} + D_{723})$
dND(522, 728)	Sonobe and Wang (2017)	$(D_{522} - D_{728})/(D_{522} + D_{728})$
dND (516,744)	This study	$(D_{516} - D_{744})/(D_{516} + D_{744})$

Note: *, **, ***, ****, and ***** indicate dHVIs based on the second, third, fourth, fifth, and sixth derivative spectra, respectively. dRE refers to the maximum of the first derivative of reflectance in the red edge, while dG is the maximum of the first derivative of reflectance in the green domain.

REFERENCES

Al-Amri, S. S., N. V. Kalyankar and S. D. Khamitkar. 2010. "A comparative study of removal noise from remote sensing image." *International Journal of Computer Science Issues* 7(12): 32–36.

Bannari, A., D. Morin, F. Bonn and A. R. Huete. 1995. "A review of vegetation indices." *Remote Sensing Reviews* 13(1–2): 95–120.

Baret, F. and G. Guyot. 1991. "Potentials and limits of vegetation indices for LAI and APAR assessment." *Remote Sensing of Environment* 35(2): 161–173.

Ben-Dor, E., D. Schlapfer, A. J. Plaza and T. Malthus. 2013. *Hyperspectral Remote Sensing*. In M. Wendisch and J.-L. Brenguier (Eds.), Airborne Measurements for Environmental Research: Methods and Instruments, Weinheim, Germany: Wiley-VCH Verlag & Co. KGaA: 413–456.

Benoist, D., Y. Tourbier and S. Germain-Tourbier. 1994. *Plans d'Experience: Construction et Analyse*. London, Lavoisier.

Bioucas-Dias, J. M., A. Plaza, G. Camps-Valls, P. Scheunders, N. Nasrabadi and J. Chanussot. 2013. "Hyperspectral remote sensing data analysis and future challenges." *IEEE Geoscience and Remote Sensing Magazine* 1(2): 6–36.

Blackburn, G. A. 1998a. "Quantifying chlorophylls and carotenoids at leaf and canopy scales: An evaluation of some hyperspectral approaches." *Remote Sensing of Environment* 66(3): 273–285.

Blackburn, G. A. 1998b. "Spectral indices for estimating photosynthetic pigment concentrations: A test using senescent tree leaves." *International Journal of Remote Sensing* 19(4): 657–675.

Boochs, F., G. Kupfer, K. Dockter and W. Kuhbauch. 1990. "Shape of the red edge as vitality indicator for plants." *International Journal of Remote Sensing* 11(10): 1741–1753.

Broge, N. H. and E. Leblanc. 2001. "Comparing prediction power and stability of broadband and hyperspectral vegetation indices for estimation of green leaf area index and canopy chlorophyll density." *Remote Sensing of Environment* 76(2): 156–172.

Carter, G. A. 1994. "Ratios of leaf reflectances in narrow wavebands as indicators of plant stress." *International Journal of Remote Sensing* 15(3): 697–703.

Carter, G. A. and A. K. Knapp. 2001. "Leaf optical properties in higher plants: Linking spectral characteristics to stress and chlorophyll concentration." *American Journal of Botany* 88(4): 677–684.

Chang, C.-W., D. A. Laird, M. J. Mausbach and C. R. Hurburgh. 2001. "Near-infrared reflectance spectroscopy? Principal components regression analyses of soil properties." *Soil Science Society of America Journal* 65(2): 480–490.

Chappelle, E. W., M. S. Kim and J. E. McMurtrey. 1992. "Ratio analysis of reflectance spectra (RARS): An algorithm for the remote estimation of the concentrations of chlorophyll A, chlorophyll B, and carotenoids in soybean leaves." *Remote Sensing of Environment* 39(3): 239–247.

Chapra, S. C. and R. P. Canale. 1988. *Numerical Methods for Engineers*, McGraw-Hill, New York.

Cheng, C., Y. Wei, G. Lv and Z. Yuan. 2013. "Remote estimation of chlorophyll-a concentration in turbid water using a spectral index: A case study in Taihu Lake, China." *Journal Of Applied Remote Sensing* 7(1): 073465.

Cho, M. A., A. Skidmore, F. Corsi, S. E. van Wieren and I. Sobhan. 2007. "Estimation of green grass/herb biomass from airborne hyperspectral imagery using spectral indices and partial least squares regression." *International Journal of Applied Earth Observation and Geoinformation* 9(4): 414–424.

Cloutis, E. A. 1996. "Hyperspectral geological remote sensing: Evaluation of analytical techniques." *International Journal of Remote Sensing* 17(12): 2215–2242.

Collins, W. 1978. "Remote sensing of crop type and maturity." *Photogrammetric Engineering and Remote Sensing* 44(1): 43–55.

Combal, B., F. Baret, M. Weiss, A. Trubuil, D. Mace, A. Pragnere, R. Myneni, Y. Knyazikhin and L. Wang. 2003. "Retrieval of canopy biophysical variables from bidirectional reflectance using prior information to solve the ill-posed inverse problem." *Remote Sensing of Environment* 84(1): 1–15.

Corner, B. R., R. M. Narayanan and S. E. Reichenbach. 2003. "Noise estimation in remote sensing imagery using data masking." *International Journal of Remote Sensing* 24(4): 689–702.

Curran, P. J., J. L. Dungan and H. L. Gholz. 1990. "Exploring the relationship between reflectance red edge and chlorophyll content in slash pine." *Tree Physiology* 7(1–4): 33–48.

Datt, B. 1998. "Remote sensing of chlorophyll a, chlorophyll b, chlorophyll a+b, and total carotenoid content in eucalyptus leaves." *Remote Sensing of Environment* 66(2): 111–121.

Datt, B. 1999a. "A new reflectance index for remote sensing of chlorophyll content in higher plants: Tests using eucalyptus leaves." *Journal of Plant Physiology* 154(1): 30–36.

Datt, B. 1999b. "Visible/near infrared reflectance and chlorophyll content in eucalyptus leaves." *International Journal of Remote Sensing* 20(14): 2741–2759.

Delegido, J., J. Verrelst, C. M. Meza, J. P. Rivera, L. Alonso and J. Moreno. 2013. "A red-edge spectral index for remote sensing estimation of green LAI over agroecosystems." *European Journal of Agronomy* 46: 42–52.

Demetriades-Shah, T. H., M. D. Steven and J. A. Clark. 1990. "High resolution derivative spectra in remote sensing." *Remote Sensing of Environment* 33(1): 55–64.

Diaz, B. M. and G. A. Blackburn. 2003. "Remote sensing of mangrove biophysical properties: Evidence from a laboratory simulation of the possible effects of background variation on spectral vegetation indices." *International Journal of Remote Sensing* 24(1): 53–73.

Dorigo, W. A., R. Zurita-Milla, A. J. W. de Wit, J. Brazile, R. Singh and M. E. Schaepman. 2007. "A review on reflective remote sensing and data assimilation techniques for enhanced agroecosystem modeling." *International Journal of Applied Earth Observation and Geoinformation* 9(2): 165–193.

Elvidge, C. D. and Z. K. Chen. 1995. "Comparison of broad-band and narrow-band red and near-infrared vegetation indices." *Remote Sensing of Environment* 54(1): 38–48.

Fassnacht, F. E., S. Stenzel and A. A. Gitelson. 2015. "Non-destructive estimation of foliar carotenoid content of tree species using merged vegetation indices" *Journal of Plant Physiology* 176: 210–217.

Feret, J.-B., C. Francois, G. P. Asner, A. A. Gitelson, R. E. Martin, L. P. R. Bidel, S. L. Ustin, G. le Maire and S. Jacquemoud. 2008. "PROSPECT-4 and 5: Advances in the leaf optical properties model separating photosynthetic pigments." *Remote Sensing of Environment* 112(6): 3030–3043.

Feret, J.-B., C. Francois, A. Gitelson, G. P. Asner, K. M. Barry, C. Panigada, A. D. Richardson and S. Jacquemoud. 2011. "Optimizing spectral indices and chemometric analysis of leaf chemical properties using radiative transfer modeling." *Remote Sensing of Environment* 115(10): 2742–2750.

Filella, I., T. Amaro, J. L. Araus and J. Penuelas. 1996. "Relationship between photosynthetic radiation-use efficiency of barley canopies and the photochemical reflectance index (PRI)." *Physiologia Plantarum* 96(2): 211–216.

Filella, I., L. Serrano, J. Serra and J. Penuelas. 1995. "Evaluating wheat nitrogen status with canopy reflectance indices and discriminant analysis." *Crop Science* 35(5): 1400–1405.

Fu, Y., G. Yang, J. Wang, X. Song and H. Feng. 2014. "Winter wheat biomass estimation based on spectral indices, band depth analysis and partial least squares regression using hyperspectral measurements." *Computers and Electronics in Agriculture* 100: 51–59.

Gamon, J. A., J. Penuelas and C. B. Field. 1992. "A narrow-waveband spectral index that tracks diurnal changes in photosynthetic efficiency." *Remote Sensing of Environment* 41(1): 35–44.

Gamon, J. A. and J. S. Surfus. 1999. "Assessing leaf pigment content and activity with a reflectometer." *New Phytologist* 143(1): 105–117.

Gandia, S., G. Fernandez and J. Moreno. 2005. "Retrieval of vegetation biophysical variables from CHRIS/PROBA data in the SPARC Campaign." *The 2nd CHRIS/Proba Workshop*, Frascati, Italy, ESA/ESRIN.

Gitelson, A. A., Y. J. Kaufman and M. N. Merzlyak. 1996. "Use of a green channel in remote sensing of global vegetation from EOS-MODIS." *Remote Sensing of Environment* 58(3): 289–298.

Gitelson, A. A., G. P. Keydan and M. N. Merzlyak. 2006. "Three-band model for noninvasive estimation of chlorophyll, carotenoids, and anthocyanin contents in higher plant leaves." *Geophysical Research Letters* 33: L11402.

Gitelson, A. and M. N. Merzlyak. 1994. "Spectral reflectance changes associated with autumn senescence of *Aesculus hippocastanum* L. and *Acer platanoides* L. leaves. Spectral features and relation to chlorophyll estimation." *Journal of Plant Physiology* 143(3): 286–292.

Gitelson, A. A. and M. N. Merzlyak. 1996. "Signature analysis of leaf reflectance spectra: Algorithm development for remote sensing of chlorophyll." *Journal of Plant Physiology* 148(3–4): 494–500.

Gitelson, A. A., M. N. Merzlyak and O. B. Chivkunova. 2001. "Optical properties and nondestructive estimation of anthocyanin content in plant leaves." *Photochemistry and Photobiology* 74(1): 38–45.

Gitelson, A. A., Y. Zur, O. B. Chivkunova and M. N. Merzlyak. 2002. "Assessing carotenoid content in plant leaves with reflectance spectroscopy." *Photochemistry and Photobiology* 75(3): 272–281.

Goetz, A. F. H. 2009. "Three decades of hyperspectral remote sensing of the Earth: A personal view." *Remote Sensing of Environment* 113: S5–S16.

Gonsamo, A. and P. Pellikka. 2012. "The sensitivity based estimation of leaf area index from spectral vegetation indices." *ISPRS Journal of Photogrammetry and Remote Sensing* 70: 15–25.

Green, R. O., M. L. Eastwood, C. M. Sarture, T. G. Chrien, M. Aronsson, B. J. Chippendale, J. A. Faust, B. E. Pavri, C. J. Chovit, M. Solis, M. R. Olah and O. Williams. 1998. "Imaging spectroscopy and the airborne visible/infrared imaging spectrometer (AVIRIS)." *Remote Sensing of Environment* 65(3): 227–248.

Haboudane, D., J. R. Miller, E. Pattey, P. J. Zarco-Tejada and I. B. Strachan. 2004. "Hyperspectral vegetation indices and novel algorithms for predicting green LAI of crop canopies: Modeling and validation in the context of precision agriculture." *Remote Sensing of Environment* 90(3): 337–352.

Hansen, P. M. and J. K. Schjoerring. 2003. "Reflectance measurement of canopy biomass and nitrogen status in wheat crops using normalized difference vegetation indices and partial least squares regression." *Remote Sensing of Environment* 86(4): 542–553.

Hoffman, J. D. and S. Frankel. 2001. *Numerical Methods for Engineers and Scientists*, New York: Marcel Dekker.

Horler, D. N. H., M. Dockray and J. Barber. 1983. "The red edge of plant leaf reflectance." *International Journal of Remote Sensing* 4(2): 273–288.

Hosgood, B., S. Jacquemoud, G. Andreoli, J. Verdebout, G. Pedrini and G. Schmuck. 1994. *Leaf Optical Properties EXperiment 93 (LOPEX93)*. Ispra, Italy: European Commission—Joint Research Centre: 20.

Huang, Z., B. J. Turner, S. J. Dury, I. R. Wallis and W. J. Foley. 2004. "Estimating foliage nitrogen concentration from HYMAP data using continuum removal analysis." *Remote Sensing of Environment* 93(1–2): 18–29.

Huete, A. R. 1988. "A soil-adjusted vegetation index (SAVI)." *Remote Sensing of Environment* 25(3): 295–309.

Huete, A., K. Didan, T. Miura, E. P. Rodriguez, X. Gao and L. G. Ferreira. 2002. "Overview of the radiometric and biophysical performance of the MODIS vegetation indices." *Remote Sensing of Environment* 83(1–2): 195–213.

Im, J. and J. R. Jensen. 2008. "Hyperspectral remote sensing of vegetation." *Geography Compass* 2(6): 1943–1961.

Imanishi, J., K. Sugimoto and Y. Morimoto. 2004. "Detecting drought status and LAI of two Quercus species canopies using derivative spectra." *Computers and Electronics in Agriculture* 43(2): 109–129.

Inoue, Y., E. Sakaiya, Y. Zhu and W. Takahashi. 2012. "Diagnostic mapping of canopy nitrogen content in rice based on hyperspectral measurements." *Remote Sensing of Environment* 126: 210–221.

Jacquemoud, S. 1993. "Inversion of the PROSPECT+ SAIL canopy reflectance model from AVIRIS equivalent spectra: Theoretical study." *Remote Sensing of Environment* 44(2–3): 281–292.

Jacquemoud, S. and F. Baret. 1990. "PROSPECT: A model of leaf optical properties spectra." *Remote Sensing of Environment* 34(2): 75–91.

Jin, J. and Q. Wang. 2016. "Hyperspectral indices based on first derivative spectra closely trace canopy transpiration in a desert plant." *Ecological Informatics* 35: 1–8.

Jordan, C. F. 1969. "Derivation of leaf-area index from quality of light on the forest floor." *Ecology* 50: 663–666.

Kalacska, M. and G. A. Sanchez-Azofeifa. 2008. *Hyperspectral Remote Sensing of Tropical and Sub-Tropical Forests*, Boca Raton: CRC Press.

Kawata, S. and S. Minami. 1984. "Adaptive smoothing of spectroscopic data by a linear mean-square estimation." *Applied Spectroscopy* 38(1): 49–58.

Kochubey, S. M. and T. A. Kazantsev. 2012. "Derivative vegetation indices as a new approach in remote sensing of vegetation." *Frontiers of Earth Science* 6(2): 188–195.

Kokaly, R. F. and R. N. Clark. 1999. "Spectroscopic determination of leaf biochemistry using band-depth analysis of absorption features and stepwise multiple linear regression." *Remote Sensing of Environment* 67(3): 267–287.

Kruse, F. A., J. W. Boardman and J. F. Huntington. 2003. "Comparison of airborne hyperspectral data and EO-1 Hyperion for mineral mapping." *IEEE Transactions on Geoscience and Remote Sensing* 41(6): 1388–1400.

le Maire, G., C. Francois and E. Dufrene. 2004. "Towards universal broad leaf chlorophyll indices using PROSPECT simulated database and hyperspectral reflectance measurements." *Remote Sensing of Environment* 89(1): 1–28.

le Maire, G., C. Francois, K. Soudani, D. Berveiller, J.-Y. Pontailler, N. Breda, H. Genet, H. Davi and E. Dufrene. 2008. "Calibration and validation of hyperspectral indices for the estimation of broadleaved forest leaf chlorophyll content, leaf mass per area, leaf area index and leaf canopy biomass." *Remote Sensing of Environment* 112(10): 3846–3864.

Li, P. and Q. Wang. 2012. "Retrieval of chlorophyll for assimilating branches of a typical desert plant through inversed radiative transfer models." *International Journal of Remote Sensing* 34(7): 2402–2416.

Li, P. and Q. Wang. 2013. "Developing and validating novel hyperspectral indices for leaf area index estimation: Effect of canopy vertical heterogeneity." *Ecological Indicators* 32: 123–130.

Li, P. H. and Q. Wang. 2011. "Retrieval of leaf biochemical parameters using PROSPECT inversion: A new approach for alleviating Ill-posed problems." *IEEE Transactions on Geoscience and Remote Sensing* 49(7): 2499–2506.

Liang, S. 2005. *Quantitative Remote Sensing of Land Surfaces*. Hoboken, New Jersey: John Wiley & Sons.

Lichtenthaler, H. K., M. Lang, M. Sowinska, F. Heisel and J. A. Miehe. 1996. "Detection of vegetation stress via a new high resolution fluorescence imaging system." *Journal of Plant Physiology* 148(5): 599–612.

Madeira, A. C., A. Mentions, M. E. Ferreira and M. d. L. Taborda. 2000. "Relationship between spectroradiometric and chlorophyll measurements in green beans." *Communications in Soil Science and Plant Analysis* 31(5–6): 631–643.

Main, R., M. A. Cho, R. Mathieu, M. M. O'Kennedy, A. Ramoelo and S. Koch. 2011. "An investigation into robust spectral indices for leaf chlorophyll estimation." *ISPRS Journal of Photogrammetry and Remote Sensing* 66(6): 751–761.

Marino, G., E. Pallozzi, C. Cocozza, R. Tognetti, A. Giovannelli, C. Cantini and M. Centritto. 2014. "Assessing gas exchange, sap flow and water relations using tree canopy spectral reflectance indices in irrigated and rainfed *Olea europaea* L." *Environmental and Experimental Botany* 99: 43–52.

Marshall, M., P. Thenkabail, T. Biggs and K. Post. 2016. "Hyperspectral narrowband and multispectral broadband indices for remote sensing of crop evapotranspiration and its components (transpiration and soil evaporation)." *Agricultural and Forest Meteorology 218?* 219: 122–134.

McMurtrey, J. E., E. W. Chappelle, M. S. Kim, J. J. Meisinger and L. A. Corp. 1994. "Distinguishing nitrogen fertilization levels in field corn (*Zea mays* L.) with actively induced fluorescence and passive reflectance measurements." *Remote Sensing of Environment* 47(1): 36–44.

Miller, J. R., E. W. Hare and J. Wu. 1990. "Quantitative characterisation of the red edge reflectance 1. An inverted-Gaussian model." *International Journal of Remote Sensing* 11(10): 1755–1773.

Mutanga, O. and A. K. Skidmore. 2004. "Hyperspectral band depth analysis for a better estimation of grass biomass (*Cenchrus ciliaris*) measured under controlled laboratory conditions." *International Journal of Applied Earth Observation and Geoinformation* 5(2): 87–96.

Mutanga, O., A. K. Skidmore, L. Kumar and J. Ferwerda. 2005. "Estimating tropical pasture quality at canopy level using band depth analysis with continuum removal in the visible domain." *International Journal of Remote Sensing* 26(6): 1093–1108.

Nguyen, H. T. and B.-W. Lee. 2006. "Assessment of rice leaf growth and nitrogen status by hyperspectral canopy reflectance and partial least square regression." *European Journal of Agronomy* 24(4): 349–356.

Pearlman, J. S., P. S. Barry, C. C. Segal, J. Shepanski, D. Beiso and S. L. Carman. 2003. "Hyperion, a space-based imaging spectrometer." *IEEE Transactions on Geoscience and Remote Sensing* 41(6): 1160–1173.

Pearson, R. L. and L. D. Miller. 1972. "Remote mapping of standing crop biomass for estimation of the productivity of the shortgrass prairie." *Remote Sensing of Environment*, VIII.

Penuelas, J., I. Filella, C. Biel, L. Serrano and R. Save. 1993. "The reflectance at the 950–970 nm region as an indicator of plant water status." *International Journal of Remote Sensing* 14(10): 1887–1905.

Penuelas, J., J. A. Gamon, A. L. Fredeen, J. Merino and C. B. Field. 1994. "Reflectance indices associated with physiological changes in nitrogen- and water-limited sunflower leaves." *Remote Sensing of Environment* 48(2): 135–146.

Poorter, H., S. Pepin, T. Rijkers, Y. de Jong, J. R. Evans and C. Korner. 2006. "Construction costs, chemical composition and payback time of high- and low-irradiance leaves." *Journal of Experimental Botany* 57(2): 355–371.

Qi, J., A. Chehbouni, A. R. Huete, Y. H. Kerr and S. Sorooshian. 1994. "A modified soil adjusted vegetation index." *Remote Sensing of Environment* 48(2): 119–126.

Rady, A. M., D. E. Guyer, W. Kirk and I. R. Donis-Gonzalez. 2014. "The potential use of visible/near infrared spectroscopy and hyperspectral imaging to predict processing-related constituents of potatoes." *Journal of Food Engineering* 135: 11–25.

Ramoelo, A., A. K. Skidmore, M. A. Cho, R. Mathieu, I. M. A. Heitkonig, N. Dudeni-Tlhone, M. Schlerf and H. H. T. Prins. 2013. "Non-linear partial least square regression increases the estimation accuracy of grass nitrogen and phosphorus using *in situ* hyperspectral and environmental data." *ISPRS Journal of Photogrammetry and Remote Sensing* 82: 27–40.

Rodriguez-Perez, J. R., D. Riano, E. Carlisle, S. Ustin and D. R. Smart. 2007. "Evaluation of hyperspectral reflectance indexes to detect grapevine water status in vineyards." *American Journal of Enology and Viticulture* 58(3): 302–317.

Rouse, J., R. Haas, J. Schell, D. Deering and J. Harlan. 1974. "Monitoring the vernal advancement of retrogradation of natural vegetation, NASA/GSFC, Type III, Final Report." Greenbelt, MD.

Ryu, C., M. Suguri and M. Umeda. 2009. "Model for predicting the nitrogen content of rice at panicle initiation stage using data from airborne hyperspectral remote sensing." *Biosystems Engineering* 104(4): 465–475.

Ryu, C., M. Suguri and M. Umeda. 2011. "Multivariate analysis of nitrogen content for rice at the heading stage using reflectance of airborne hyperspectral remote sensing." *Field Crops Research* 122(3): 214–224.

Savitzky, A. and M. J. E. Golay. 1964. "Smoothing and differentiation of data by simplified least squares procedures." *Analytical Chemistry* 36(8): 1627–1639.

Schowengerdt, R. A. 2006. *Remote Sensing: Models and Methods for Image Processing*, San Diego, California: Academic Press.

Singer, R. B. and P. E. Geissler. 1988. "An independent assessment of derivative analysis of reflectance spectra." *Lunar and Planetary Science Conference*. 19.

Smith, K. L., M. D. Steven and J. J. Colls. 2004. "Use of hyperspectral derivative ratios in the red-edge region to identify plant stress responses to gas leaks." *Remote Sensing of Environment* 92(2): 207–217.

Smith, R. C. G., J. Adams, D. J. Stephens and P. T. Hick. 1995. "Forecasting wheat yield in a Mediterranean-type environment from the NOAA satellite." *Australian Journal of Agricultural Research* 46(1): 113–125.

Sonobe, R. and Q. Wang. 2016. "Assessing the xanthophyll cycle in natural beech leaves with hyperspectral reflectance." *Functional Plant Biology* 43(5): 438–447.

Sonobe, R. and Q. Wang. 2017. "Towards a universal hyperspectral index to assess chlorophyll content in deciduous forests." *Remote Sensing* 9(3): 191.

Sun, P., S. Wahbi, T. Tsonev, M. Haworth, S. Liu and M. Centritto. 2014. "On the use of leaf spectral indices to assess water status and photosynthetic limitations in *Olea europaea* L. during water-stress and recovery." *PloS ONE* 9(8): e105165.

Tanaka, S., K. Kawamura, M. Maki, Y. Muramoto, K. Yoshida and T. Akiyama. 2015. "Spectral index for quantifying leaf area index of winter wheat by field hyperspectral measurements: A case study in Gifu prefecture, Central Japan." *Remote Sensing* 7(5): 5329–5346.

Thenkabail, P. S., J. G. Lyon and A. Huete. 2012. *Hyperspectral Remote Sensing of Vegetation*, Boca Raton, Florida: CRC Press.

Thenkabail, P. S., R. B. Smith and E. De Pauw. 2000. "Hyperspectral vegetation indices and their relationships with agricultural crop characteristics." *Remote Sensing of Environment* 71(2): 158–182.

Tsai, F. and W. Philpot. 1998. "Derivative analysis of hyperspectral data." *Remote Sensing of Environment* 66(1): 41–51.

Vane, G., R. O. Green, T. G. Chrien, H. T. Enmark, E. G. Hansen and W. M. Porter. 1993. "The airborne visible/infrared imaging spectrometer (AVIRIS)." *Remote Sensing of Environment* 44(2): 127–143.

Vogelmann, J. E., B. N. Rock and D. M. Moss. 1993. "Red edge spectral measurements from sugar maple leaves." *International Journal of Remote Sensing* 14(8): 1563–1575.

Wang, Q. and J. Jin. 2015. "Leaf transpiration of drought tolerant plant can be captured by hyperspectral reflectance using PLSR analysis." *iForest—Biogeosciences and Forestry* 9: 30–37.

Wang, Q. and P. Li. 2012a. "Hyperspectral indices for estimating leaf biochemical properties in temperate deciduous forests: Comparison of simulated and measured reflectance data sets." *Ecological Indicators* 14(1): 56–65.

Wang, Q. and P. Li. 2012b. "Identification of robust hyperspectral indices on forest leaf water content using PROSPECT simulated dataset and field reflectance measurements." *Hydrological Processes* 26(8): 1230–1241.

Wang, Q. and P. Li. 2013. "Canopy vertical heterogeneity plays a critical role in reflectance simulation." *Agricultural And Forest Meteorology* 169: 111–121.

Wang, Y., X. Li, Z. Nashed, F. Zhao, H. Yang, Y. Guan and H. Zhang. 2007. "Regularized kernel-based BRDF model inversion method for ill-posed land surface parameter retrieval." *Remote Sensing of Environment* 111(1): 36–50.

Watanabe, S. 2010. "Asymptotic equivalence of Bayes cross validation and widely applicable information criterion in singular learning theory." *Journal of Machine Learning Research* 11: 3571–3594.

Westoby, M., D. S. Falster, A. T. Moles, P. A. Vesk and I. J. Wright. 2002. "Plant ecological strategies: Some leading dimensions of variation between species." *Annual Review of Ecology and Systematics* 33: 125–159.

Wiegand, C. L., A. J. Richardson, D. E. Escobar and A. H. Gerbermann. 1991. "Vegetation indices in crop assessments." *Remote Sensing of Environment* 35(2–3): 105–119.

Williams, P. C. 1987. *Variable Affecting Near Infrared Reflectance Spectroscopic Analysis. Near-Infrared Technology in the Agriculture and Food Industries.* P. Williams and K. Norris. St. Paul, MN, American Association of Cereal Chemists Inc.: 143–167.

Yao, X., H. Ren, Z. Cao, Y. Tian, W. Cao, Y. Zhu and T. Cheng. 2014. "Detecting leaf nitrogen content in wheat with canopy hyperspectrum under different soil backgrounds." *International Journal of Applied Earth Observation and Geoinformation* 32: 114–124.

Yebra, M. and E. Chuvieco. 2009. "Linking ecological information and radiative transfer models to estimate fuel moisture content in the Mediterranean region of Spain: Solving the ill-posed inverse problem." *Remote Sensing of Environment* 113(11): 2403–2411.

Yi, Q., G. Jiapaer, J. Chen, A. Bao and F. Wang. 2014. "Different units of measurement of carotenoids estimation in cotton using hyperspectral indices and partial least square regression." *Journal of Photogrammetry and Remote Sensing* 91: 72–84.

Yoder, B. J. and R. E. Pettigrew-Crosby. 1995. "Predicting nitrogen and chlorophyll content and concentrations from reflectance spectra (400–2500 nm) at leaf and canopy scales." *Remote Sensing of Environment* 53(3): 199–211.

Zarco-Tejada, P. J., A. Berjon, R. Lopez-Lozano, J. R. Miller, P. Martin, V. Cachorro, M. R. Gonzalez and A. de Frutos. 2005. "Assessing vineyard condition with hyperspectral indices: Leaf and canopy reflectance simulation in a row-structured discontinuous canopy." *Remote Sensing of Environment* 99(3): 271–287.

Zarco-Tejada, P. J., J. Miller, A. Morales, A. Berjon and J. Aguera. 2004. "Hyperspectral indices and model simulation for chlorophyll estimation in open-canopy tree crops." *Remote Sensing of Environment* 90(4): 463–476.

Zarco-Tejada, P. J., J. R. Miller, D. Haboudane, N. Tremblay and S. Apostol. 2003b. "Detection of chlorophyll fluorescence in vegetation from airborne hyperspectral CASI imagery in the red edge spectral region." *IGARSS 2003: IEEE International Geoscience and Remote Sensing Symposium, Vols I—Vii, Proceedings: Learning from Earth's Shapes and Sizes*: 598–600.

Zarco-Tejada, P. J., J. C. Pushnik, S. Dobrowski and S. L. Ustin. 2003a. "Steady-state chlorophyll a fluorescence detection from canopy derivative reflectance and double-peak red-edge effects." *Remote Sensing of Environment* 84(2): 283–294.

Zhang, L., Z. Zhou, G. Zhang, Y. Meng, B. Chen and Y. Wang. 2012. "Monitoring the leaf water content and specific leaf weight of cotton (*Gossypium hirsutum* L.) in saline soil using leaf spectral reflectance." *European Journal of Agronomy* 41: 103–117.

Zhang, X., F. Liu, Y. He and X. Gong. 2013. "Detecting macronutrients content and distribution in oilseed rape leaves based on hyperspectral imaging." *Biosystems Engineering* 115(1): 56–65.

Zhao, D. H., J. L. Li and J. G. Qi. 2005. "Identification of red and NIR spectral regions and vegetative indices for discrimination of cotton nitrogen stress and growth stage." *Computers and Electronics in Agriculture* 48(2): 155–169.

Zhao, J., M. Feng, C. Wang, W. Yang, Z. Li, Z. Zhu, P. Ren, T. Liu and H. Wang. 2014. "Simulating the Content of Chlorophyll in Winter Wheat Based on Spectral Vegetation Index." *Journal of Shanxi Agricultural University (Natural Science Edition)* 34(5): 391–396. (*in Chinese*)

Zheng, S., C. X. Cao, Y. F. Dang, H. B. Xiang, J. Zhao, Y. X. Zhang, X. J. Wang and H. W. Guo. 2014. "Retrieval of forest growing stock volume by two different methods using Landsat TM images." *International Journal of Remote Sensing* 35(1): 29–43.

Section II

Hyperspectral Image Classification
Methods and Approaches

3 Advances in Hyperspectral Image Classification Methods for Vegetation and Agricultural Cropland Studies

Edoardo Pasolli, Saurabh Prasad, Melba M. Crawford, and James C. Tilton

CONTENTS

3.1 INTRODUCTION

Hyperspectral data are becoming more widely available via sensors on airborne and unmanned aerial vehicle (UAV) platforms, as well as proximal platforms. While space-based hyperspectral data continue to be limited in availability, multiple spaceborne Earth-observing missions on traditional platforms are scheduled for launch, and companies are experimenting with small satellites for constellations to observe the Earth, as well as for planetary missions. Land cover mapping via classification is one of the most important applications of hyperspectral remote sensing and will increase in significance as time series of imagery are more readily available.

However, while the narrow bands of hyperspectral data provide new opportunities for chemistry-based modeling and mapping, challenges remain. Hyperspectral data are high dimensional, and many bands are highly correlated or irrelevant for a given classification problem. For supervised classification methods, the quantity of training data is typically limited relative to the dimension of the input space. The resulting Hughes phenomenon [1], often referred to as the curse of dimensionality, increases potential for unstable parameter estimates, overfitting, and poor generalization of classifiers [2]. This is particularly problematic for parametric approaches such as Gaussian maximum likelihood–based classifiers that have been the backbone of pixel-based multispectral classification methods. This issue has motivated investigation of alternatives, including regularization of the class covariance matrices [3], ensembles of weak classifiers [4,5], development of feature selection and extraction methods [6], adoption of nonparametric classifiers, and exploration of methods to exploit unlabeled samples via semi-supervised [7] and active learning [8,9]. Data sets are also quite large, motivating computationally efficient algorithms and implementations.

This chapter provides an overview of the recent advances in classification methods for mapping vegetation using hyperspectral data. Three data sets that are used in the hyperspectral classification literature (e.g., Botswana Hyperion satellite data and AVIRIS airborne data over both Kennedy Space Center and Indian Pines) are described in Section 3.2 and used to illustrate methods described in the chapter. An additional high-resolution hyperspectral data set acquired by a SpecTIR sensor on an airborne platform over the Indian Pines area is included to exemplify the use of new deep learning approaches, and a multiplatform example of airborne hyperspectral data is provided to demonstrate transfer learning in hyperspectral image classification.

Classical approaches for supervised and unsupervised feature selection and extraction are reviewed in Section 3.3. In particular, nonlinearities exhibited in hyperspectral imagery have motivated development of nonlinear feature extraction methods in manifold learning, which are outlined in Section 3.3.1.4. Spatial context is also important in classification of both natural vegetation with complex textural patterns and large agricultural fields with significant local variability within fields. Approaches to exploit spatial features at both the pixel level (e.g., co-occurrence–based texture and extended morphological attribute profiles [EMAPs]) and integration of segmentation approaches (e.g., HSeg) are discussed in this context in Section 3.3.2.

Recently, classification methods that leverage nonparametric methods originating in the machine learning community have grown in popularity. An overview of both widely used and newly emerging approaches, including support vector machines (SVMs), Gaussian mixture models, and deep learning based on convolutional neural networks is provided in Section 3.4. Strategies to exploit unlabeled samples, including active learning and metric learning, which combine feature extraction and augmentation of the pool of training samples in an active learning framework, are outlined in Section 3.5. Integration of image segmentation with classification to accommodate spatial coherence typically observed in vegetation is also explored, including as an integrated active learning system. Exploitation of multisensor strategies for augmenting the pool of training samples is investigated via a transfer learning framework in Section 3.5.1.2. Finally, we look to the future, considering opportunities soon to be provided by new paradigms, as hyperspectral sensing is becoming common at multiple scales from ground-based and airborne autonomous vehicles to manned aircraft and space-based platforms.

3.2 HYPERSPECTRAL TESTBED FOR VEGETATION CLASSIFICATION

Five publically available hyperspectral benchmark data sets, which have been used to evaluate classification algorithms for vegetation-based studies in the literature, are included to illustrate the methodology presented in this chapter. The data were acquired by space-based and airborne sensors covering the visible and short-wave infrared portions of the spectrum at spatial resolutions ranging from 2 to 30 m. The higher spatial resolution acquisition allows both discrimination of smaller objects and utilization of texture information by the classification algorithms. Spectral signatures of the classes are complex and often overlapping, and spatial patterns include agricultural fields with regular boundaries and natural vegetation where classes are often fragmented or mixed. Characteristics of the data sets are listed in Table 3.1.

3.2.1 Botswana Hyperion Data (BOT)

The NASA EO-1 satellite was launched in November 2000, with Hyperion as an "auxiliary" hyperspectral sensor. The EO-1 platform was designed for one year of operation, but was finally decommissioned nearly two decades later in 2017. Hyperion data were acquired in 7.7-km strips at 30 m spatial resolution for a multiyear study of flooding in the Okavango Delta, Botswana. Uncalibrated and noisy bands were removed, leaving 145 bands as candidate features for classification. Nine classes of complex natural vegetation were identified by researchers at the Okavango Research Center. Class groupings include seasonal swamps, occasional swamps, and woodlands, and are distributed in fragmented patterns over a large area. RGB images of the area, maps of the ground reference data, and a class legend are included in Figure 3.1. As shown in the figure, the pointing angle of the satellite changed after the May acquisition, necessitating development of knowledge transfer models. Signatures of several classes overlap spectrally, resulting in a challenging data set for classification. Class 3 (riparian) and Class 6 (woodlands) are particularly difficult to discriminate. After removing water absorption and noisy and overlapping spectral bands in the visible and near-infrared (VNIR) and short-wave infrared (SWIR) sensors, 145 bands of May and July 2001 images were used for classification experiments.

3.2.2 Kennedy Space Center AVIRIS Data (KSC)

Airborne hyperspectral data were acquired by NASA AVIRIS at 18 m spatial resolution and 10 nm spectral resolution over a natural wetland/upland environment adjacent to the Kennedy Space Center, Florida, in March 1996, to evaluate the impact of drainage management practices on the incursion of invasive species into an endangered species habitat. Figure 3.2 includes an RGB image of the area, ground reference data, and class reference information. Noisy and water absorption bands

TABLE 3.1

Testbed Data Sets and Associated Characteristics

Data Set	BOT	KSC	Indian Pine 1992	Indian Pine 2010	Galveston, Texas	Galveston, Texas
Scene description	Vegetation and flooding	Wetland/upland vegetation	Early season agriculture	Early season agriculture	Wetland vegetation	Wetland vegetation
Sensor	Hyperion	AVIRIS	AVIRIS	SpecTIR	SpecTIR	Headwall nano
Platform	Satellite	Airborne	Airborne	Airborne	Airborne	Terrestrial
Spectral range	0.4–2.5 μm	0.4–2.5 μm	0.4–2.5 μm	0.4–2.5 μm	0.4–2.5 μm	0.4–1.0 μm
Spatial resolution	30 m	18 m	18 m	2 m	1 m	Variable
No. of bands	220	176	176	360	360	274
No. of classes	9	13	16	12	12	12

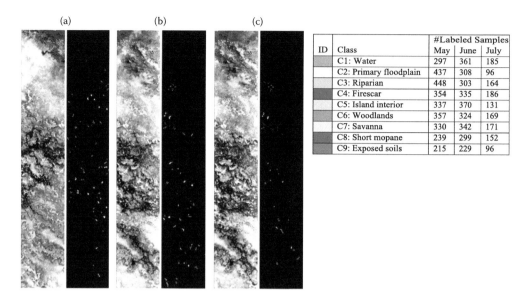

ID	Class	#Labeled Samples		
		May	June	July
	C1: Water	297	361	185
	C2: Primary floodplain	437	308	96
	C3: Riparian	448	303	164
	C4: Firescar	354	335	186
	C5: Island interior	337	370	131
	C6: Woodlands	357	324	169
	C7: Savanna	330	342	171
	C8: Short mopane	239	299	152
	C9: Exposed soils	215	229	96

FIGURE 3.1 Botswana (BOT) Hyperion data. True color composites with corresponding ground reference data in (a) May, (c) June, and (e) July 2001, class labels and # labeled samples. The pointing angle changed after the May acquisition, and then remained the same for subsequent dates.

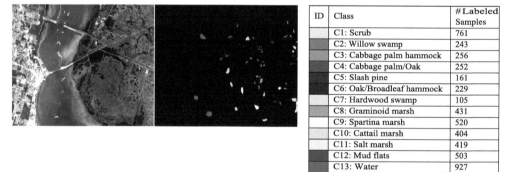

ID	Class	#Labeled Samples
	C1: Scrub	761
	C2: Willow swamp	243
	C3: Cabbage palm hammock	256
	C4: Cabbage palm/Oak	252
	C5: Slash pine	161
	C6: Oak/Broadleaf hammock	229
	C7: Hardwood swamp	105
	C8: Graminoid marsh	431
	C9: Spartina marsh	520
	C10: Cattail marsh	404
	C11: Salt marsh	419
	C12: Mud flats	503
	C13: Water	927

FIGURE 3.2 Kennedy Space Center (KSC) AVIRIS data. RGB true-color composite and corresponding ground reference map, class labels, and # labeled samples.

were removed from the reflectance data, leaving 176 features for 13 wetland and upland classes. The spectral signatures of multiple classes are mixed and often exhibit only subtle differences. Cabbage Palm Hammock (Class 3) and Broad Leaf/Oak Hammock (Class 6) are upland trees; Willow Swamp (Class 2), Hardwood Swamp (Class 7), Graminoid Marsh (Class 8), and Spartina Marsh (Class 9) are trees and grasses in transition wetlands. Classification results for all 13 classes and for these difficult classes are reported in several publications.

3.2.3 INDIAN PINE AVIRIS 1992 AND SPECTIR 2010 DATA

The historical data acquired by the NASA AVIRIS sensor in June 1992 over a central Indiana farming area have been widely used to evaluate classification methods that exploit spatial information. After removing 20 water absorption bands, 200 bands are used for analysis. The scene is composed of agricultural fields with regular geometry, providing an opportunity to evaluate the impact of within-class and between-class variability at medium spatial resolution. The spectral signatures of corn and soybean fields, which were planted only a short time prior to the acquisition, illustrate the impact of tillage management practices.

The 16 classes of labeled reference data are reported at the field scale for crops, but significant within-field variability resulted in heterogeneous spectral responses. Although labeled as vegetation, the spectral responses of many classes are dominated by soil and residue signatures from the previous year. Figure 3.3 includes an RGB image of the area, class legends, and the corresponding labeled data.

Additional hyperspectral imagery was acquired by the airborne ProSpecTIR VS2 VNIR/SWIR in June 2010 for a study of residue cover estimates over an area near the location of the original Indian Pine AVIRIS data. The data were collected at 2 m spatial resolution in 360 channels at 5 nm spectral resolution over the range of 390–2450 nm. Bands were aggregated to 10 nm, and 178 spectral bands were used for analysis. Nineteen classes of crops, residue, and buildings were identified. A true color composite is shown in Figure 3.4 with associated ground reference data and class information.

3.2.4 TEXAS COAST SPECTIR DATA

A heterogeneous hyperspectral data set composed of airborne imagery and a unit collected over wetland vegetation are included to illustrate domain adaptation (transfer learning) (Figure 3.5).

ID	Class	# Labeled Samples
	C1: Corn-no till	1434
	C2: Corn-min till	834
	C3: Corn	234
	C4: Soybeans-no till	968
	C5: Soybeans-min till	2468
	C6: Soybeans-clean till	614
	C7: Alfalfa	54
	C8: Grass/Pasture	497
	C9: Grass/Trees	747
	C10: Grass/Pasture-mowed	26
	C11: Hay-windrowed	489
	C12: Oats	20
	C13: Wheat	212
	C14: Woods	1294
	C15: Building/Grass/Trees/Drive	380
	C16: Stone/Steel towers	95

FIGURE 3.3 Indian Pine 1992 AVIRIS data. RGB true-color composite and corresponding ground reference map, class labels, and # labeled samples.

ID	Class	# Labeled Samples
	C1: Corn-high residue	47,168
	C2: Corn-mid residue	183,530
	C3: Corn-low residue	356
	C4: Soybean-high residue	226,130
	C5: Soybean-mid residue	120,400
	C6: Soybean-low residue	29,210
	C7: Other residues	5,795
	C8: Wheat	3,387
	C9: Hay	63,135
	C10: Grass/Pasture	6,512
	C11: Grass	26,853
	C12: Wood-uniform	64,947
	C13: Wood-rugged	288,500
	C14: Highway	10,637
	C15: Local road	6,570
	C16: Power station	6,929
	C17: Power towers	411
	C18: Houses/Buildings	2,128
	C19: Urban areas	1,532

FIGURE 3.4 Indian Pine 2010 SpecTIR data. RGB true-color composite and corresponding ground reference map, class labels, and # labeled samples.

FIGURE 3.5 Galveston, Texas, data. True-color images of the aerial view (target domain) and street view (source domain) wetland data. The location of the study site is indicated by the red box in a Google Earth screenshot image.

Changes in distribution of wetland vegetation species can have profound impacts on the coastal economy and ecology; hence, studying wetland vegetation through remote sensing is of great importance, particularly over extended geographic areas.. Marshes in Misson-Aransas estuary, which were previously dominated by smooth cordgrass (*Spartina alterniflora*), have been replaced by black mangroves (*Avicennia germinans*).

Hyperspectral imagery was acquired by the airborne ProSpecTIR VS sensor with a spatial coverage of 3462 × 5037 pixels at a 1 m spatial resolution. A field survey was undertaken on September 16, 2016. Figure 3.6 depicts the mean spectral signatures of 12 identified species/classes (Table 3.2) As part of this field survey, side-looking hyperspectral imagery (called "street view" imagery in this chapter) was collected using a Headwall Nano-Hyperspec sensor of the same area. It has 274 bands spanning 400–1000 nm at a 3 nm spectral resolution. This resulted in a unique domain adaptation problem where models were trained from very high-resolution street-view (terrestrial) imagery and transferred to aerial imagery. It is also very challenging because the two domains (street view and aerial) are different in many ways, including the viewpoints, illumination conditions, atmospheric conditions, and so on.

3.3   MANAGING THE FEATURE SPACE

3.3.1   Dimensionality Reduction via Feature Selection and Extraction

3.3.1.1   Feature Selection

Identifying features that are effective for modeling data class characteristics is a critical preprocessing step for hyperspectral image classification. Apart from computational

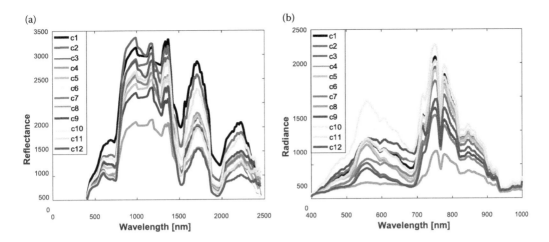

FIGURE 3.6 Galveston, Texas, SpecTIR airborne and Headwall handheld camera data. Mean spectral signatures of the (a) aerial view (target domain) and (b) street view (source domain) wetland data.

TABLE 3.2

Galveston, Texas, Data

Class	# Labeled Samples (Aerial View)	# Labeled Samples (Street View)
C1: Upland grass	794	1463
C2: St. Augustine grass	100	1009
C3: Sesbania	294	1021
C4: Upland tree	426	1040
C5: *Phragmites australis*	780	1029
C6: *Sabal mexicana*	74	1189
C7: *Spartina alterniflora*	733	1152
C8: *Juncus roemerianus*	202	1264
C9: *Batis maritima/Distichlis spicata*	596	1106
C10: *Distichlis spicata*	1197	1087
C11: *Baccharis halimifolia*	360	1017
C12: *Avicennia germinans*	1663	1119

requirements, classifiers tend to have low generalization capabilities when data are characterized by high dimensionality, especially when the number of training samples is limited with respect to the number of features. A traditional solution to deal with this problem is represented by feature selection (FS), which aims to reduce the dimensionality of the original feature space by choosing the best—and ideally the minimum—subset of features. Numerous FS approaches have been proposed in the last few decades [6] and can be grouped in two main categories: *filter* and *wrapper* methods. Filter methods perform FS as a preprocessing step where the selection criterion is independent of the classifiers used to subsequently perform classification of the data [10]. Wrapper strategies perform FS based on the performance of a given classifier [11]. These techniques are generally applied to the original spectral bands, although they can be extended to newly generated [12] or extracted features [13]. Selection from the original feature space is advantageous in the sense that the resulting features retain a physical relationship to the original process. We focus in the following on filter methods, which can be categorized as supervised and unsupervised approaches.

Supervised FS can be further subdivided in two major groups, that is, parametric and nonparametric methods. *Parametric supervised FS* involves class modeling using training data. A widely adopted method is based on the Jeffries-Matusita (J-M) distance, which measures the separability of two class density functions [14]. When classes are assumed to be Gaussian distributed [15], computation of the J-M distance is based on the Bhattacharyya distance. This approach performs well when the Gaussian distribution assumption is valid. It is used to find a subset of features to best accommodate class data variations at multiple sites/locations and generate a visual representation of the separation capability provided by each band, which then leads to quantitative band selection [16]. Other distance measures can also be used, including spectral angle, Euclidean distance, and Mahalanobis distance. Apart from distances between classes, the ratio of within-class and between-class variance can be used as well to define separability [17]. *Nonparametric supervised FS* considers the information provided by the training data directly, without requiring class data modeling. As an example, overlapping and noisy bands can be removed using a canonical correlation measure to obtain an optimal subset of features that provide the best estimate of the center of classes [18]. Information theory can also be applied to perform nonparametric supervised FS. Mutual information (MI) provides a measure of linear and nonlinear dependency between two variables [19] and is suitable for general cases since no assumptions about the shape of the class data density functions must be made.

Unsupervised FS aims at reducing feature redundancy. For example, features that are dissimilar to those already selected can be chosen one by one via linear prediction error analysis [20]. Other methods use similarity measures to partition the original set of features into a number of homogeneous clusters and then select a representative feature from each cluster [21]. MI is used to find the subset of features with minimum dependency [22]. Subsets of representative features can also be selected simultaneously through geometry-based FS methods [23].

After defining the selection criterion, adopting the appropriate *searching strategy* is a challenging task. A first solution is represented by *exhaustive search* in which all the possible combinations are evaluated. Exhaustive search is often impractical, so alternative suboptimal approaches are usually adopted and defined as *greedy search*. These optimization problems are usually not convex, and heuristic strategies are needed. In the case of a monotonic criterion, the branch and bound method [24] can be applied to avoid exhaustive search in a moderate-sized searching space. Sequential forward selection and backward elimination are fast approaches, but do not allow feedback to revise previously selected features. Improvements are represented by sequential forward floating selection and sequential backward floating selection [25] in which the selected features are reconsidered for inclusion or deletion at each iteration. Combinatorial optimization approaches use heuristic methods to reduce the number of features. Proposed solutions include methods based on genetic algorithms [26], particle swarm optimization [27], and clonal selection [28].

3.3.1.2 Spectral Indices

Spectral indices computed as ratios of broad-band spectral bands or of normalized differences between pairs of bands based on multispectral radiance and reflectance data have been used for nearly half a century in studies of vegetation. More recently, narrow-band indices based on spectral bands and derivatives of the reflectance spectrum have been explored in vegetation studies for their value in characterizing biophysical properties, predicting variables of interest, and mapping or unmixing land cover. The continuous spectrum provided by hyperspectral sensors and the relationship of spectral absorption features in spectral signatures to chemistry-based properties provides robust capability to target explicit characteristics compared to what can be achieved by broad-band multispectral indices. Thenkabail et al. [29] conducted a comprehensive review of spectral indices derived from the EO-1 Hyperion hyperspectral sensor using data acquired for a wide range of agricultural applications and geographic sites. Recently, narrow-band indices have also become a focus of high-throughput phenotyping by plant breeders seeking to map and relate phenotypic traits to genotypes [30].

Spectral indices also provide an appealing physically-based capability for reducing the dimensionality and redundancy that are inherent in hyperspectral signatures over vegetation.

Bridging band-specific feature selection and extraction approaches that seek to represent relevant information in the spectrum via global transformations, hyperspectral indices provide robust, local transformations that are useful for a wide range of applications in remote sensing–based studies of vegetation, including classification.

3.3.1.3 Linear Transformation–Based Approaches

Feature extraction (FE) seeks to project the original feature space onto a small number of features. While extracted features lose their relationship to the original physical phenomena, they provide a compressed version of the original feature set. Each feature in the original hyperspectral data is characterized by a contribution determined by the transformation matrix associated with the chosen extraction method. FE techniques can be categorized as unsupervised (global data oriented) and supervised (class data oriented) or as linear and nonlinear.

In *supervised FE* methods, new discriminative features can be obtained by combining the original features into groups. As an example, adjacent correlated features can be combined into a smaller number of features retaining the original spectral interpretation [31]. B-dis can be adopted as a grouping criterion before creating new features as a weighted sum of the features in each group [32]. In [33], contiguous groups of features are averaged based on JM distance. Linear discriminant analysis (LDA) and canonical analysis are traditional parametric FE techniques that are based on the mean vector and covariance matrix of each class. The ratio of within-class to between-class scatter matrices is used to formulate an effective criterion of class separability. Limitations of LDA include its dependence on the distributions of classes being approximately Gaussian and its inability to handle cases where class data do not form a single cluster. Further, the maximum rank of the between-class scatter matrix is the number of classes (M) minus one. Decision boundary FE (DBFE), an early method developed specifically for hyperspectral data [34], aims at finding new features that are normal to class decision boundaries. Nonparametric FE via regularization techniques have also been proposed to overcome the limitations of LDA and obtain more stable results [35]. Nonparametric discriminant analysis (NDA) [36] defines a nonparametric between-class scatter matrix based on a critical finding that samples close to the boundary are more relevant than those far from it. Nonparametric weighted FE (NWFE) was developed in light of NDA, introducing regularization techniques to achieve better performance for hyperspectral image classification than LDA and NDA. Double nearest proportion (DNP) FE builds new scatter matrices based on a double nearest proportion structure [37].

Unsupervised FE is usually obtained by combining the original group of features via an average or weighted sum operation. For example, top-down and bottom-up decompositions can be adopted to merge highly correlated adjacent features and then project them onto their Fisher directions [31]. Apart from combining groups of contiguous features, a more general FE approach consists of mapping the original high-dimensional space into a low-dimensional one via a data transform. Two typical data transformation methods are principal component analysis (PCA), which reduces the original set of features into a smaller set of orthogonal ones computed as linear combinations of the original features with maximum variance [38], and independent component analysis (ICA), a statistical technique for separating the independent signals from overlapping signals [39]. Other techniques seek projections via different optimization models. Projection pursuit (PP) methods search for projections that optimize certain projection indexes [40]. In both supervised and unsupervised FE, the kernel trick is an easy way to extend linear models to nonlinear ones, as we describe in more detail in the next section. In [41–43], angular distance–based supervised, unsupervised, and semisupervised discriminant analysis and their kernel variants were proposed and shown to outperform Euclidean distance–based variants of such algorithms for hyperspectral classification.

3.3.1.4 Manifold Learning

Although hyperspectral data are typically modeled assuming that the data originate from linear processes, and linear feature extraction approaches are simple and straightforward to implement,

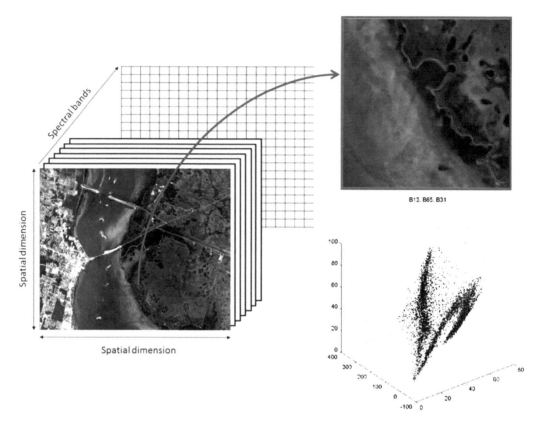

FIGURE 3.7 Nonlinearity in the spectral data is exhibited in a plot of bands 13, 65, and 31 for the Kennedy Space Center data set. (Adapted from L. Bruzzone et al. *Hyperspectral Data Exploitation: Theory and Applications*, pp. 275–311, 2007. [45])

nonlinearities associated with physical processes are often exhibited in the narrow-band data [44]. Nonlinear feature extraction techniques assume that the high-dimensional data inherently lie on a lower-dimensional manifold, as shown in Figure 3.7.

The machine learning community initially demonstrated the potential of manifold-based approaches for nonlinear dimensionality reduction and modeling of nonlinear structures [46–50], and its application to classification of hyperspectral data has been exhibited over the past decade for unsupervised, supervised, and semisupervised learning [45,51], as well as for multitemporal scenarios [52,53]. Nonlinear manifold learning methods are broadly characterized as globally or locally based approaches. Global manifold methods seek to maintain the fidelity of the overall topology of the data set at multiple scales of the data, while local methods preserve local geometry and are computationally efficient because they only require sparse matrix computations. Global manifolds are generally less susceptible to overfitting the data, which is beneficial for generalization in classification, but local methods potentially provide opportunities for better representation of heterogeneous data with submanifolds. Traditional manifold learning methods whose theoretical foundation is associated with the eigenspectrum and kernel framework include isometric feature mapping (ISOMAP) [46], kernel principal component analysis (KPCA) [47], and locally linear embedding (LLE) [48].

Manifold learning is classically represented in a graph-embedding approach, both for efficiency and computational simplicity. A Laplacian regularizer is employed to constrain the classification function to be smooth with respect to the data manifold, and the resulting coordinates are obtained from the eigenvectors of a graph Laplacian matrix. Different manifold learning methods correspond to specific graph structures. Compared to global methods that represent distances across the full manifold, local manifold learning evaluates the local geometry in each neighborhood, and data points that are included in neighborhood are connected in the graph. In the graph-embedding framework, manifold coordinates are obtained from the eigenvectors of the graph Laplacian matrix. The graph can either be developed from the combined training and testing data (unsupervised), or manifold learning can be applied to the training data and an out-of-sample extension method employed to incorporate the testing data (supervised) [54,55]. The first strategy can provide more accurate manifold coordinates, while the latter is advantageous when the quantity of testing data is large.

In the general formulation of the graph Laplacian-based framework, samples are defined in a data matrix $X = [x_1, x_2, \ldots, x_n]$, $x_i \in \mathbb{R}^m$, where n is the number of samples and m is the feature dimension. The dimensionality reduction problem seeks a set of manifold coordinates $Z = [z_1, z_2 \ldots, z_n]$, $z_i \in \mathbb{R}^p$, where typically, $m \gg p$, through feature mapping $\Phi : x \to y$, which may be analytical (explicit) or data driven (implicit), and linear or nonlinear. Assuming an undirected weighted graph $G = \{X, W\}$ with data samples X and algorithm-dependent similarity matrix W, the graph Laplacian $L = D - W$, with diagonal degree matrix $D_{ii} = \Sigma_j W_{ij}$, $\forall_i$. Given labeled data $X_l = [x_1, x_2, \ldots, x_l]$ and unlabeled data $X_u = [x_{l+1}, x_{l+2}, \ldots, x_{l+u}]$, $\Omega = [\Omega_1, \ldots, \Omega_C]$, where C is the number of classes. Class labels of X_l are denoted as $Y_l \in \mathbb{R}^{C \times l}$ with $Y_{ij} = 1$ if $x_j \in \Omega_i$. The data points $X = [X_l, X_u]$ produce a weighted graph $G = \{X, W\}$, where X consists of $N = l + u$ data points, and $W \in \mathbb{R}^{N \times N}$ is the adjacency matrix of the connected edges between data points. For a directed graph, a symmetric adjacency matrix $W = W + W^T$ can be assumed. The dimensionality reduction criterion can be represented as $Y^* = \arg \min_{Y BY^T = I} tr(Y LT^T)$, where B is a constraint matrix that depends on the dimensionality reduction method.

Plots of coordinates obtained using PCA, Isomap, and LLE for the nine-class NASA Hyperion BOT data, with optimal parameters for each manifold embedding, are shown in Figure 3.8.

Differences in the separation of classes, which relate to potential discrimination via classification, are shown for the three projections. The objective functions for neither linear projections such as PCA nor manifold learning are related to the classification objective, so the resulting projections may or may not provide improved separation of classes.

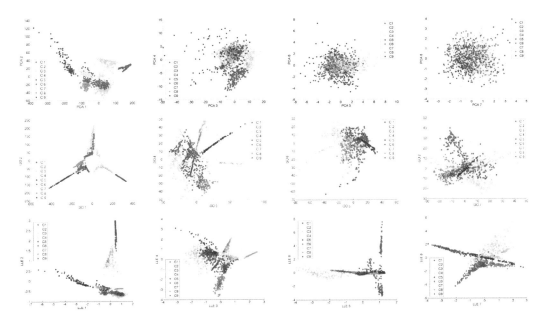

FIGURE 3.8   Plots of coordinates obtained from (first line) PCA, (second line) ISOMAP, and (third line) LLE for the nine Botswana classes (C1-C9). (a) bands 1–2, (b) bands 3–4, (c) bands 5–6, (d) bands 7–8.

3.3.2 Incorporation of Spatial Context

For many applications, a key discriminative component that makes hyperspectral imaging appealing is the richness of the spectral reflectance profiles. There is much to be gained for most applications, including vegetation classification, by leveraging the spatial context in the imagery. A common approach to incorporating spatial context for classification is to extract meaningful morphological and textural features [56,57] and then learn the classifier in the resulting feature space. Within this space, a choice that has been demonstrated to be particularly successful for hyperspectral classification tasks is extended morphological attribute profile (EMAP) features, that represent profiles created by removing connected components that do not meet some criteria specified a priori. For situations where the criteria are satisfied, the regions are kept intact; otherwise, they are set to the gray level of a darker or brighter surrounding region. Attributes (features) can represent the morphology/geometric properties of the objects (e.g., image moments and shape) or the textural information about the objects (e.g., range, standard deviation, and entropy). EMAP features have found significant success in hyperspectral image analysis—in the context of multichannel images, such features are often computed from a subset of features extracted from the hypercube (e.g., selected bands or the first few principal components of the cube). They have been applied to a wide variety of remote sensing applications to extract spatial context for image analysis [58,59].

3.4 IMAGE CLASSIFICATION STRATEGIES

3.4.1 Classical Pixel (Sample)–Based Classification for Hyperspectral Image Analysis

Within the realm of geospatial image analysis in general and hyperspectral image analysis in particular, a significant focus within the research community has been on the design of feature reduction and analysis algorithms (classification, change and anomaly detection, target recognition, spectral unmixing, etc.) [60–63]. Similar to feature extraction and dimension reduction, classifiers that utilize domain-specific properties of hyperspectral data have emerged as popular choices [64–68]. Over the past decade, the choice of classifiers for hyperspectral classification has shifted from traditional approaches such as K-nearest neighbors, Gaussian maximum-likelihood classification, and random forests to more advanced approaches that offer greater capacity in modeling nonlinear decision surfaces in the feature space. K-nearest neighbor–based classifiers assume that data can be classified by surveying the K training data points nearest the test point in the feature space—an assumption that is effective when coupled with local manifold learning approaches, but is far too simplistic as a classification scheme in itself. Likewise, Gaussian maximum-likelihood classifiers assume that class-conditional likelihoods are modeled as Gaussian distributions in the spectral reflectance parameter space, and the underlying parametrization is then learned via maximum likelihood—the approach has several shortcomings, including the necessity to undertake feature reduction as a preprocessing step prior to inferring the high-dimensional parameters and departure of class-conditional distributions from unimodal Gaussian behavior under a variety of practical remote sensing scenarios (such as the same class being present in a well-illuminated part of the scene and under cloud shadows). More advanced classification approaches have hence been proposed and utilized in recent years for hyperspectral classification. These include multikernel learning and variants of support vector machines [69,70], and other model-based classifiers, spectrally constrained implementations of Gaussian mixture models (GMMs) [66], sparse representation classification (SRC) [71,72], local approaches [73], and so on.

3.4.1.1 Single-Kernel Support Vector Machines

Given a training data set $\mathbf{X} = \left\{\mathbf{x}_i\right\}_{i=1}^{n}$ in $\mathbb{R}^d$ with class labels $y_i \in \{+1, -1\}$ and a nonlinear kernel function $\phi(\cdot)$, an SVM [74] classifies data by learning the optimal hyperplane in the Hilbert space induced by the kernel function

$$\min_{\omega, \xi_i, b} \left\{ \frac{1}{2}\|\omega\|^2 + \varsigma \sum_{i=1}^{n} \xi_i \right\} \tag{3.1}$$

subject to the constraints

$$y_i(\langle \phi(\omega, \mathbf{x}_i) \rangle + b) \geq 1 - \xi_i, \tag{3.2}$$

for $\xi_i \geq 0$ and $i = 1, \dots, n$, where ω is normal to the optimal decision hyperplane [i.e., $\langle \omega, \phi(\mathbf{x}) \rangle + b = 0$], n denotes the number of samples, b is the bias term, ς is the regularization parameter that controls the generalization capacity of the machine, and ξ_i is the positive slack variable allowing appropriate accommodation of permitted errors. The above problem is solved by maximizing its Lagrangian dual form,

$$\max_{\alpha} \left\{ \sum_{i=1}^{n} \alpha_i - \frac{1}{2} \sum_{i,j=1}^{n} \alpha_i \alpha_j y_i y_j K(\mathbf{x}_i, \mathbf{x}_j) \right\}, \tag{3.3}$$

where $\alpha_1, \alpha_2, \dots, \alpha_n$ are nonzero Lagrange multipliers constrained to $0 \leq \alpha_i \leq \varsigma$, and $\Sigma_i \alpha_i y_i = 0$, for $i = 1, \dots, n$. Some commonly implemented kernel functions are the linear kernel, the polynomial kernel, and the radial-basis-function (RBF) kernel [74].

$$K(\mathbf{x}_i, \mathbf{x}_j) = \exp\left(-\frac{\|\mathbf{x}_i - \mathbf{x}_j\|^2}{2\sigma^2} \right), \tag{3.4}$$

where σ is a width parameter of the RBF that controls the generalization capacity of the machine. The decision function is then given as

$$f(\mathbf{x}) = \mathrm{sgn}\left(\sum_{i=1}^{n} y_i \alpha_i K(\mathbf{x}_i, \mathbf{x}) + b \right). \tag{3.5}$$

SVM has become one of the most widely used nonparametric classifiers for hyperspectral data, in part due to its relative independence from data dimensionality, and it is included now in many software libraries.

3.4.2 Bayesian Parametric and Nonparametric Classification

Bayesian classification entails modeling class-conditional likelihoods $[p(\mathbf{x}|y_i), i \in \{1, \dots, c\}]$ through an underlying probability model that is learned in the feature space (or some other appropriate subspace) from the training data. The state of the art in such methods for hyperspectral classification assumes that the class-conditional likelihoods are best modeled as mixtures (weighted linear combination) of "basis" Gaussian density functions. Depending upon whether one assumes there are finite or infinite Gaussian components in the mixture, the resulting model is either a traditional Gaussian mixture model or infinite Gaussian mixture model (IGMM), respectively.

3.4.2.1 Finite Gaussian Mixture Model

A GMM is a weighted linear combination of a finite number (K) of Gaussian components, such that the resulting likelihood function of $\mathbf{X} = \{\mathbf{x}_i\}_{i=1}^n$ in $\mathbb{R}^d$ is

$$p(\mathbf{x}) = \sum_{k=1}^{K} \alpha_k \mathcal{N}(\mathbf{x}, \mu_k, \Sigma_k) \tag{3.6}$$

where

$$\mathcal{N}\left(\mathbf{x},\mu_k,\Sigma_k\right) = \frac{1}{(2\pi)^{d/2}\left|\Sigma_k\right|^{1/2}}\exp\left[-\frac{1}{2}(\mathbf{x}-\mu_k)^{\top}\Sigma_k^{-1}(\mathbf{x}-\mu_k)\right]. \tag{3.7}$$

GMMs are hence a parametric representation of the underlying probability model of the data, where the parametrization is formed by the parametric representations of each of the K Gaussian distributions (mean vector, μ_k, and covariance matrix, Σ_k), as well as the weights of each component in the mixture α_k, $\Theta = \{\alpha_k,\mu_k,\Sigma_k\}_{k=1}^{K}$.

Training GMM models (i.e., given $\mathbf{X}$, estimating $\Theta = \{\alpha_k,\mu_k,\Sigma_k\}_{k=1}^{K}$) is typically undertaken using the expectation maximization (EM) approach. A crucial design choice when training such GMM models is then determining the number of mixtures/components, K. This is often learned empirically by making use of an information theoretic measure, such as the Akaike information criterion (AIC) [75].

Such a Bayesian model can be used for both unsupervised and supervised learning. In an unsupervised learning framework, GMMs are often used as a robust Bayesian paradigm for clustering data—for instance, given an imagery data set (a set of pixels, $\mathbf{X}$) with no labels, cluster the data into its -K constituent clusters, where K is the underlying number of components (e.g., number of classes in the scene, assumed to be known a priori). In a supervised learning framework, GMMs can be used to learn class-conditional likelihood functions, with the assumption that each class may have a likelihood function that requires a mixture of multiple Gaussians to represent it effectively. In the context of remotely sensed imagery, such multimodal class-conditional distributions may arise due to practical effects (e.g., the same class in the image under different illumination conditions or variations in spectral response of vegetation due to varying vegetation stress within the same class in different parts of the scene).

3.4.2.2 Infinite Gaussian Mixture Models

A fundamental challenge of traditional GMMs is that the number of mixture components, K, may be difficult to ascertain a priori, and empirical approaches to estimating K may result in under- or overestimation of the required number of components within the mixture model. A recent development in the field of Bayesian nonparametrics, the IGMM, addresses this issue very effectively. Unlike traditional GMMs, the number of mixture components in IGMMs can be ascertained as part of the Bayesian inference process. IGMMs are a specific variant [76] of Dirichlet process mixture models (DPMMs). In DPMM formulations, often a stick-breaking construction process [76] is used to generate mixture/component weights that have a Dirichlet process prior placed on them. Unlike traditional GMMs, IGMMs do not assume that the number of mixture components is known a priori and instead infer the mixture components via Bayesian inference strategies, such as Markov chain Monte Carlo sampling [76].

3.4.2.3 Practical Issues: Dimensionality

A key point to be made here with respect to Bayesian inference approaches is that the success of such approaches is contingent on the training sample size relative to the dimensionality of the feature space. For example, with a K-component GMM, assuming full covariance matrices, one needs to estimate $K(1 + d(d - 1)/2) + Kd$ parameters from the training data. Learning such high-dimensional parameters with limited training data is highly impractical—hence, feature extraction (dimensionality reduction) is often a critical preprocessing step to such Bayesian inference strategies. Figure 3.9 illustrates the performance of GMM-based classification coupled with a locality preserving feature reduction approach for classification of the Indian Pine 1992 data set. For this data set, LFDA-SVM and LFDA-GMM consistently resulted in the highest overall accuracy. In our recent work [77], we found that for hyperspectral image analysis, including for vegetation, locality-preserving approaches

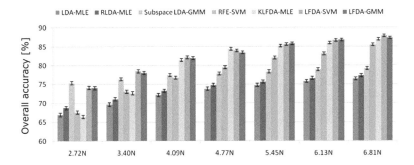

FIGURE 3.9 Classification accuracy as a function of training sample size using GMM classifiers coupled with LFDA-based feature reduction for the Indian Pine 1992 data set. Comparisons to baseline methods are also provided. Linear discriminant analysis-max likelihood estimate (LDA-MLE); regularized linear discriminant analysis-max likelihood estimate (RLDA-MLE); subspace LDA-Gaussian mixture model (Subspace LDA-GMM); recursive feature elimination-SVM (RFE-SVM); kernel local Fisher discriminant analysis-SVM (KLFDA-SVM); local Fisher discriminant analysis-SVM (LFDA-SVM); local Fisher discriminant analysis-Gaussian mixture models. (Adapted from W. Li et al. *IEEE Transactions on Geoscience and Remote Sensing*, vol. 50, no. 4, pp. 1185–1198, 2012. [78])

to subspace learning are the most effective feature extraction/preprocessing for Bayesian inference, as they preserve locality (neighborhood structure) of the feature space in the embedded subspace.

3.4.3 Deep Learning of Hyperspectral Imagery Data

Deep learning has emerged as a very effective approach in contemporary computer vision and signal analysis tasks. This has been driven in part by increased availability of high-performance computing infrastructures and rich libraries for training models. Deep learning algorithms include deep convolutional neural networks (CNNs), recurrent neural networks (RNNs), and deep belief networks (DBNs) that have been successfully deployed for speech recognition, computer vision, natural language processing, and, more recently, remote sensing applications [79–81]. CNNs and their variants have been successfully used for tasks such as large-scale object detection, transfer learning/domain adaptation, and so on. RNNs and their variants have also been demonstrated to be useful for temporal modeling for applications such as speech recognition [82,83].

Within the context of pixel-level hyperspectral image analysis, we have observed that the spectral reflectance characteristics of the data can be modeled quite effectively through what we call convolutional recurrent neural networks (CRNNs). The approach entails a series of convolutional and pooling layers followed by recurrent neural network layers, as shown in Figure 3.10. The convolutional layers are adept at extracting stable, locally invariant features from the spectral reflectance features. Pooling layers help minimize effects of overfitting. The recurrent layers extract the interchannel relationship in spectral reflectance profiles. Each convolutional layer has multiple 1-D convolutional filters where the filter support is a data-dependent parameter. The pooling layers are used for subsampling to reduce the dimensionality of the network, which can help reduce computation and control overfitting. A recurrent layer has a recursive function f that takes as input one input vector x_t and the previous hidden state h_{t-1}, and returns the new hidden state as:

$$h_t = f(x_t, h_{t-1}) = tanh(Wx_t + Uh_{t-1}) \tag{3.8}$$

and the outputs are calculated as:

$$o_t = softmax(Vh_t) \tag{3.9}$$

where W, U, and V are the weight matrices that are shared across all steps, and the activation function *tanh* is the hyperbolic tangent function. In our framework (CRNN), the last hidden state (node)

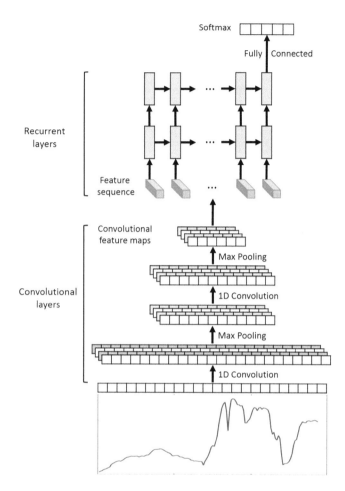

FIGURE 3.10 CRNN architecture for hyperspectral data classification. (Adapted from H. Wu and S. Prasad. *Remote Sensing*, vol. 9, no. 3, p. 298, 2017. [86])

of the RNN is fully connected to the classification layer that is based on the softmax activation function. Cross-entropy is used as a loss function, and mini-batch gradient descent is used to find the best network parameters. Gradients in the convolutional layers are calculated by traditional back-propagation, while the gradients in the recurrent layers are computed by back-propagation through time (BPTT) [84]. As an illustration of the potential for pixel-level deep learning based on CRNN for vegetation classification, classification maps with the Indian Pine 2010 data set as well as comparisons to baselines are provided in Table 3.3.

3.4.4 SEGMENTATION-BASED APPROACHES TO SUPPORT IMAGE CLASSIFICATION

Segmentation is often an effective precursor to classification, providing the capability to identify homogeneous regions that are classified as objects. A myriad of image segmentation approaches have been proposed and developed over the years (see, e.g., [87]). These approaches can be adapted to segmenting hyperspectral imagery as a preprocessing stage to classification by first reducing the data dimensionality through feature selection and extraction (see previous section) or by utilizing an appropriate region similarity (or dissimilarity) criterion.

A dissimilarity criterion designed for hyperspectral data is the spectral angle mapper (SAM) criterion [88]. An important property of the SAM criterion is that poorly illuminated and more

TABLE 3.3

Classification Results (Overall Accuracy and Standard Deviations) Obtained by Different Approaches with Different Numbers of Training Samples on the Indian Pine 2010 Data Set

# Training Samples	1900	3800	5700
RBF-SVM	92.82($\pm$1.07)	94.36($\pm$0.93)	95.13($\pm$0.64)
CNN	93.11($\pm$0.95)	94.53($\pm$0.39)	95.84($\pm$0.31)
RNN	84.83($\pm$1.62)	89.74($\pm$0.98)	91.86($\pm$0.77)
CRNN	94.43($\pm$1.01)	96.24($\pm$0.60)	96.83($\pm$0.47)

Source: Adapted from H. Wu and S. Prasad. *Remote Sensing*, vol. 9, no. 3, p. 298, 2017. [86]

brightly illuminated pixels of the same color will be mapped to the same spectral angle despite the difference in illumination.

The most successful approaches to spatial-spectral image segmentation are based on best merge region growing. An early version of best merge region growing, hierarchical step-wise optimization (HSWO), is an iterative form of region growing in which the iterations consist of finding the most optimal or best segmentation with one region less than the current segmentation [89]. The HSWO approach can be summarized as follows:

1. Initialize the segmentation by assigning each image pixel a region label. If a presegmentation is provided, label each image pixel according to the presegmentation. Otherwise, label each image pixel as a separate region.
2. Calculate the dissimilarity criterion value, d, between all pairs of spatially adjacent regions, find the smallest dissimilarity criterion value, T_{merge}, and merge all pairs of regions with $d = T_{merge}$.
3. Stop if no more merges are required. Otherwise, return to step 2.

HSWO naturally produces a segmentation hierarchy consisting of the entire sequence of segmentations. For practical applications, however, a subset of segmentation needs to be selected from this exhaustive segmentation hierarchy. A portion of such a segmentation hierarchy is illustrated in Figure 3.11 (the selection of a single segmentation from a segmentation hierarchy is discussed in a later section).

A unique feature of the segmentation hierarchy produced by HSWO and related region growing segmentation approaches is that the segment or region boundaries are maintained at the full image spatial resolution for all levels of the segmentation hierarchy. This is important for classification problems.

Many variations on best merge region growing have been described in the literature. As early as 1994, [90] described an implementation of the HSWO form of best merge region growing that utilized a heap data structure [91] for efficient determination of best merges and a dissimilarity criterion based on minimizing the mean squared error between the region mean image and original image. The main differences between most of these region growing approaches are the dissimilarity criterion employed and, perhaps, some control logic designed to remove small regions or otherwise tailor the segmentation output. In complex scenes, such as remotely sensed images of the Earth, objects with similar spectral signatures (e.g., lakes, agricultural fields, buildings, etc.) appear in spatially separated locations. In such cases, it is useful to aggregate these spectrally similar but spatially disjoint region objects into groups of region objects, or region classes. This is the basis of the hybridization of HSWO best merge region growing with spectral clustering [92,93] called HSeg (hierarchical segmentation). HSeg adds to HSWO a step following each step of adjacent region merges in which all pairs of spatially nonadjacent regions are merged that have *dissimilarity* $\leq S_w T_{merge}$,

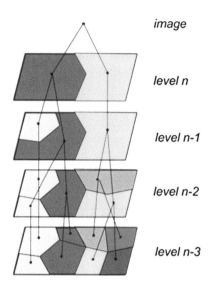

FIGURE 3.11 The last four levels of an *n*-level segmentation hierarchy produced by a region growing segmentation process. Note that when depicted in this manner, the region growing process is a "bottom-up" approach. (Adapted from J. C. Tilton et al. "Image segmentation algorithms for land categorization," in *Remotely Sensed Data Characterization, Classification, and Accuracies*, 2015, pp. 317–342. [87])

where $0.0 \leq S_w \leq 1.0$ is a factor that sets the priority between spatially adjacent and nonadjacent region merges. Note that when $S_w = 0.0$, HSeg reduces to HSWO.

A recursive divide-and-conquer approximation of HSeg (called RHSeg) [94] was developed to enable a straightforward parallel implementation. The computational requirements of HSeg were further reduced by refinements discussed in [93]. HSeg segmentation results are also compared in [93] with other segmentation approaches based on region growing.

Determining the optimal level of segmentation detail for a particular application is a challenge for all image segmentation approaches. The level of segmentation detail produced by a segmentation approach can usually be specified by adjusting one or more of the approach's parameters. Additionally, segmentation approaches based on region growing can also be easily designed to output a hierarchical set of image segmentations that can be examined later to select an optimal level of segmentation detail. In the case of HSeg, a particular set of hierarchical segmentations can be specified by providing a set of merge thresholds or number of region classes. HSeg can also automatically select a hierarchical set of image segmentations over a specified range of the number of region classes. This automatic approach outputs the image segmentation result at the region growing iteration prior to the point where any region classes are involved in a second merge since the last segmentation results output.

The following subsections describe various approaches for selecting an optimal level of segmentation detail out of an image segmentation hierarchy, such as that produced by HSeg:

1. *A Semisupervised Approach for Selecting Segmentations from a Segmentation Hierarchy*: [95] describes a semisupervised approach for adaptively selecting segmentations from the region class segmentation hierarchy produced by HSeg (or RHSeg). With the HSegLearn tool described therein, an analyst selects region classes (groups of region objects) as positive or negative examples of a specific ground cover class. Based on these selections, HSegLearn searches the HSeg segmentation hierarchy for the coarsest level of segmentation at which the submitted positive example region classes do not conflict with the submitted negative example region classes and displays the submitted positive example region classes at

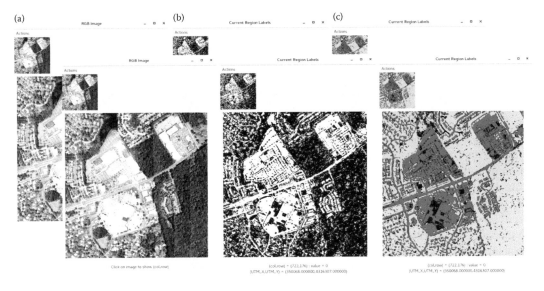

FIGURE 3.12 Example of using HSegLearn to generate a ground reference map of vegetation. (a) RGB image panel displaying a subsection of a hyperspectral image over a portion of downtown Bowie, MD. (b) HSegLearn Current Region Labels Panel after selecting some positive example region classes (vegetation, yellow) and some negative example region classes (nonvegetation, white). (c) HSegLearn Current Region Labels Panel: the positive example regions (vegetation, green) and negative regions (nonvegetation, red); region labeling by finding the coarsest levels in the HSeg segmentation hierarchy that do not conflict with the analyst labeling. These hyperspectral data were obtained by NASA Goddard's LiDAR, Hyperspectral and Thermal (G-LiHT) Airborne Imager on June 28, 2012, at a nominal 1-m spatial resolution. The data set has 114 spectral bands over the spectral range of 418–918 nm.

that level of segmentation detail. HSegLearn was successfully used to generate ground reference data from 1- to 2-m resolution satellite imagery (mainly WorldView 2 imagery) for training of the classification of Landsat TM and ETM+ data [96] for a global mapping of human-made impervious surfaces. This tool can also be used with hyperspectral data, as demonstrated by the example shown in Figure 3.12.

2. *Selecting Segmentations from a Segmentation Hierarchy by Analyzing Classification Error Rates*: [97] developed an approach for selecting the best level of segmentation detail from the segmentation hierarchy produced by HSeg (or RHSeg) for fusion of the region object labeling from HSeg with a pixel-based random forest classification of Landsat TM or ETM+ data covering the North American continent. This approach was used to determine the merge threshold for the best level of segmentation detail in the HSeg segmentation in each agricultural zone. This was done by analyzing the error rates across all the segmentation hierarchies in all image tiles containing reference objects. The merge threshold was determined as the threshold for the segmentation hierarchy with the lowest error rate in a stable portion of a graph of error rate versus segmentation hierarchy level.

3. *Selecting Segmentations from a Segmentation Hierarchy Utilizing Active Learning*: [98] describes an approach utilizing an active learning framework for selecting the best segmentation obtained by pruning the segmentation hierarchy produced by HSeg from a hyperspectral image. By pruning we mean adaptively selecting the level of segmentation detail from varying levels of the segmentation hierarchy throughout the image. Since it utilizes active learning, we defer discussion of this approach to the section on active learning.

3.5 CHALLENGES AND ADVANCED APPROACHES FOR CLASSIFICATION OF VEGETATION

3.5.1 MULTISOURCE/MULTITEMPORAL/MULTISCALE CHALLENGES AND APPROACHES

3.5.1.1 Multiple Kernel Learning

It is well understood that with traditional SVM classifiers, the choice of kernel function and the associated kernel parameter (e.g., the width of the RBF kernel) have a significant impact on the classification performance. A key practical aspect of using SVM classifiers with hyperspectral data is the need to "tune" the kernel parameters to find the appropriate (data-dependent) parameters that result in optimal decision boundaries. Additionally, when dealing with multisource data (e.g., data from different sensors), traditional single-kernel SVMs do not offer a compelling approach to fuse such data into a unified classification product. In recent work, multikernel learning methods have been shown to fill in these gaps that traditional SVMs struggle to classify. Multikernel learning can be thought of as learning a traditional SVM, with the exception that instead of traditional Mercer's kernels, one uses a mixture of "basis" kernels, where the mixing weights are learned as part of the SVM optimization.

In the multisource scenario, for a specific source p, the combined kernel function K between two pixels $\mathbf{x}_i^p$ and $\mathbf{x}_j^p$ can be represented as

$$K(\mathbf{x}_i^p, \mathbf{x}_j^p) = \sum_{m=1}^{M} d_m K_m(\mathbf{x}_i^p, \mathbf{x}_j^p)$$

$$\text{s.t. } d_m \geq 0, \text{ and } \sum_{m=1}^{M} d_m = 1,$$

(3.10)

where M is the number of candidate basis kernels representing different kernel parameters, K_m is the mth basis kernel, and d_m is the associated weight for it. The SimpleMKL [99] algorithm is an implementation of the MKL framework, where the optimization problem is posed as:

$$\min_{d} J(d), \quad \text{s.t. } d_m \geq 0, \text{ and } \sum_{m=1}^{M} d_m = 1$$

$$J(d) = \begin{cases} \min_{\mathbf{w},b,\xi} \frac{1}{2} \sum_{m=1}^{M} \frac{1}{d_m} \|\mathbf{w}_m\|^2 + C \sum_{i=1}^{N} \xi_i \\ \text{s.t. } y_i \left(\sum_{m=1}^{M} \langle \mathbf{w}_m, \Phi_m(\mathbf{x}_i^p) \rangle + b \right) \geq 1 - \xi_i \\ \xi_i \geq 0, \quad \forall i = 1, 2, \ldots, N, \end{cases}$$

(3.11)

where $\Phi_m(\mathbf{x}^p)$ is the kernel mapping function of $\mathbf{x}_i^p$, $\mathbf{w}_m$ is the weight vector of the mth decision hyperplane, C is the regularization parameter controlling the generalization capabilities of the classifier, and ξ_i is a positive slack variable.

As with traditional SVMs, SimpleMKL also has a dual representation:

$$
\max \left\{ L(\alpha_i, \alpha_j) = \sum_{i=1}^{N} \alpha_i \right.
$$

$$
\left. - \frac{1}{2} \sum_{i=1}^{N} \sum_{j=1}^{N} \alpha_i \alpha_j y_i y_j \sum_{m=1}^{M} d_m K_m(\mathbf{x}_i^p, \mathbf{x}_j^p) \right\}
$$

$$
\text{s.t.} \begin{cases} \sum_{i=1}^{N} \alpha_i y_i = 0 \\ \alpha_i, \alpha_j \in [0, C], \quad \forall i, j = 1, 2, \ldots, N \\ d_m \geq 0, \text{ and } \sum_{m=1}^{M} d_m = 1, \end{cases} \tag{3.12}
$$

where α_i and α_j are Lagrange multipliers. Kernel weights d_m are learned via gradient descent by updating it along the direction of the negative gradient of $L(\alpha_i, \alpha_j)$. The outcome of SimpleMKL yields a predicted label per test sample, and by using Platt's approach, a class-conditional posterior probability $P(y = 1|x)$ is estimated, obtaining the posterior probability that allows quantification of uncertainty associated with the underlying classification.

An alternative approach to use the MKL framework entails learning source-specific optimal MKL classifiers, which are then fused through decision fusion strategies, such as majority voting or linear opinion pools. In this alternative construction, the mixture of kernels per SVM yields a much more discriminative and linearly separable Hilbert space than a traditional single kernel–based SVM. We used multikernel SVMs to systematically fuse multisensor (hyperspectral and LiDAR) geospatial data in an active learning paradigm [100].

3.5.1.2 Transfer Learning and Domain Adaptation

A common image analysis scenario when dealing with multisource data involves transferring knowledge/information between sources. For instance, given a source data set rich in labeled data where one can optimally learn a robust classifier, can this knowledge/model be transferred to a (different) target data set where there may not otherwise be sufficient training data to train classifiers for effective classification of target data? It is common to encounter such scenarios when undertaking multiscale, multisensor, and multitemporal data sets. Here, we refer to recent efforts [53,101,102] aimed at transferring knowledge across sources (e.g., across different sensors or different viewpoints or times corresponding to the same sensor).

Denote the source domain as $\mathcal{D}_S$ with samples $\{x_1, \ldots, x_n\} \in \mathbb{R}^{d_s}$ and corresponding class labels $\{l_{x_1}, \ldots, l_{x_n}\}$. Similarly, denote the target domain as $\mathcal{D}_T$ with labeled samples $\{y_1, \ldots, y_m\} \in \mathbb{R}^{d_t}$, $d_s \neq dt$, and corresponding class labels $\{l_{y_1}, \ldots, l_{y_m}\}$, $m \ll n$. Given vectors $x \in \mathcal{D}_S$ and $y \in \mathcal{D}_T$, the goal is to project these data to a latent space $\mathbb{R}^{d_c}$ that is discriminative for the underlying classification task.

In recent work, we developed mappings **A** and **AB** that maximize the overlap of within-class samples in the latent space by ensuring that within-class samples in $\mathcal{D}_S$ and $\mathcal{D}_T$ are located in the same cluster/region of the latent feature space. Table 3.4 depicts results of this domain adaptation based on transformation learning (DATL) approach with the Botswana data and compares it to traditional approaches (including semisupervised transfer component analysis (SSTCA) and kernel principal component analysis) to domain adaptation. Class specific accuracies reported in Table 3.5 result from training on a large pool of source data (May Botswana Hyperion image) and transferring

TABLE 3.4

Overall KNN Classification Accuracies (%) and Standard Deviations (in Parentheses) for Botswana Data Set from May (Source) and July (Target)

Algorithm	# Target Domain Training Samples per Class							
	1	3	5	7	9	11	13	15
KPCA	34.0 (4.5)	35.1 (4.6)	37.5 (4.8)	38.5 (4.5)	40.5 (4.6)	40.8 (4.8)	41.2 (4.6)	42.0 (5.5)
SSTCA	47.7 (8.4)	47.5 (7.2)	50.5 (9.1)	49.6 (7.6)	52.4 (7.5)	53.0 (7.8)	52.6 (7.3)	51.7 (6.5)
DATL	46.6 (12.2)	63.6 (13.4)	68.2 (8.5)	71.3 (10.0)	73.3 (7.9)	76.9 (5.9)	76.8 (7.9)	77.6 (6.7)

Source: Adapted from X. Zhou and S. Prasad. *IEEE Transactions on Computational Imaging*, vol. 3, no. 4, pp. 822–836, December 2017. [102]

TABLE 3.5

Class-Dependent Accuracies (%) for Botswana Data from May (Source) and July (Target) with Five Training Samples per Class from the Target Domain

Algorithm	Class Index								
	1	2	3	4	5	6	7	8	9
KPCA	100.0	13.4	50.1	97.9	0.6	14.0	0.5	1.8	15.1
SSTCA	100.0	54.4	61.1	99.8	10.0	32.0	15.4	18.1	37.8
DATL	68.5	50.4	47.2	94.4	77.3	42.8	59.2	89.5	84.5

TABLE 3.6

Overall Classification Accuracies (%) and Standard Deviations (in Parentheses) for the Aerial and Street View Wetland Hyperspectral Data Set

Algorithm	# Target Domain Training Samples per Class							
	1	3	5	7	9	11	13	15
KPCA	26.4 (7.6)	46.8 (9.5)	59.9 (3.3)	64.4 (3.4)	67.0 (5.0)	70.0 (5.3)	72.5 (5.6)	73.6 (6.1)
SSTCA	25.4 (9.8)	51.0 (8.3)	65.1 (3.2)	70.0 (3.0)	74.3 (2.4)	77.9 (3.2)	80.9 (3.5)	82.7 (4.2)
DATL	42.0 (5.4)	59.8 (6.6)	68.2 (7.8)	70.6 (8.0)	74.5 (7.0)	75.6 (4.7)	78.7 (5.9)	81.0 (4.9)

Source: Adapted from X. Zhou and S. Prasad. *IEEE Transactions on Computational Imaging*, vol. 3, no. 4, pp. 822–836, December 2017. [102]

the models learned to classify a different image acquired over the same region in July (target image) using between 1 and 15 target training samples per class to facilitate the alignment.

Accuracies in Tables 3.6 and 3.7 represent the coastal wetland ecosystem monitoring application where the source domain refers to street-view (side-looking, terrestrial) hyperspectral imagery captured during our field campaigns to identify the vegetation cover. The target domain refers to aerial imagery.

The goal of such a framework is to learn models from the rich source domain data (due to being imaged at close range, there is an abundance of pixels per class to train models in the source domain) and transfer this knowledge to undertake classification in the target domain. For this data, as few as 5–15 samples per class from the target domain, very effective classification accuracies are achieved leveraging domain information. Results are particularly promising because of different

TABLE 3.7

Class-Dependent KNN Classification Accuracies (%) for the Aerial and Street View Wetland Hyperspectral Data Set with Five Training Samples per Class from the Target Domain (Aerial View)

Algorithm	Class Index											
	1	2	3	4	5	6	7	8	9	10	11	12
KPCA	64.7	48.5	57.5	29.7	67.5	23.5	45.8	72.5	88.0	78.7	75.3	67.3
SSTCA	65.0	63.3	59.8	30.8	75.0	32.7	49.2	73.2	91.8	86.5	81.2	72.5
DATL	71.0	66.5	69.3	41.6	67.4	25.2	47.5	71.5	82.0	91.2	70.6	62.6

Source: Adapted from X. Zhou and S. Prasad. *IEEE Transactions on Computational Imaging*, vol. 3, no. 4, pp. 822–836, December 2017. [102]

sun-sensor-canopy geometry but also different sensors (source spans the VNIR range, while the target spans the VNIR-SWIR range).

3.5.2 LIMITED TRAINING SAMPLES: EXPLOITING UNLABELED SAMPLES

3.5.2.1 Semi-Supervised and Active Learning

Semisupervised learning (SSL) approaches, where unlabeled samples are classified and some are incorporated into the training set, have been investigated by many researchers for classification of high-dimensional data. SSL directly utilizes the unlabeled data to facilitate the learning process without requiring any human-labeling efforts. In addition to semisupervised learning approaches, many researchers have explored active learning (AL) as an alternative strategy for leveraging unlabeled samples to mitigate the impact of limited training samples for supervised classification of remotely sensed hyperspectral data. Unlike SSL approaches, AL heuristics focus on identifying the most "informative" unlabeled samples, then obtaining the corresponding labels and incorporating the newly labeled data into the training pool.

In traditional supervised classification, as shown in the box in Figure 3.13, labeled samples/features are presented to the classifier for training, and unlabeled samples from the data pool are then classified using the resulting model. AL involves an iterative training/evaluation phase, whereby an initial model is developed using a small number of samples, and the classification results are evaluated. Samples are then selected from the unlabeled pool, evaluated according to "query" criteria, and ranked. Sample(s) are chosen from the list for labeling according to specified selection criteria, labeled, and incorporated

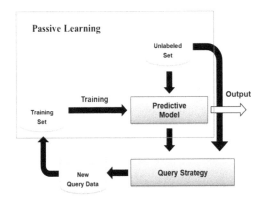

FIGURE 3.13 Flow chart for active and passive learning.

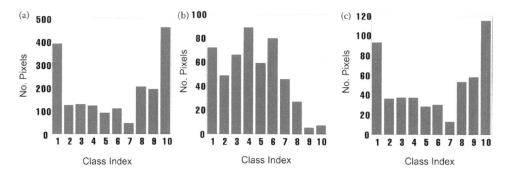

FIGURE 3.14   Distribution of labeled class samples for Kennedy Space Center data. (a) True distribution, (b) AL sampling, (c) random sampling.

into the training set. The remaining candidates are returned to the unlabeled pool. The classifier is updated using the augmented training set, and the process is repeated until a stopping criterion (e.g., number of iterations, number of samples, classification accuracy) is achieved.

AL strategies typically result in biased sampling of classes, with more samples being selected from classes that are more difficult to discriminate, as shown in the example of Figure 3.14. AL strategies are generally characterized by the criteria used to select candidate samples for labeling. The approaches include (1) margin sampling (MS)–based approaches, where the samples closest to the separating hyperplane of the classifier, such as support vector machines, are considered as the most uncertain ones [103,104]; (2) committee of learners–based approaches, where those samples exhibiting maximal disagreement between the different classification models are selected [8,105,106]; and (3) class probability distribution–based approaches, where breaking ties (BT) is a representative method in which the difference between the two highest posterior probabilities is used to quantify the uncertainty of a pixel [107,108]. Algorithms are typically implemented with spectral data as the single-source inputs, although multiple-source inputs have also been studied [109,110]. Active learning for classification has been used in multiple applications [111,112] including large-scale scenarios [113]. We refer to [9,114] for surveys on AL methods.

Results from [106] are included to illustrate AL based on a committee of learners. Views can be composed of multiple classifiers resulting from different methods, different inputs, and so on. Here, multiple views of the problem were derived by segmenting the spectral data into disjoint contiguous subband sets generated by (1) correlation of contiguous bands, (2) k-means–based band clustering, (3) deterministic selection of every kth band (band slicing), and (4) random sampling. Views generated by correlation and clustering are diverse, but may differ in their discriminative ability for individual classes, so there is a risk of insufficiency for the classification, while views obtained from band slicing may be redundant, but are sufficient for covering the full space of inputs. Finally, random sampling provides diverse views, although they are not guaranteed to be either sufficient or accurate. Figure 3.15 illustrates multiview subsetting of the input space of the KSC data based on interband correlation. A two-stage query and sample selection process was used, where the first ranked samples by the maximum disagreement of the classification results across views, and the second invoked an entropy criterion to increase diversity of the samples. Figures 3.16 and 3.17 show the accuracy curves as AL progresses. Margin sampling had the highest overall accuracy, although multiview methods converged to approximately the same overall accuracy.

3.5.2.2   Segmentation-Based Active Learning

Most of the AL methods proposed in the literature deal solely with spectral information, but improvements in terms of classification accuracies can be obtained by also exploiting the spatial dimension. Only recently, researchers have started to integrate spatial information with spectral features in the AL framework [109,115–117]. A recent strategy [98] that incorporates AL,

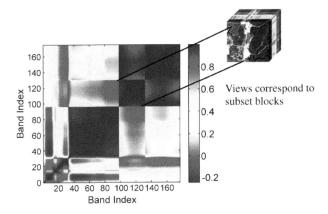

FIGURE 3.15   Correlation matrix for Kennedy Space Center AVIRIS data. Each block corresponds to a view.

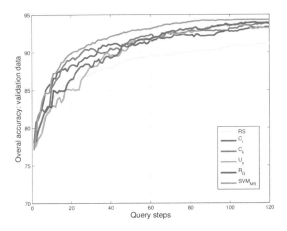

FIGURE 3.16   Overall classification accuracy for SVM classification of Kennedy Space Center with multiview active learning based on band correlation C_r, k-means clustering C_k, uniform band slicing U_s, and random view generation R_U compared to random sampling RS and margin sampling SVM$_{MS}$. (Adapted from W. Di and M. M. Crawford, *IEEE Transactions on Geoscience and Remote Sensing*, vol. 50, no. 5, pp. 1942–1954, 2012. [106])

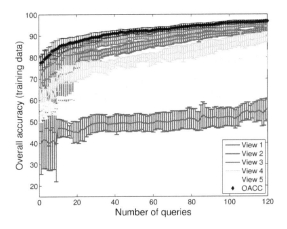

FIGURE 3.17   The overall classification accuracy and view performance derived from correlation-based view generation C_r for Kennedy Space Center data. (Adapted from W. Di and M. M. Crawford, *IEEE Transactions on Geoscience and Remote Sensing*, vol. 50, no. 5, pp. 1942–1954, 2012. [106])

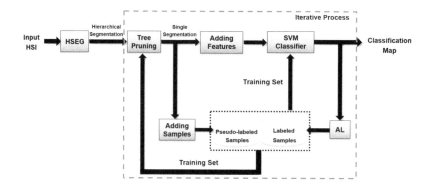

FIGURE 3.18 Flow chart illustrating the segmentation-based AL framework that incorporates AL, semisupervised learning, and segmentation into a unique framework. (Adapted from Z. Zhang et al. *IEEE Journal of Selected Topics in Applied Earth Observations and Remote Sensing*, vol. 9, no. 2, pp. 640–654, 2016. [98])

semisupervised learning, and segmentation into a unique framework is shown in Figure 3.18 and detailed in the following.

The approach relies on the HSeg algorithm introduced earlier in this chapter, whose output is a segmentation hierarchy that contains different levels of detail of the image. The appropriate level can be selected in a supervised way by quantitatively evaluating the segmentation results at each hierarchical level (as discussed previously in this chapter and in [97]) or in an unsupervised way by interactively using the HSegViewer tool. Both approaches assume that the best segmentation corresponds to a single specific level of the hierarchy. However, coherent objects may be found at different levels [118] and therefore, the best segmentation should be defined by selecting regions at different levels. This process is usually referred to as pruning, in which subtrees of the hierarchy that are homogeneous with respect to some defined homogeneity criteria are removed. Although different pruning strategies based on supervised [119], semisupervised [120], and unsupervised [121] strategies have been proposed, they are not suitable to be used within an AL framework. While AL and SSL have different workflows, they both aim to make the best use of unlabeled data and reduce human labeling efforts. It is natural to combine the two strategies to take advantage of both paradigms in the classification task. The main idea of the strategy is to find an optimal segmentation map from the segmentation hierarchy by considering a novel supervised pruning strategy. This strategy aims at removing redundant subtrees composed of nodes that are homogeneous with respect to some criteria of interest from the HSeg output. As a result, it generates an optimal segmentation map that can provide spatial information for the classification. The best segmentation does not represent one of the actual levels of the hierarchy, but incorporates regions selected from potentially different levels. Two merging criteria based on the node size and the Bhattacharyya coefficient are considered. The whole pruning process is integrated within the AL framework. Compared to the unsupervised pruning strategy proposed in [121], the new method can exploit the labeled information provided by the user (which increases through the AL process) and thus update the best segmentation map at each AL iteration. At the end of the pruning, a single segmentation map is obtained, which is further exploited to incorporate spatial information into the framework in two different ways: (1) spatial features (e.g., mean and standard deviation) are extracted from each segment and concatenated with the original spectral ones into a single stacked feature vector. An SVM is adopted as the back-end classifier and trained on the enriched feature space. (2) The continuously updated segmentation map is considered to expand the training set by employing both AL and self-learning–based SSL approaches. At each iteration, both labeled and pseudo-labeled samples are added to the training set and used jointly to train the classifier. The most uncertain samples are selected using the BT criterion and then labeled by the human expert. Pseudo-labels are assigned automatically by taking advantage of spatial information. Such pseudo-labeled samples help increase the size of the training set when the available labeled training set is small. For this purpose, first, the set of candidate unlabeled samples is defined as the samples that belong to regions with identically labeled samples based on the current segmentation map. The framework is validated experimentally on the Indian Pine 1992 data set. Improvements in terms of overall classification accuracies are exhibited by the new framework in comparison with other spectral-spatial strategies, as reported in Figures 3.19 through 3.21.

3.5.2.3 Active Metric Learning

Feature extraction and AL problems for hyperspectral image classification have usually been investigated independently. Considering a traditional AL-based hyperspectral image classification chain, feature extraction is usually executed first in the original high-dimensional feature space as a preprocessing step to obtain an optimally reduced feature space. This can be accomplished in an unsupervised way using manifold learning strategies or in a supervised way by exploiting the limited labeled information available at the beginning of the process. An AL algorithm is then applied in the reduced feature space to increase the number of points in the training set. However, the feature space extracted earlier may be suboptimal relative to the resulting training set. The unsupervised feature extraction step lacks connection with the classification problem or is performed using the few potentially nonrepresentative initial labeled samples. In both cases, the extracted feature space is fixed and does not interact in any way with the additional information provided by the user during the AL process. Therefore, even an optimal AL strategy cannot guarantee maximization of the classification accuracy since it is applied to a suboptimal feature space.

Novel solutions have recently been proposed in the literature with the aim of combining feature extraction and AL into a unique framework [122–124], as summarized in the flowchart of Figure 3.22. The overall idea is to learn and update a reduced feature space in a supervised way at each iteration of the AL process, thus exploiting the increasing labeled information provided by the user. In particular, the computation of the reduced feature space is based on the large-margin nearest

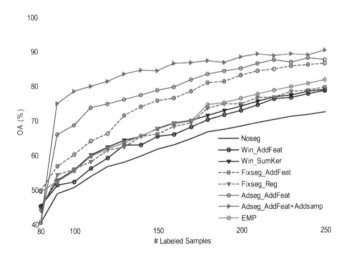

FIGURE 3.19 Learning curve (overall accuracy vs. number of labeled samples) on the Indian Pine 1992 data set. The methods *Adseg_AddFeat* and *Adseg_AddFeat+Addsamp* are compared with other spectral-spatial strategies. *Win_AddFeat*: spectral and spatial features extracted from a 3 × 3 window; *Win_SumKer*: spectral and spatial features extracted from a 3 × 3 window with a kernel-composite approach classifier; *Fixseg_AddFeat*: spatial features extracted from a fixed segmentation map; *Fixseg_Reg*: the fixed segmentation map is used as regularizer; *EMP*: morphological features extracted from the first two PCs. All methods adopt BT as the AL query criterion. (Adapted from Z. Zhang et al. *IEEE Journal of Selected Topics in Applied Earth Observations and Remote Sensing*, vol. 9, no. 2, pp. 640–654, 2016. [98])

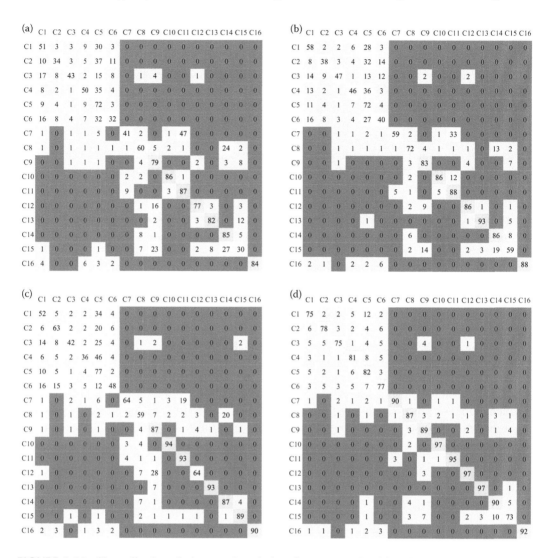

FIGURE 3.20 Normalized confusion matrices (values in percentage) achieved on the Indian Pine 1992 data set. (a) *Noseg*, (b) *Win_AddFeat*, (c) *EMP*, (d) *Adseg_AddFeat + AddSamp*. (Adapted from Z. Zhang et al. *IEEE Journal of Selected Topics in Applied Earth Observations and Remote Sensing*, vol. 9, no. 2, pp. 640–654, 2016. [98])

neighbor (LMNN) metric learning principle [125]. The metric learning strategy is applied in conjunction with k-NN classification and novel sample selection criteria.

Specifically, consider a hyperspectral image $X = \{\mathbf{x}_i\}_{i=1}^n$, where $x_i = [x_{i,1}, \ldots, x_{i,d}]$ is a sample in the original high d-dimensional feature space and n is the total number of pixels in I. Define $\mathbf{x}_i' = [x_{i,1}', \ldots x_{i,r}']$ as the same sample in the reduced r-dimensional feature space obtained by adopting a feature extraction strategy. A training set $L = x_i, y_i \,_{i=1}^l$ is constructed by selecting l samples from X and assigning corresponding discrete labels y_i (where $y_i \in 1, \ldots, \Omega$, and Ω is the number of thematic classes). $U = x_i \,_{i=l+1}^{l+u}$ is defined as the set of u remaining labeled samples, that is, $U = X - L$ and $n = u + l$. The supervised dimensionality reduction strategy exploits the training set L to generate a low-dimensional feature space X' from X, and an AL strategy, where the most uncertain samples are selected and labeled, is then applied on X'. The process is iterated until a convergence criterion is satisfied.

(a)

(b)

(c)

(d)

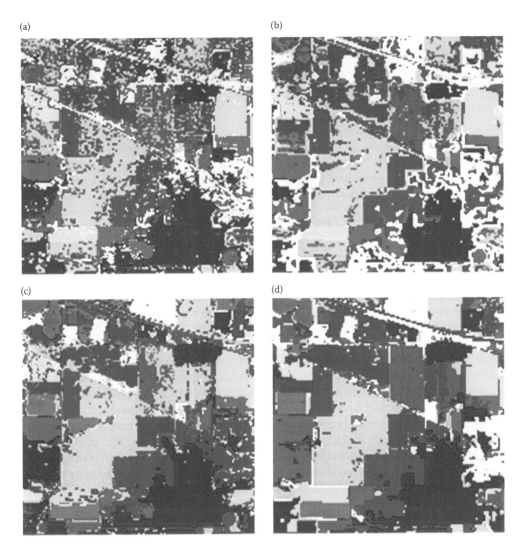

FIGURE 3.21 Classification maps achieved on the Indian Pine 1992 data set. (a) *Noseg*, (b) *Win_AddFeat*, (c) *EMP*, (d) *Adseg_AddFeat + AddSamp*. (Adapted from Z. Zhang et al. *IEEE Journal of Selected Topics in Applied Earth Observations and Remote Sensing*, vol. 9, no. 2, pp. 640–654, 2016. [98])

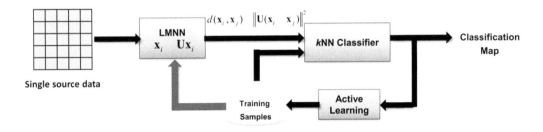

FIGURE 3.22 Flow chart illustrating the active-metric learning approach for supervised classification. (Adapted from E. Pasolli et al. *IEEE Transactions on Geoscience and Remote Sensing*, vol. 54, no. 4, pp. 1925–1939, 2016. [122])

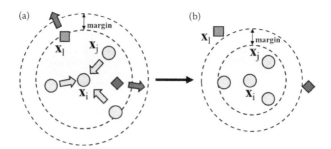

FIGURE 3.23 Schematic representation of LMNN metric learning method. (a) Original feature space. (b) Feature space after training. After training, *K* similar samples (in green) are separated from the dissimilar ones by a unit margin. Local neighborhood in gray. (Adapted from E. Pasolli et al. *IEEE Transactions on Geoscience and Remote Sensing*, vol. 54, no. 4, pp. 1925–1939, 2016. [122])

The dimensionality reduction step associated with the LMNN algorithm [125], which is schematically represented in Figure 3.23, is implemented in conjunction with the Mahalanobis distance [126] and extended to improve classification performance, in terms of accuracy and computational time: (1) dimensionality reduction is incorporated directly into the objective function optimization process, (2) the optimization process is iterated through a multipass approach, and (3) *k*-NN search is accelerated through ball tree formulation.

The original method, which can only handle single input features (e.g., pure spectral features) and does not specifically accommodate multiple feature scenarios, is extended in [123], where multiple feature types are concatenated by extending LMNN to heterogeneous multimetric learning (HMML) [127]. The reduced feature space is obtained for each feature type by adopting a modified version of the HMML algorithm, and AL is then applied in the resulting single feature space in conjunction with *k*-NN classification. Further improvements were obtained in [124] via a regularized multimetric AL framework to jointly learn distinct metrics for different feature types. The regularizer incorporates unlabeled data based on the neighborhood relationship, which also helps avoid overfitting at early AL stages. As the iterative process proceeds, the regularizer is updated through similarity propagation, thus taking advantage of informative labeled samples. Finally, multiple features are projected into a common feature space, in which a new batch-mode selection strategy that incorporates uncertainty and diversity criteria is used. Comparison on the Indian Pine 1992 data set among the different active-metric learning methods is reported in Figure 3.24.

3.6 SUMMARY AND FUTURE DIRECTIONS FOR CLASSIFICATION OF HYPERSPECTRAL IMAGES

This chapter has addressed key issues in classification of hyperspectral data and provided an overview of strategies to address these problems. As noted in the introduction, availability of hyperspectral imagery and associated ground reference data, including class labels, has been a significant hurdle to advancing classification methods focused on vegetation and agriculture croplands. This problem will be addressed in part by the upcoming launches of combinations of traditional near polar orbiting satellite missions and constellations of small satellites carrying hyperspectral cameras. The resulting improvement in temporal resolution of data will be particularly relevant to applications in agriculture and vegetation. Hyperspectral camera and global navigation satellite system (GNSS) technologies are also advancing, including miniaturization, making them viable sensors for UAVs and ground-based platforms that provide higher spatial resolution data that can be collected on demand at reduced cost. The resulting data sets will be enormous, motivating the need for advances in data processing and management, including on-board processing and analysis.

The high dimensionality of hyperspectral data, coupled with band redundancy, is a well-known obstacle for traditional parametric classifiers, particularly when the quantity of labeled data for

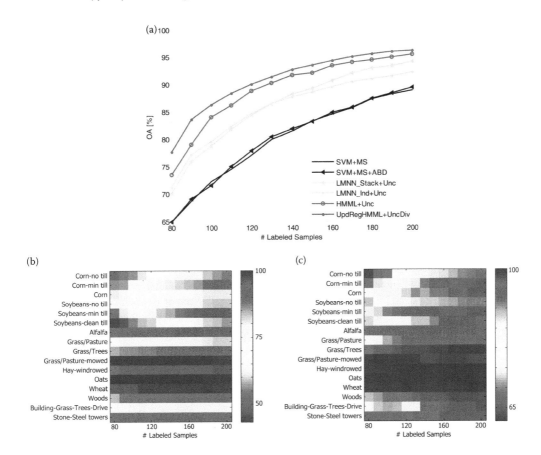

FIGURE 3.24 Comparisons among different active-metric learning methods on the Indian Pine 1992 data set. (a) Learning curve (overall accuracy in function of the number of labeled samples) for four strategies *UpdRegHMML+Unc* [124], *HMML+Unc* [123], *LMNN_Stack+Unc* [122], and *LMNN_Stack+Unc* [122] in addition to *SVM + MS* and *SVM + MS + ABD* [8]. Class-specific accuracies as a function of the number of labeled samples for (b) *SVM + MS* and (c) *UpdRegHMML + UncDiv*. (Adapted from Z. Zhang and M. M. Crawford, *IEEE Transactions on Geoscience and Remote Sensing*, vol. 55, no. 11, pp. 6594–6609, 2017. [124])

training is limited. Dimensionality reduction as a front-end processing step is both the most direct and most widely used approach to address this problem. Traditional methods that include direct feature selection as an optimization problem as either a supervised and unsupervised strategy (Section 3.3.1.1) and use of narrow band indices (Section 3.3.1.2) continue to be popular for classification, disease detection, and yield prediction in agriculture, as they are often correlated with plant vigor or specific chemistry-based responses at various stages of crop development. Both feature selection and vegetation indices have the advantage of being interpretable, but potentially ignore useful information in the rest of the spectrum. Alternatively, feature extraction approaches, which can exploit the full set of bands and inputs from other sources, are more popular for classification of single images, but suffer from interpretability and generalizability across multiple images. Traditional linear feature extraction methods (Section 3.3.1.3) are still widely used as inputs for classification and for other applications of hyperspectral data such as unmixing, in part because of their availability in commercial software. However, as discussed and illustrated in Section 3.3.1.4, nonlinear extraction approaches, particularly local graph-based methods, are promising for increasing class separation and thereby classification accuracy, although at increased computational cost.

Inclusion of spatial information can be extremely beneficial for classification of scenes containing natural vegetation or agriculture. Combining high spectral resolution data with spatial information has made it possible to discriminate classes that are spectrally quite similar, even at the relatively coarse spatial resolution of AVIRIS and Hyperion, as the impact of within-class spatial variability is reduced. Higher spatial resolution data obtained by low-altitude aircraft, UAVs, and ground-based systems provide the capability to actually exploit higher-frequency spatial components in the data, which may be useful for improved class discrimination. Incorporation of texture measures as inputs (Section 3.3.2) and use of hierarchical segmentation approaches (Section 3.4.4) and new classification strategies such as deep learning (Section 3.4.3) all leverage multiscale spatial information for classification, as illustrated in Section 3.4.3 using the 2010 Indian Pine data. While no single strategy dominates as the best approach for inclusion of spatial information in a classification problem, example results in this chapter, as well as the literature, clearly demonstrate the importance of spatial context in classification of natural vegetation and agricultural images.

New models of inputs, as well as nonparametric data-driven approaches, are also receiving significant attention for classification of hyperspectral data. SVM classifiers (Section 3.4.1.1), which directly avoid the issue of non-Gaussian class conditional density functions and high-dimensional inputs, are now widely implemented, although proper parameter tuning is challenging and requires adequate quantities of training data. Multikernel extensions of traditional single-kernel SVMs (Section 3.5.1.1) are especially promising for multisource inputs, as demonstrated by the examples in this chapter and the literature. When coupled with feature extraction for dimensionality reduction, Gaussian mixture models implemented in a Bayesian framework can also provide an effective strategy for classifying hyperspectral data (Section 3.4.2). Deep learning approaches (Section 3.4.3), which are now being widely explored for classification of hyperspectral data, have the capability to exploit nonlinear relationships and interactions, but require very large data sets for training. The potential of deep learning for classification of hyperspectral data is really in its infancy in this domain. Early results have stimulated significant research for both algorithm development and applications, including approaches for leveraging the structure of hyperspectral spectra and new strategies for addressing the limited training data issue. The chapter also included examples of newly developed approaches for addressing challenges and new opportunities in hyperspectral data classification. Relative to training and application of classifiers, methods to tackle domain adaptation and transfer learning will be necessary as more hyperspectral data become available. We presented one strategy for feature alignment that yielded promising results, and referenced others (Section 3.5.1.2). We also included a novel example where training data acquired by a ground-based system were used to train a classifier that was applied over an extended area. It provided not only an opportunity for multiscale learning, but also a strategy for expanding the limited training data set.

Finally, we addressed the issue of limited training data through active and metric learning in Section 3.5.2. Active learning provides a flexible framework for initiating the classification process with a small number of training samples and augmenting the pool with unlabeled data. The framework can be implemented with front-end feature extraction from single or multiple sources and with appropriate backend classifiers. An example based on the Kennedy Space Center AVIRIS data illustrates the potential of the approach, including targeted learning related to classes that are difficult to discriminate (Section 3.5.2.1). One limitation of active learning is that the true labels of pixels identified for inclusion in the training set must be determined. One strategy for addressing this problem is illustrated using the original Indian Pine data, where active learning is coupled with hierarchical segmentation to leverage spatially homogeneous areas and semisupervised learning (Section 3.5.2.2). As shown in the quantitative and qualitative results, this is a particularly effective approach for the testbed data. Another potential problem of active learning is that features that are extracted from the initial small training set may not be optimal as learning progresses, necessitating updates that can be computationally intensive, particularly for nonlinear feature extractors. Metric learning (Section 3.5.2.3) provides a new strategy that naturally integrates updates to the reduced feature space into the active learning framework using the concept of the large-margin nearest neighbor principle, which naturally couples with a k-NN classifier (which can also be considered a limitation). Early results from active metric learning are promising for both single- and multiple-source input data.

While the current strategies for classification of hyperspectral data build on a rich foundation, significant advances are still needed. New classification algorithms whose data structures and architecture exploit hyperspectral data for multitemporal, multiscale studies of agriculture and vegetation are particularly needed to effectively utilize hyperspectral imagery, both as standalone methods and in conjunction with biophysical models. Classification methods have traditionally been developed in a stove-pipe approach by algorithm developers working in isolation from the application domain and with limited understanding of this domain. The next generation of classification systems must leverage the knowledge of collaborative, multidisciplinary research composed of both methodologically focused researchers and partners from the basic and applied earth and life sciences.

ACKNOWLEDGMENTS

The authors would like to acknowledge the graduate students and postdoctoral scholars at Purdue University and University of Houston for setting up the experiments and generating results with methods reviewed in this overview chapter. We would also like to thank Farideh Foroozandeh Shahraki at the University of Houston for her help with formatting the chapter.

REFERENCES

1. G. Hughes. "On the mean accuracy of statistical pattern recognizers," *IEEE Transactions on Information Theory*, vol. 14, no. 1, pp. 55–63, 1968.
2. L. David. "Hyperspectral image data analysis as a high dimensional signal processing problem," *IEEE Signal Processing Magazine*, vol. 19, no. 1, pp. 17–28, 2002.
3. S. Tadjudin and D. A. Landgrebe. "Covariance estimation with limited training samples," *IEEE Transactions on Geoscience and Remote Sensing*, vol. 37, no. 4, pp. 2113–2118, 1999.
4. L. Breiman. "Random forests," *Machine Learning*, vol. 45, no. 1, pp. 5–32, 2001.
5. J. Ham, Y. Chen, M. M. Crawford, and J. Ghosh. "Investigation of the random forest framework for classification of hyperspectral data," *IEEE Transactions on Geoscience and Remote Sensing*, vol. 43, no. 3, pp. 492–501, 2005.
6. X. Jia, B.-C. Kuo, and M. M. Crawford. "Feature mining for hyperspectral image classification," *Proceedings of the IEEE*, vol. 101, no. 3, pp. 676–697, 2013.
7. L. Bruzzone, M. Chi, and M. Marconcini. "Semisupervised support vector machines for classification of hyperspectral remote sensing images," in *Hyperspectral Data Exploitation: Theory and Applications*, New York, John Wiley & Sons, pp. 275–311, 2007.
8. D. Tuia, F. Ratle, F. Pacifici, M. Kanevski, and W. Emery. "Active learning methods for remote sensing image classification," *IEEE Transactions on Geoscience and Remote Sensing*, vol. 47, no. 7, pp. 2218–2232, 2009.
9. M. M. Crawford, D. Tuia, and H. L. Yang. "Active learning: Any value for classification of remotely sensed data?" *Proceedings of the IEEE*, vol. 101, no. 3, pp. 593–608, 2013.
10. H. Liu and L. Yu. "Toward integrating feature selection algorithms for classification and clustering," *IEEE Transactions on Knowledge and Data Engineering*, vol. 17, no. 4, pp. 491–502, 2005.
11. Q. Cheng, P. K. Varshney, and M. K. Arora. "Logistic regression for feature selection and soft classification of remote sensing data," *IEEE Geoscience and Remote Sensing Letters*, vol. 3, no. 4, pp. 491–494, 2006.
12. X. Chen, T. Fang, H. Huo, and D. Li. "Graph-based feature selection for object-oriented classification in VHR airborne imagery," *IEEE Transactions on Geoscience and Remote Sensing*, vol. 49, no. 1, pp. 353–365, 2011.
13. X. Jia and J. A. Richards. "Segmented principal components transformation for efficient hyperspectral remote-sensing image display and classification," *IEEE Transactions on Geoscience and Remote Sensing*, vol. 37, no. 1, pp. 538–542, 1999.
14. S. M. Davis, D. A. Landgrebe, T. L. Phillips, P. H. Swain, R. M. Hoffer, J. C. Lindenlaub, and L. F. Silva. *Remote Sensing: The Quantitative Approach*, New York, McGraw-Hill, 1978.
15. A. Paoli, F. Melgani, and E. Pasolli. "Clustering of hyperspectral images based on multiobjective particle swarm optimization," *IEEE Transactions on Geoscience and Remote Sensing*, vol. 47, no. 12, pp. 4175–4188, 2009.

16. A. Ifarraguerri and M. W. Prairie. "Visual method for spectral band selection," *IEEE Geoscience and Remote Sensing Letters*, vol. 1, no. 2, pp. 101–106, 2004.
17. K. Fukunaga, *Introduction to Statistical Pattern Recognition*. Academic Press, 2013.
18. B. Paskaleva, M. M. Hayat, Z. Wang, J. S. Tyo, and S. Krishna. "Canonical correlation feature selection for sensors with overlapping bands: Theory and application," *IEEE Transactions on Geoscience and Remote Sensing*, vol. 46, no. 10, pp. 3346–3358, 2008.
19. T. M. Cover and J. A. Thomas, *Elements of Information Theory*. Wiley-Interscience, 2012.
20. Q. Du and H. Yang. "Similarity-based unsupervised band selection for hyperspectral image analysis," *IEEE Geoscience and Remote Sensing Letters*, vol. 5, no. 4, pp. 564–568, 2008.
21. P. Mitra, C. Murthy, and S. K. Pal. "Unsupervised feature selection using feature similarity," *IEEE Transactions on Pattern Analysis and Machine Intelligence*, vol. 24, no. 3, pp. 301–312, 2002.
22. J. M. Sotoca, F. Pla, and J. S. Sanchez. "Band selection in multispectral images by minimization of dependent information," *IEEE Transactions on Systems, Man, and Cybernetics, Part C (Applications and Reviews)*, vol. 37, no. 2, pp. 258–267, 2007.
23. L. Wang, X. Jia, and Y. Zhang. "A novel geometry-based feature-selection technique for hyperspectral imagery," *IEEE Geoscience and Remote Sensing Letters*, vol. 4, no. 1, pp. 171–175, 2007.
24. P. M. Narendra and K. Fukunaga. "A branch and bound algorithm for feature subset selection," *IEEE Transactions on Computers*, vol. 9, no. C-26, pp. 917–922, 1977.
25. P. Pudil, J. Novovičová, and J. Kittler. "Floating search methods in feature selection," *Pattern Recognition Letters*, vol. 15, no. 11, pp. 1119–1125, 1994.
26. H. Yao and L. Tian. "A genetic-algorithm-based selective principal component analysis (GA-SPCA) method for high-dimensional data feature extraction," *IEEE Transactions on Geoscience and Remote Sensing*, vol. 41, no. 6, pp. 1469–1478, 2003.
27. A. Paoli, F. Melgani, and E. Pasolli. "Swarm intelligence for unsupervised classification of hyperspectral images," in *IEEE International Symposium on Geoscience and Remote Sensing*, vol. 5, no. XX, 2009, pp. V-96–V-99.
28. L. Zhang, Y. Zhong, B. Huang, J. Gong, and P. Li. "Dimensionality reduction based on clonal selection for hyperspectral imagery," *IEEE Transactions on Geoscience and Remote Sensing*, vol. 45, no. 12, pp. 4172–4186, 2007.
29. P. S. Thenkabail, I. Mariotto, M. K. Gumma, E. M. Middleton, D. R. Landis, and K. F. Huemmrich. "Selection of hyperspectral narrowbands (HNBS) and composition of hyperspectral two-band vegetation indices (HVIS) for biophysical characterization and discrimination of crop types using field reflectance and Hyperion/EO-1 data," *IEEE Journal of Selected Topics in Applied Earth Observations and Remote Sensing*, vol. 6, no. 2, pp. 427–439, 2013.
30. J. W. White, P. Andrade-Sanchez, M. A. Gore, K. F. Bronson, T. A. Coffelt, M. M. Conley, K. A. Feldmann, A. N. French, J. T. Heun, D. J. Hunsaker et al. "Field-based phenomics for plant genetics research," *Field Crops Research*, vol. 133, pp. 101–112, 2012.
31. S. Kumar, J. Ghosh, and M. M. Crawford. "Best-bases feature extraction algorithms for classification of hyperspectral data," *IEEE Transactions on Geoscience and Remote Sensing*, vol. 39, no. 7, pp. 1368–1379, 2001.
32. S. De Backer, P. Kempeneers, W. Debruyn, and P. Scheunders. "A band selection technique for spectral classification," *IEEE Geoscience and Remote Sensing Letters*, vol. 2, no. 3, pp. 319–323, 2005.
33. S. B. Serpico and G. Moser. "Extraction of spectral channels from hyperspectral images for classification purposes," *IEEE Transactions on Geoscience and Remote Sensing*, vol. 45, no. 2, pp. 484–495, 2007.
34. D. A. Landgrebe, *Signal Theory Methods in Multispectral Remote Sensing*. John Wiley & Sons, 2005.
35. B.-C. Kuo and D. A. Landgrebe. "Nonparametric weighted feature extraction for classification," *IEEE Transactions on Geoscience and Remote Sensing*, vol. 42, no. 5, pp. 1096–1105, 2004.
36. K. Fukunaga and J. Mantock. "Nonparametric discriminant analysis," *IEEE Transactions on Pattern Analysis and Machine Intelligence*, vol. 5, no. 6, pp. 671–678, 1983.
37. H.-Y. Huang and B.-C. Kuo. "Double nearest proportion feature extraction for hyperspectral-image classification," *IEEE Transactions on Geoscience and Remote Sensing*, vol. 48, no. 11, pp. 4034–4046, 2010.
38. I. T. Jolliffe. "Principal component analysis and factor analysis," in *Principal Component Analysis*. Springer, 1986, pp. 115–128.
39. H. Oja and K. Nordhausen. "Independent component analysis," *Encyclopedia of Environmetrics*, 2001.
40. L. O. Jimenez-Rodriguez, E. Arzuaga-Cruz, and M. Velez-Reyes. "Unsupervised linear feature-extraction methods and their effects in the classification of high-dimensional data," *IEEE Transactions on Geoscience and Remote Sensing*, vol. 45, no. 2, pp. 469–483, 2007.

41. M. Cui and S. Prasad. "Angular discriminant analysis for hyperspectral image classification," *IEEE Journal of Selected Topics in Signal Processing*, vol. 9, no. 6, pp. 1003–1015, 2015.

42. M. Cui and S. Prasad. "Spectral-angle-based discriminant analysis of hyperspectral data for robustness to varying illumination," *IEEE Journal of Selected Topics in Applied Earth Observations and Remote Sensing*, vol. 9, no. 9, pp. 4203–4214, 2016.

43. S. Mukherjee, M. Cui, and S. Prasad. "Spatially constrained semisupervised local angular discriminant analysis for hyperspectral images," *IEEE Journal of Selected Topics in Applied Earth Observations and Remote Sensing*, vol. 11, no. 4, pp. 1203–1212, 2018.

44. C. M. Bachmann, T. L. Ainsworth, and R. A. Fusina. "Improved manifold coordinate representations of large-scale hyperspectral scenes," *IEEE Transactions on Geoscience and Remote Sensing*, vol. 44, no. 10, pp. 2786–2803, 2006.

45. D. Lunga, S. Prasad, M. Crawford, and O. Ersoy. "Manifold-learning-based feature extraction for classification of hyperspectral data: A review of advances in manifold learning," *IEEE Signal Processing Magazine*, vol. 31, no. 1, pp. 55–66, 2014.

46. J. B. Tenenbaum, V. De Silva, and J. C. Langford. "A global geometric framework for nonlinear dimensionality reduction," *Science*, vol. 290, no. 5500, pp. 2319–2323, 2000.

47. B. Schölkopf, A. Smola, and K.-R. Müller. "Nonlinear component analysis as a kernel eigenvalue problem," *Neural Computation*, vol. 10, no. 5, pp. 1299–1319, 1998.

48. S. T. Roweis and L. K. Saul. "Nonlinear dimensionality reduction by locally linear embedding," *Science*, vol. 290, no. 5500, pp. 2323–2326, 2000.

49. V. D. Silva and J. B. Tenenbaum. "Global versus local methods in nonlinear dimensionality reduction," in *Advances in Neural Information Processing Systems*, 2003, pp. 721–728.

50. L. K. Saul and S. T. Roweis. "Think globally, fit locally: Unsupervised learning of low dimensional manifolds," *Journal of Machine Learning Research*, vol. 4, pp. 119–155, 2003.

51. L. Ma, M. M. Crawford, X. Yang, and Y. Guo. "Local-manifold-learning-based graph construction for semisupervised hyperspectral image classification," *IEEE Transactions on Geoscience and Remote Sensing*, vol. 53, no. 5, pp. 2832–2844, 2015.

52. H. L. Yang and M. M. Crawford. "Spectral and spatial proximity-based manifold alignment for multitemporal hyperspectral image classification," *IEEE Transactions on Geoscience and Remote Sensing*, vol. 54, no. 1, pp. 51–64, 2016.

53. H. L. Yang and M. M. Crawford. "Domain adaptation with preservation of manifold geometry for hyperspectral image classification," *IEEE Journal of Selected Topics in Applied Earth Observations and Remote Sensing*, vol. 9, no. 2, pp. 543–555, 2016.

54. S. Yan, D. Xu, B. Zhang, and H.-J. Zhang. "Graph embedding: A general framework for dimensionality reduction," in *IEEE Computer Society Conference on CVPR*, vol. 2, 2005, pp. 830–837.

55. S. Yan, D. Xu, B. Zhang, H.-J. Zhang, Q. Yang, and S. Lin. "Graph embedding and extensions: A general framework for dimensionality reduction," *IEEE Transactions on Pattern Analysis and Machine Intelligence*, vol. 29, no. 1, pp. 40–51, 2007.

56. R. M. Haralick and K. Shanmugam. "Textural features for image classification," *IEEE Transactions on Systems, Man, and Cybernetics*, vol. SMC-3, no. 6, pp. 610–621, 1973.

57. A. Baraldi and F. Parmiggiani. "An investigation of the textural characteristics associated with gray level cooccurrence matrix statistical parameters," *IEEE Transactions on Geoscience and Remote Sensing*, vol. 33, no. 2, pp. 293–304, 1995.

58. J. Li, P. R. Marpu, A. Plaza, J. M. Bioucas-Dias, and J. A. Benediktsson. "Generalized composite kernel framework for hyperspectral image classification," *IEEE Transactions on Geoscience and Remote Sensing*, vol. 51, no. 9, pp. 4816–4829, 2013.

59. M. Dalla Mura, J. Atli Benediktsson, B. Waske, and L. Bruzzone. "Extended profiles with morphological attribute filters for the analysis of hyperspectral data," *International Journal of Remote Sensing*, vol. 31, no. 22, pp. 5975–5991, 2010.

60. D. Landgrebe. "Hyperspectral image data analysis," *IEEE Signal Processing Magazine*, vol. 19, no. 1, pp. 17–28, 2002.

61. G. Shaw and D. Manolakis. "Signal processing for hyperspectral image exploitation," *IEEE Signal Processing Magazine*, vol. 19, no. 1, pp. 12–16, 2002.

62. N. Keshava and J. Mustard. "Spectral unmixing," *IEEE Signal Processing Magazine*, vol. 19, no. 1, pp. 44–57, 2002.

63. D. Manolakis and G. Shaw. "Detection algorithms for hyperspectral imaging applications," *IEEE Signal Processing Magazine*, vol. 19, no. 1, pp. 29–43, 2002.

64. S. Prasad, H. Wu, and J. Fowler. "Compressive data fusion for multi-sensor image analysis," in *Proceedings of the IEEE International Conference on Image Processing*, October 2014, pp. 5032–5036.

65. W. Li, S. Prasad, E. W. Tramel, J. E. Fowler, and Q. Du. "Decision fusion for hyperspectral image classification based on minimum-distance classifiers in the wavelet domain," in *Proceedings of the Signal and Information Processing China Summit & International Conference*, 2014, pp. 162–165.

66. S. Prasad, M. Cui, W. Li, and J. Fowler. "Segmented mixture of Gaussian classification for robust sub-pixel hyperspectral ATR," *IEEE Geoscience and Remote Sensing Letters*, vol. 11, no. 1, pp. 138–142, 2014.

67. S. Prasad, W. Li, J. E. Fowler, and L. M. Bruce. "Information fusion in the redundant-wavelet-transform domain for noise-robust hyperspectral classification," *IEEE Transactions on Geoscience and Remote Sensing*, vol. 50, no. 99, pp. 3474–3486, 2012.

68. W. Li, S. Prasad, and J. E. Fowler. "Classification and reconstruction from random projections for hyperspectral imagery," *IEEE Transactions on Geoscience and Remote Sensing*, vol. 51, no. 2, pp. 1–11, 2012.

69. D. Tuia, G. Camps-Valls, G. Matasci, and M. Kanevski. "Learning relevant image features with multiple-kernel classification," *IEEE Transactions on Geoscience and Remote Sensing*, vol. 48, no. 10, pp. 3780–3791, 2010.

70. D. Tuia, F. Ratle, A. Pozdnoukhov, and G. Camps-Valls. "Multisource composite kernels for urban-image classification," *IEEE Geoscience and Remote Sensing Letters*, vol. 7, no. 1, pp. 88–92, January 2010.

71. M. Cui and S. Prasad. "Class dependent sparse representation classifier for robust hyperspectral image classification," *IEEE Transactions on Geosciences and Remote Sensing*, vol. 53, no. 5, pp. 2683–2695, May 2015.

72. M. Cui and S. Prasad. "Multiscale sparse representation classification for robust hyperspectral image analysis," in *Proceedings of the Global Conference on Signal and Information Processing*, 2013, pp. 969–972.

73. N. Segata, E. Pasolli, F. Melgani, and E. Blanzieri. "Local SVM approaches for fast and accurate classification of remote-sensing images," *International Journal of Remote Sensing*, vol. 33, no. 19, pp. 6186–6201, 2012.

74. G. Camps-Valls and L. Bruzzone. "Kernel-based methods for hyperspectral image classification," *IEEE Transactions on Geoscience and Remote Sensing*, vol. 43, no. 6, pp. 1351–1362, June 2004.

75. H. Akaike. "A new look at the statistical model identification," *IEEE Transactions on Automatic Control*, vol. 19, no. 6, pp. 716–723, 1974.

76. C. E. Rasmussen. "The infinite Gaussian mixture model," in *Advances in Neural Information Processing Systems*, vol. 12, 1999, pp. 554–560.

77. H. Wu and S. Prasad. "Dirichlet process based active learning and discovery of unknown classes for hyperspectral image classification," *IEEE Transactions on Geoscience and Remote Sensing*, vol. 54, no. 8, pp. 4882–4895, 2016.

78. W. Li, S. Prasad, J. Fowler, and L. Bruce. "Locality-preserving dimensionality reduction and classification for hyperspectral image analysis," *IEEE Transactions on Geoscience and Remote Sensing*, vol. 50, no. 4, pp. 1185–1198, 2012.

79. A. Krizhevsky, I. Sutskever, and G. E. Hinton. "Imagenet classification with deep convolutional neural networks," in *Advances in Neural Information Processing Systems*, 2012, pp. 1097–1105.

80. R. Girshick, J. Donahue, T. Darrell, and J. Malik. "Rich feature hierarchies for accurate object detection and semantic segmentation," in *Proceedings of the IEEE Conference on Computer Vision and Pattern Recognition*, 2014, pp. 580–587.

81. K. Simonyan and A. Zisserman. "Very deep convolutional networks for large-scale image recognition," *arXiv preprint arXiv:1409.1556*, 2014.

82. A. Graves, A.-R. Mohamed, and G. Hinton. "Speech recognition with deep recurrent neural networks," in *IEEE International Conference on Acoustics, Speech and Signal Processing*, 2013, pp. 6645–6649.

83. H. Sak, A. Senior, K. Rao, and F. Beaufays. "Fast and accurate recurrent neural network acoustic models for speech recognition," *arXiv preprint arXiv:1507.06947*, 2015.

84. P. J. Werbos. "Backpropagation through time: What it does and how to do it," *Proceedings of the IEEE*, vol. 78, no. 10, pp. 1550–1560, 1990.

85. Z. Zuo, B. Shuai, G. Wang, X. Liu, X. Wang, B. Wang, and Y. Chen. "Convolutional recurrent neural networks: Learning spatial dependencies for image representation," in *Proceedings of the IEEE Conference on Computer Vision and Pattern Recognition Workshops*, 2015, pp. 18–26.

86. H. Wu and S. Prasad. "Convolutional recurrent neural networks for hyperspectral data classification," *Remote Sensing*, vol. 9, no. 3, p. 298, 2017.

87. J. C. Tilton, S. Aksoy, and Y. Tarabalka. "Image segmentation algorithms for land categorization," in *Remotely Sensed Data Characterization, Classification, and Accuracies*, 2015, pp. 317–342.

88. F. A. Kruse, A. Lefkoff, J. Boardman, K. Heidebrecht, A. Shapiro, P. Barloon, and A. Goetz. "The spectral image processing system (SIPS), interactive visualization and analysis of imaging spectrometer data," *Remote Sensing of Environment*, vol. 44, no. 2–3, pp. 145–163, 1993.

89. J.-M. Beaulieu and M. Goldberg. "Hierarchy in picture segmentation: A stepwise optimization approach," *IEEE Transactions on Pattern Analysis and Machine Intelligence*, vol. 11, no. 2, pp. 150–163, 1989.

90. T. Kurita. "An efficient agglomerative clustering algorithm for region growing," in *IAPR Workshop on Mach. Vis. Appl. Citeseer*, 1994.

91. J. Williams. "Algorithm 232—Heapsort," *Communications of the ACM*, vol. 7, no. 6, pp. 347–348, 1964.

92. J. C. Tilton. "Image segmentation by region growing and spectral clustering with a natural convergence criterion," in *IEEE International Geoscience and Remote Sensing Symposium*, vol. 4, 1998, pp. 1766–1768.

93. J. C. Tilton, Y. Tarabalka, P. M. Montesano, and E. Gofman. "Best merge region-growing segmentation with integrated nonadjacent region object aggregation," *IEEE Transactions on Geoscience and Remote Sensing*, vol. 50, no. 11, pp. 4454–4467, 2012.

94. J. C. Tilton, *Parallel Implementation of the Recursive Approximation of an Unsupervised Hierarchical Segmentation Algorithm*, New York, Chapman & Hall, 2007.

95. J. C. Tilton, E. B. de Colstoun, R. E. Wolfe, B. Tan, and C. Huang. "Generating ground reference data for a global impervious surface survey," in *IEEE International Geoscience and Remote Sensing Symposium*, 2012, pp. 5993–5996.

96. E. C. B. de Colstoun, C. Huang, P. Wang, J. C. Tilton, B. Tan, J. Phillips, S. Niemczura, P.-Y. Ling, and R. Wolfe. "Documentation for the global man-made impervious surface (GMIS) dataset from LANDSAT," 2017.

97. R. Massey, T. T. Sankey, K. Yadav, R. G. Congalton, and J. C. Tilton. "Integrating cloud-based workflows in continental-scale cropland extent classification," submitted to the *Remote Sensing of Environment*, 2017.

98. Z. Zhang, E. Pasolli, M. M. Crawford, and J. C. Tilton. "An active learning framework for hyperspectral image classification using hierarchical segmentation," *IEEE Journal of Selected Topics in Applied Earth Observations and Remote Sensing*, vol. 9, no. 2, pp. 640–654, 2016.

99. A. Rakotomamonjy, F. R. Bach, S. Canu, and Y. Grandvalet. "SimpleMKL," *Journal of Machine Learning Research*, vol. 9, November, pp. 2491–2521, 2008.

100. Y. Zhang, L. Yang, S. Prasad, E. Pasolli, J. Jung, and M. Crawford. "Ensemble multiple kernel active learning for classification of multisource remote sensing data," *IEEE Journal of Selected Topics in Applied Earth Observations and Remote Sensing*, vol. 8, no. 2, 2015.

101. D. Tuia, J. Munoz-Mari, L. Gómez-Chova, and J. Malo. "Graph matching for adaptation in remote sensing," *IEEE Transactions on Geoscience and Remote Sensing*, vol. 51, no. 1, pp. 329–341, 2013.

102. X. Zhou and S. Prasad. "Domain adaptation for robust classification of disparate hyperspectral images," *IEEE Transactions on Computational Imaging*, vol. 3, no. 4, pp. 822–836, December 2017.

103. G. Schohn and D. Cohn. "Less is more: Active learning with support vector machines," in *ICML. Citeseer*, 2000, pp. 839–846.

104. P. Mitra, B. Uma Shankar, and S. Pal. "Segmentation of multispectral remote sensing images using active support vector machines," *Pattern Recognition Letters*, vol. 25, no. 9, pp. 1067–1074, 2004.

105. L. Copa, D. Tuia, M. Volpi, and M. Kanevski. "Unbiased query-by-bagging active learning for VHR image classification," in *Image and Signal Processing for Remote Sensing XVI, vol. 7830. International Society for Optics and Photonics*, 2010, p. 78300K.

106. W. Di and M. M. Crawford. "View generation for multiview maximum disagreement based active learning for hyperspectral image classification," *IEEE Transactions on Geoscience and Remote Sensing*, vol. 50, no. 5, pp. 1942–1954, 2012.

107. T. Luo, K. Kramer, D. B. Goldgof, L. O. Hall, S. Samson, A. Remsen, and T. Hopkins. "Active learning to recognize multiple types of plankton," *Journal of Machine Learning Research*, vol. 6, no. April, pp. 589–613, 2005.

108. S. Rajan, J. Ghosh, and M. Crawford. "An active learning approach to hyperspectral data classification," *IEEE Transactions on Geoscience and Remote Sensing*, vol. 46, no. 4, pp. 1231–1242, 2008.

109. E. Pasolli, F. Melgani, D. Tuia, F. Pacifici, and W. J. Emery. "SVM active learning approach for image classification using spatial information," *IEEE Transactions on Geoscience and Remote Sensing*, vol. 52, no. 4, pp. 2217–2233, 2014.

110. Q. Shi, B. Du, and L. Zhang. "Spatial coherence-based batch-mode active learning for remote sensing image classification," *IEEE Transactions on Image Processing*, vol. 24, no. 7, pp. 2037–2050, 2015.

111. D. Tuia, E. Pasolli, and W. J. Emery. "Using active learning to adapt remote sensing image classifiers," *Remote Sensing of Environment*, vol. 115, no. 9, pp. 2232–2242, 2011.

112. E. Pasolli, F. Melgani, N. Alajlan, and N. Conci. "Optical image classification: A ground-truth design framework," *IEEE Transactions on Geoscience and Remote Sensing*, vol. 51, no. 6, pp. 3580–3597, 2013.

113. N. Alajlan, E. Pasolli, F. Melgani, and A. Franzoso. "Large-scale image classification using active learning," *IEEE Geoscience and Remote Sensing Letters*, vol. 11, no. 1, pp. 259–263, 2014.

114. D. Tuia, M. Volpi, L. Copa, M. Kanevski, and J. Munoz-Mari. "A survey of active learning algorithms for supervised remote sensing image classification," *IEEE Journal of Selected Topics in Signal Processing*, no. 99, pp. 1–1, 2011.

115. J. Li, J. M. Bioucas-Dias, and A. Plaza. "Hyperspectral image segmentation using a new Bayesian approach with active learning," *IEEE Transactions on Geoscience and Remote Sensing*, vol. 49, no. 10, pp. 3947–3960, 2011.

116. J. Li, J. M. Bioucas-Dias, and A. Plaza. "Spectral–spatial classification of hyperspectral data using loopy belief propagation and active learning," *IEEE Transactions on Geoscience and Remote Sensing*, vol. 51, no. 2, pp. 844–856, 2013.

117. A. Stumpf, N. Lachiche, J.-P. Malet, N. Kerle, and A. Puissant. "Active learning in the spatial domain for remote sensing image classification," *IEEE Transactions on Geoscience and Remote Sensing*, vol. 52, no. 5, pp. 2492–2507, 2014.

118. P. Salembier and L. Garrido. "Binary partition tree as an efficient representation for image processing, segmentation, and information retrieval," *IEEE Transactions on Image Processing*, vol. 9, no. 4, pp. 561–576, 2000.

119. S. Valero, P. Salembier, and J. Chanussot. "Hyperspectral image representation and processing with binary partition trees," *IEEE Transactions on Image Processing*, vol. 22, no. 4, pp. 1430–1443, 2013.

120. J. Munoz-Mari, D. Tuia, and G. Camps-Valls. "Semisupervised classification of remote sensing images with active queries," *IEEE Transactions on Geoscience and Remote Sensing*, vol. 50, no. 10, pp. 3751–3763, 2012.

121. J. Jung, E. Pasolli, S. Prasad, J. C. Tilton, and M. M. Crawford. "A framework for land cover classification using discrete return LiDAR data: Adopting pseudo-waveform and hierarchical segmentation," *IEEE Journal of Selected Topics in Applied Earth Observations and Remote Sensing*, vol. 7, no. 2, pp. 491–502, 2014.

122. E. Pasolli, H. L. Yang, and M. M. Crawford. "Active-metric learning for classification of remotely sensed hyperspectral images," *IEEE Transactions on Geoscience and Remote Sensing*, vol. 54, no. 4, pp. 1925–1939, 2016.

123. Z. Zhang, E. Pasolli, H. L. Yang, and M. M. Crawford. "Multimetric active learning for classification of remote sensing data," *IEEE Geoscience and Remote Sensing Letters*, vol. 13, no. 7, pp. 1007–1011, 2016.

124. Z. Zhang and M. M. Crawford. "A batch-mode regularized multimetric active learning framework for classification of hyperspectral images," *IEEE Transactions on Geoscience and Remote Sensing*, vol. 55, no. 11, pp. 6594–6609, 2017.

125. K. Q. Weinberger and L. K. Saul. "Distance metric learning for large margin nearest neighbor classification," *Journal of Machine Learning Research*, vol. 10, no. February, pp. 207–244, 2009.

126. A. Bellet, A. Habrard, and M. Sebban. "A survey on metric learning for feature vectors and structured data," *arXiv preprint arXiv:1306.6709*, 2013.

127. H. Zhang, T. S. Huang, N. M. Nasrabadi, and Y. Zhang. "Heterogeneous multi-metric learning for multi-sensor fusion," in *Proceedings of the 14th International Conference on Information Fusion*, IEEE, 2011, pp. 1–8.

4 Big Data Processing on Cloud Computing Architectures for Hyperspectral Remote Sensing

Zebin Wu, Jin Sun, and Yi Zhang

CONTENTS

4.1 INTRODUCTION

4.1.1 WHAT IS CLOUD COMPUTING?

The well-known definition of cloud computing is given by the National Institute of Standards and Technology (NIST) and can be described as follows. Cloud computing is a model for enabling ubiquitous, convenient, on-demand network access to a shared pool of configurable computing resources (e.g., networks, servers, storage, applications, and services) that can be rapidly provisioned and released with minimal management effort or service provider interaction [1]. This model provides five essential characteristics, which are introduced in Table 4.1.

As an important part of cloud computing, Infrastructure as a Service (IaaS) becomes very popular as the foundation for higher-level services such as Platform as a Service (PaaS) and Software as a Service (SaaS) [2]. The three service models of cloud computing are illustrated in Figure 4.1.

The capability provided by the IaaS provider is to provision processing, storage, networks, and other fundamental computing resources where users are able to deploy and run selected software (including operating systems and applications) by virtualization technologies. Users do not manage

TABLE 4.1

Five Essential Characteristics of Cloud Computing

On-demand self-service	Users can unilaterally achieve computing capabilities (e.g., server time and network storage) from service providers as needed automatically without involving any human interaction.
Broad network access	Capabilities are delivered over the network and accessed by standard mechanisms that promote use by heterogeneous thin or thick client devices, (e.g., mobile phones, tablets, laptops, and workstations).
Resource pooling	Service providers' computing resources are pooled to serve multiple users using a multitenant model, with different physical and virtual resources dynamically scheduled according to users' demands. Users do not care about where the resources really are.
Rapid elasticity	Capabilities can be elastically provisioned and released, in some cases automatically, to scale rapidly outward and inward commensurate with users' demands. From the perspectives of users, the available capabilities provided by service providers often appear to be unlimited and can be achieved in any quantity at any time.
Measured service	Cloud systems automatically control and optimize resource use by employing a metering capability at some level of abstraction appropriate to the type of service (e.g., storage, processing, bandwidth, and active user accounts). Resource usage can be monitored, controlled, and reported, providing transparency for both the provider and users.

or control the underlying cloud infrastructure but have control over operating systems, storage, and deployed applications, and possibly have limited control of select networking components (e.g., host firewalls).

The capability provided by the PaaS provider is to deploy onto the cloud infrastructure user-created or acquired applications created using programming languages, libraries, services, and tools supported by the provider. Users do not manage or control the underlying cloud infrastructure, including network, servers, operating systems, or storage, but have control over the deployed applications and possibly configuration settings for the application-hosting environment.

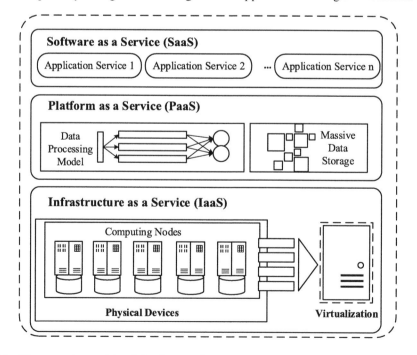

FIGURE 4.1 Three service models of cloud computing.

The capability provided by the SaaS provider is to use the provider's applications running on a cloud platform. The applications are accessible from various client devices through either a thin client interface, such as a web browser (e.g., web-based email), or a program interface. Users do not manage or control the underlying cloud infrastructure, including network, servers, operating systems, storage, or even individual application capabilities, with the possible exception of limited user-specific application configuration settings.

Among the three service models, the most widely accepted one is IaaS. IaaS providers (e.g., Amazon EC2) always offer different virtual machine (VM) types that are characterized by machine configuration, quality of service (QoS), and a pricing model. Users select VM types according to their business requirements and pay for the rented instances of their selected VM types. QoS promised by providers is described in service-level agreements (SLAs).

In general, there are four deployment models that are generally used: private cloud, public cloud, community cloud, and hybrid cloud. The cloud resources of the private cloud are provisioned for exclusive use by a single organization comprising multiple consumers (e.g., business units). The cloud resources of the public cloud are provisioned for open use by the general public. The cloud resources of the community cloud are provisioned for exclusive use by a specific community of users from organizations that have shared concerns. The cloud resources of the hybrid cloud are a composition of two or more distinct cloud infrastructures (private, community, or public) that remain unique entities, but are bound together by standardized or proprietary technology that enables data and application portability.

4.1.2 WHY DO WE NEED TO USE CLOUD COMPUTING IN HYPERSPECTRAL DATA ANALYSIS?

Hyperspectral imaging instruments collect hundreds of narrow spectral bands spanning any portion of the electromagnetic spectrum (e.g., visible, near-infrared, short-wave infrared, thermal). In hyperspectral image cubes, each pixel can be represented by a vector whose entries correspond to the spectral bands, providing a representative spectral signature of the underlying materials within the pixel [12,13]. Hyperspectral images comprise hundreds of contiguous spectral bands, thus imposing significant requirements in terms of storage and data processing. Moreover, recent technological advances in spaceborne and airborne remote sensing have exponentially increased the volume of remote sensing image data that are collected and stored in hyperspectral data repositories [8]. The availability of new hyperspectral missions producing large amounts of data on a daily basis has posed important challenges for scalable and efficient processing of hyperspectral data in the context of various applications.

For example, the data collection rate of the NASA Jet Propulsion Laboratory's Airborne Visible/Infrared Imaging Spectrometer (AVIRIS) [3] is 2.55 MB/sec, which means that it can obtain nearly 9 GB of data in 1 hour. The Chinese Pushbroom Hyperspectral Imager (PHI) has a data collection rate of 7.2 MB/sec, and it can collect more than 25 GB data in 1 hour. The space-borne Hyperion instrument collects a hyperspectral data cube with 256×6925 pixels, 242 spectral bands, and radiometric resolution of 12 bits in 30 seconds, collecting almost 71.9 GB data in 1 hour (over 1.6 TB daily). Other satellite missions that will be soon in operation, such as the environmental analysis and mapping program (EnMAP), present higher data collection ratios [8]. Since hyperspectral data volumes are becoming increasingly massive, it is hard to meet the storage and computational requirements of large-scale hyperspectral data processing applications without resorting to distributed computing facilities. Considering the high volume and fast generation velocity of remotely sensed data, the processing of hyperspectral data can be regarded as a big data problem [4,5].

Distributed computing technologies are in high demand for dynamic, on-demand processing of large remote sensing datasets. Cloud computing offers the potential to tackle massive data processing workloads by means of its distributed parallel architecture. The use of cloud computing for the analysis of large remote sensing data repositories represents a natural solution, as well as an evolution of previously developed techniques. With the continuously increasing demand for massive data processing in remote sensing applications, there have been several efforts in the literature oriented toward exploiting the cloud computing infrastructure for processing remote sensing big data [6–10]. For example, the literature [8] provides a distributed parallel implementation of a principal component analysis (PCA) algorithm for hyperspectral dimensionality reduction based on the MapReduce parallel model [11]. The literature [9] proposes a new distributed architecture for supervised classification of large volumes of earth observation data. The experimental results in these studies have demonstrated the effectiveness of the proposed methods, not only in terms of accuracy but also in terms of computational performance.

4.1.3   Advances and Advantages of Cloud Computing

Considering hyperspectral data processing as a big data problem, cloud computing has shown the following key advantages on the parallel and distributed processing of large-scale hyperspectral data.

- *Cost Efficiency*: Cloud computing is probably the most cost-efficient method to use, maintain, and upgrade. Traditional desktop software costs companies a lot in terms of finance. Adding up the licensing fees for multiple users is very expensive for the establishment concerned. On the other hand, the cloud is available at much cheaper rates and can thus reduce the company's IT expenses significantly. Besides, cloud computing technology enables delivery of large-scale computational resources in the form of a pay-as-you-go model and brings low total cost of ownership (TCO) and good return on investment (ROI) due to the economy of scale.
- *Unlimited Resource Capacity*: Cloud service providers always provide users the option of unlimited resource capacity. Accordingly, users do not need to worry about running out of resource capacity.
- *High Reliability*: It is relatively much easier to store data on clouds than to store them on a physical device. Furthermore, most cloud service providers are usually competent enough to handle data backup and recovery. Hence, this makes the entire process of storing data on clouds much more reliable than other traditional methods of data storage.
- *Automatic Software Integration*: Software can be integrated on clouds automatically. As a result, users do not need to make additional efforts to customize and integrate their applications as per their preferences.
- *Easy Access*: Once registered on clouds, users can access their data from anywhere as long as there is an Internet connection. This convenient feature allows users to move beyond time zone and geographic location issues.
- *Quick Deployment*: Applications can be deployed on clouds quickly, since cloud providers provide lots of easy tools. The required time is relatively much shorter than that required to deploy applications on platforms on premise.

4.2   CLOUD COMPUTING TECHNOLOGIES

4.2.1   Distributed Programming Model

The most well-known and widely used model to process large amounts of data on a distributed cluster is MapReduce (MR), which was proposed by researchers from Google [11]. This model takes full advantage of the high-performance capabilities provided by cloud computing architectures. Specifically, in the MapReduce model, the processing of a computation task consists of two distributed operations (i.e., "map" and "reduce"). With all datasets organized as key/value pairs, the "map" function processes a key/value pair to generate a set of intermediate pairs. In this manner, a task has been divided into several independent subtasks that can be executed in parallel. On the other hand, the "reduce" function is in charge of processing all those intermediate pairs that have an identical intermediate key, and further merges all subtask results to obtain the final result of the original task. Lots of real computation jobs can be expressed in this model.

Programs implementing both interfaces can be automatically executed on a large cluster of machines in parallel. The runtime system implementing the MR model is responsible for partitioning the input data, scheduling execution of the program's tasks across over multiple machines, handling exceptions, and managing the required communication. All these good features hide those issues occurring in parallel and distributed systems, thus freeing the developers. As a result, the MR model can help developers to use the resources of a large cluster in an easy and efficient way.

The details of the MR execution process are illustrated in Figure 4.2. As seen in the figure, map invocations are distributed on multiple machines (i.e., workers) by partitioning the input data files into M splits automatically. Reduced invocations are distributed by partitioning the intermediate key space into R pieces using a partitioning function [e.g., hash(key) mod R]. These two partitioning procedures can also be processed on multiple machines in parallel.

The runtime of the MR model includes two important components, a master and multiple workers. The master manages all the workers and takes care of assigning all map and reduce tasks to workers, controlling lifecycles of tasks. Accordingly, the master should maintain the state (idle, in-progress, or completed) of all map and reduce tasks and the state of all workers (idle and busy). Meanwhile, the master should record the locations of the input files required by each map task

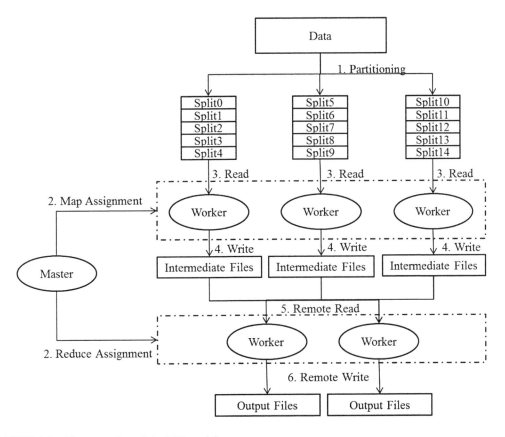

FIGURE 4.2   The execution of the MR model.

and the locations of the generated intermediate data required by each reduce task. The six steps in Figure 4.2 are as follows.

1. The input files are first partitioned into M pieces. The size per piece can be controlled by setting some parameters. The Google researchers recommend 16–64 MB per piece. Meanwhile, many copies of the program implementing both the map and reduce interfaces are generated and distributed on a cluster of machines.

2. When assigning map or reduce tasks, the master should take into account the states of all workers. In order to balance workloads, the master always tries to pick up an idle worker for each map or reduce task.

3. Each worker assigned a map task reads its corresponding splits of the input files. Multiple key/value pairs are obtained from the input files and passed to the user-defined map function. During the execution of the map task, each worker generates lots of intermediate key/value pairs that are buffered in the memory first.

4. Periodically, the buffered pairs are written into local disks. The master acquires the locations of these buffered pairs so that it will send the locations to those workers assigned reduce tasks.

5. Once a worker assigned a reduce task receives the locations of the intermediate pairs, the worker reads all these data remotely. After this remote reading procedure is finished, all obtained pairs should be sorted first in order to group those pairs having the same key.

6. All the sorted intermediate pairs are passed to the reduce tasks that include the user-defined reduce functions. The output of the reduce tasks is written to a remote final file.

Despite its great computation efficiency of distributed processing, there are also several essential issues associated with this distributed programming model.

- *High Reliability*: The system implementing the MR model should have high reliability, as the system is designed to process large amounts of data on lots of machines. In other words, the system is able to ensure reliability automatically and appropriately.In terms of the reliability of workers, the master checks the living state of each worker periodically. If a worker does not give a response in a specified amount of time, all the map tasks finished by this worker and those map or reduce tasks assigned to this worker should be marked as idle and re-executed. Completed reduce tasks do not need to be reprocessed, as the output data are written into a remote file in a remote shared file system.Regarding the reliability of the master, a feasible way is to make the master record some checkpoints. If the master does not work, a backup master can be activated and use the checkpoint for recovery.

- *Data Locality*: As mentioned above, map tasks read data from the local disks. Obviously, it is difficult to always find an idle worker that contains the data required by map tasks. Accordingly, if a map task is not assigned to a worker containing its required data, the required data should be transferred to the worker from the original location via the network. In other words, network bandwidth is a relatively scarce resource in the system. In order to improve the system's efficiency, if there are no available workers containing the data required by a map task, the master attempts to schedule the map task near its required data. Accordingly, we need a shared file system to support the system implementing the MR model. Google's system uses the Google file system(GFS), while the Hadoop and Spark systems employ the HDFS, which is an open-source system similar to GFS. In GFS, each file is portioned into multiple blocks with the size of 64 MB, and each block has three replicas that are stored on different machines.

- *Backup Tasks*: According to the mechanism ensuring the reliability of workers, if one worker does not work, its map tasks are reassigned to another worker. In other words, those tasks should be reprocessed. As a result, the new worker becomes a "straggler" or a worker completing one of the last few map or reduce tasks in the computation with an unusually long time. A feasible solution recommended by Google's researchers [11] to alleviate the problem of stragglers is to use a backup task of the remaining in-process tasks when tasks are approaching completion. The task is regarded as completed when either the primary or backup task is finished. According to the report presented by Google researchers, this mechanism is able to reduce the completion time of large MapReduce operations significantly. As a result, the efficiency of the entire system implementing the MR model can be improved dramatically.

- *Combiner Function*: A combiner function is designed to allow users to merge data with the same keys before writing the data into the output files. The combiner function can be executed on the worker performing map and/or reduce worker tasks. According to the report presented by Google researchers [11], this combiner function can greatly improve the system's effectiveness.

4.2.2 COMPUTING ENGINES

Regarding the distributed computing engine, a possible solution is Apache Hadoop* due to its reliability and scalability, as well as its completely open-source nature. It has been successfully applied in industrial fields [8]. However, Apache Hadoop only supports simple one-pass computations (e.g., aggregation or database SQL queries) and is generally not appropriate for multipass algorithms. Google Earth Engine (GEE)† is a public cloud that offers Google's massive computational capabilities to solve issues in the geographical field. It empowers not only traditional remote sensing scientists, but also a much wider audience that lacks the technical capacity needed to utilize traditional supercomputers or large-scale commodity cloud computing resources. GEE provides lots of functions in terms of libraries, whose application programming interfaces (APIs) are exposed and can be used by users to make programs (the Python and JavaScript languages are currently supported) to solve problems. However, although users can use those APIs provided by GEE, the platform exerts multiple restrictions on the aspects by which users deploy and apply their own functions. Additionally, users use GEE resources for free with some restrictions given by GEE; for example, a cached object cannot exceed 200 MB. Amazon EC2‡ is a well-known general public IaaS platform providing variety of VM types. Users can select VM types according to their business requirements and create instances of the selected types to complete their work. As EC2 only offers VM instances, users can perform actions they want to do. ArcGIS Online (AO)§ is a public SaaS mapping platform provided by ESRI. AO provides those capabilities in the fields of applications, analytics, administration, and collaboration involving maps. As AO is a SaaS platform, users can only use their provided functions. Additionally, AO is not a free platform, and users should make payments for what they use.

Apache Spark¶ is a newly developed computing engine for large-scale data processing on cloud computing architectures, which implements a fault-tolerant abstraction for in-memory cluster computing and provides fast and general data processing on large clusters [8]. One of the main differences between Hadoop and Spark is that the output data of map tasks in Spark are written into memory, whereas those data in Hadoop are first written into memory and then written onto local disks. Obviously, Spark requires more memory and has higher efficiency.

Spark supports the multipass algorithms that are required for complicated data analysis. It extends the MR model to include primitives for data sharing named resilient distributed datasets (RDDs), and offers an API based on coarse-grained transformation that allows efficient data

---

* https://hadoop.apache.org

† https://earthengine.google.com

‡ https://aws.amazon.com/

§ http://www.esri.com/software/arcgis/arcgisonline.

¶ https://spark.apache.org/

recovery. In addition, Spark has an advanced directed acyclic graph (DAG) execution engine that supports cyclic data flow and an in-memory computing mechanism. Owing to this advanced engine, it achieves tremendous speedups in program execution when compared with traditional computing engines. To use Spark, the user is required to prepare a driver program that defines RDDs, invokes actions on RDDs, and tracks the lineage of RDDs. The driver program usually connects to a cluster of workers, which are long-lived processes that can store RDD partitions in random access memory (RAM) across operations. During the runtime, the driver program launches multiple workers to load RDD data from a distributed file system and holds the loaded data partitions in memory.

4.2.3 DYNAMIC STORAGE

The distributed processing of big data application requires a dynamic storage mechanism for distributing the data across several geographic locations. This requirement is well fulfilled by HDFS, which can be deployed on low-cost hardware and can provide high-throughput access to large-scale application datasets. Spark provides common interfaces to use the data files stored in HDFS as the input RDDs. HDFS is a highly reliable distributed file system that implements Google's GFS and can be deployed on a large cluster. HDFS aims for storing large files whose sizes are in granularity of gigabytes or terabytes. Each large file is partitioned into multiple blocks with the same size, 64 MB (the size can be set via an optional parameter, and the default value is 64 MB). Each block has three replicas (this value can be set via an optional parameter; the default value is 3), which are distributed across three different machines. As a result, HDFS achieves high fault tolerance. HDFS provides common operating interfaces like creating, writing, closing, and deleting files. Note that HDFS is designed for those applications that process large amounts of data (e.g., OLAP applications) rather than the ones involving lots of user interactions (e.g., OLTP applications). Accordingly, the efficiency of updating data is not high. HDFS is developed in a master/slave architecture. Each cluster consists of a NameNode (i.e., the master) and multiple DataNodes (i.e., the slaves).

4.3 FRAMEWORK OF BIG DATA PROCESSING ON CLOUD COMPUTING ARCHITECTURES

In order to implement parallel and distributed processing of massively big remote sensing data, three essential issues have to be well developed: (1) the distributed programming model, (2) the underlying engine that supports parallel computing, and (3) dynamic storage of distributed remote sensing data.

As discussed in the previous section, the appropriate and effective solutions to the above-mentioned three issues are the MapReduce model, Apache Spark, and HDFS, respectively. Taking into account all the above-mentioned issues, Figure 4.3 [8] provides a graphical overview of the distributed parallel framework for massive hyperspectral data processing using Apache Spark and HDFS. Built upon a cloud environment, this framework can be decomposed into three levels: the engine level, platform level, and commodity hardware level. The hardware level includes all computing resources, for example, physical/virtual servers, storage devices, and network devices. A virtualization process is necessary to convert all physical resources into virtual resources in order to build a high-performance, flexible, and shared computing resource pool that provides the distributed processing capacities. The platform level usually includes a master node and a cluster of slave nodes processing RDD instances. By executing the driver program, the master communicates with the slave nodes to complete the map and reduce operations in the distributed programming model. On the engine level, Apache Spark provides common interfaces with HDFS data blocks and the driver program, and HDFS is responsible for the dynamic storage of large-scale data distributed across the cloud environment.

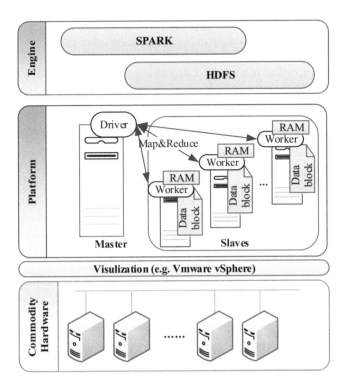

FIGURE 4.3   An overview of the distributed parallel framework for big hyperspectral data processing.

By taking advantage of the distributed parallel framework and the MapReduce model, a massive hyperspectral data processing application can be implemented in a parallel and distributed way with the following key steps:

First, we store the original hyperspectral datasets on HDFS in a distributed and flexible way. Since the original datasets are divided into many spatial-domain partitions, we read every partition of the data on HDFS as a key/value pair in which the key is the offset of this partition in the original dataset and the value is the hyperspectral data partition. Thus, the Apache driver program is able to read the data partitions of the original hyperspectral dataset.

Subsequently, a map operation is performed to convert the loaded data partitions into RDD format. The obtained RDD instances after conversion can then be processed in parallel on Apache Spark. Driven by the driver program on the master node, all slave nodes work simultaneously to process the partitioned data on individual RDD instances. The slave nodes submit the results to the master node after they finish data processing.

The driver program performs the reduce operation to merge the results produced by all slave nodes. The reduce operation analyzes all key/value pairs and combines the values that shared the same key following the user-specified driver program. The final results will be broadcast to all slave nodes and stored into HDFS.

4.4 CASE STUDY

In this section, we use hyperspectral image classification, one of the most popular and important techniques for hyperspectral remotely sensed image interpretation, as a case study to demonstrate the applicability and efficiency of utilizing cloud computing technologies to efficiently perform distributed parallel processing of big hyperspectral data and accelerate hyperspectral data computations. To that end, we develop a parallel and distributed implementation of a hyperspectral image classification algorithm (i.e., spatial correlation regularized sparse representation classification [SCSRC] [12]) on the proposed framework using advanced cloud computing technologies such as HDFS and Apache Spark, as well as MapReduce methodology.

4.4.1 A COMPUTATION- AND DATA-INTENSIVE TASK OF HYPERSPECTRAL IMAGE CLASSIFICATION

The task of image classification has wide applications in various fields of Earth observation and space exploration. It amounts to labeling the pixels into a set of predefined classes [12,13]. Over the last decade, various supervised and unsupervised methods have been developed for hyperspectral image classification [14,15], including support vector machines (SVMs) [16,17], artificial neural networks (ANNs) [18], multinomial logistic regression (MLR) [19], orthogonal matching pursuit (OMP) [20], composite kernels (CKs) [21–23], and so on.

Among this extensive set of techniques, sparse representation-based classification (SRC) methods often lead to state-of-the-art classification accuracy in hyperspectral imaging [25,26]. This approach relies on the fact that a single pixel in a hyperspectral image can be sparsely represented by a linear combination of a few training samples from a structured dictionary. The sparse representation of an unknown pixel is expressed as a sparse vector whose nonzero entries correspond to the weights of the selected training samples [24]. The class label of the test sample can be easily determined via the sparse vector that is recovered by solving a sparse optimization problem.

SRC methods have been shown to be highly effective for solving the supervised classification problem from the viewpoint of classification accuracy. However, their application to real problems is compromised by the fact that they often suffer from high complexity of the model solution, which often limits their application in time-critical scenarios. Taking into consideration that SRC-based hyperspectral image classification is generally a computation- and data-intensive task, it is important to employ cloud computing technologies and our proposed framework to accelerate its computation and make them suitable for massive data processing.

For an SRC method, a test spectral sample in hyperspectral imaging can be approximately represented by a few training samples among a given over-complete training dictionary, and the class label of a test pixel is determined by the characteristics of the sparse vector recovered by a sparsity-constrained optimization problem. The fundamental assumption is that the spectral signals in the same class usually lie in a low-dimensional subspace.

Moreover, it is expected that neighboring pixels usually include materials with similar spectral characteristics, especially in homogeneous regions of a hyperspectral image. Previous works indicate that by taking advantage of the spatial-contextual information and spectral characteristics, classification accuracy can be significantly improved [27–33]. For instance, the method of spatial correlation regularized sparse representation classification [12], which introduces spatial smoothness regularization and spectral fidelity terms to the sparse representation model, has successfully promoted the stability of classification results and achieved very high classification accuracy.

Assume there are K distinct classes, including J training samples in total. The kth class has N_k training samples, denoted as $\boldsymbol{A}^k = \left[\boldsymbol{a}_1^k, \boldsymbol{a}_2^k, \ldots, \boldsymbol{a}_{N_k}^k\right] \in \boldsymbol{R}^{L \times N_k}$, where L is the number of spectral bands. $\boldsymbol{A} = [\boldsymbol{A}^1, \boldsymbol{A}^2, \ldots, \boldsymbol{A}^K] \in \boldsymbol{R}^{L \times J}$ is a structured dictionary formed by the concatenation of several class-wise subdictionaries consisting of training samples, $J = \sum_{k=1}^{K} N_k$, and $\hat{\boldsymbol{s}} = [\boldsymbol{s}^1 \ldots \boldsymbol{s}^K]^{\mathrm{T}} \in \boldsymbol{R}^J$ is a sparse vector formed by concatenating the sparse vectors $\{\boldsymbol{s}^k\}_{k=1,2,\ldots,K}$. Let $\boldsymbol{x} \in \boldsymbol{R}^L$ be an L-dimensional hyperspectral pixel vector. The hyperspectral image with L bands and I pixels is denoted by $\boldsymbol{X} \in \boldsymbol{R}^{L \times I}$, in which a pixel is represented by a column vector $\{\boldsymbol{X}_{:,i}\}_{i=1}^{I} \in \boldsymbol{R}^L$. The problem of spatial-spectral classification can be modeled as

$$\min_{\boldsymbol{S}} \frac{1}{2} \| \boldsymbol{X} - \boldsymbol{AS} \|_F^2 + \lambda_1 \| \boldsymbol{S} \|_{1,1} + \lambda_2 \| \boldsymbol{HS} \|_F^2 + l_0(\boldsymbol{S}_\Lambda - \boldsymbol{I}),$$

where $\|\cdot\|_F$ is the Frobenius norm; $\|\cdot\|_{1,1}$ denotes the sum of absolute values of all the matrix elements; $\lambda_1 > 0$ and $\lambda_2 > 0$ are the regularization parameters; $\mathbf{S} \in \mathbf{R}^{J \times I}$ is the recovered sparse matrix whose row vectors $\{\mathbf{S}_{j,:}\}_{j=1}^{J} \in \mathbf{R}^I$ are the corresponding weights associated with the training samples (atoms) of the dictionary; and $\mathbf{HS} = [(\nabla \mathbf{S}_{1,:})^T, (\nabla \mathbf{S}_{2,:})^T, \dots, (\nabla \mathbf{S}_{J,:})^T]^T$, whose row column $\nabla \mathbf{S}_{j,:}$ denotes the discrete gradient of $\mathbf{S}_{j,:}$ in a two-dimensional sense. $l_0(x)$ is an indicative function [if $x = 0$, $l_0(x) = 0$; $x \neq 0$, $l_0(x) = \infty$], $\mathbf{S}_\Lambda$ denotes the coefficient matrix of training sample set A in X ($X_\Lambda = A$), and $\mathbf{I}$ is the identity matrix.

This model can be efficiently solved by employing the alternating direction method of multipliers [34,35] as Algorithm 4.1 [12] and Figure 4.4 show. Then, the class label of test sample x can be determined by the minimum residual [25] between x and its approximation from each class-wise subdictionary as follows:

$$Class(\mathbf{x}) = \min_{k=1,\dots K} \| \mathbf{x} - \mathbf{A}^k \hat{s}^k \|_2,$$

where $\hat{s}^k$ is the portion of the recovered sparse coefficients corresponding to subdictionary $\mathbf{A}^k$.

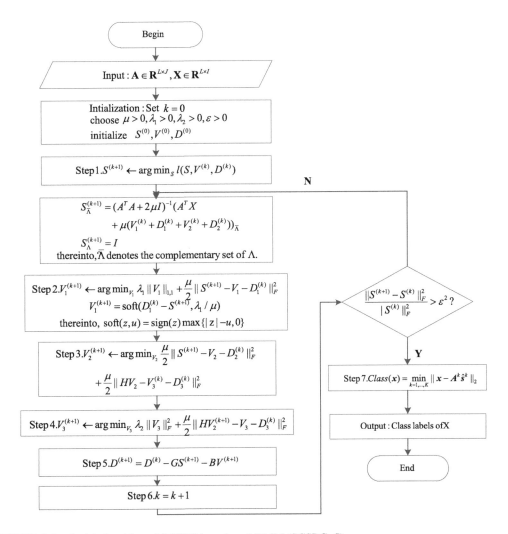

FIGURE 4.4 Serial algorithm of SCSRC based on ADMM (SCSRC_S).

Algorithm 4.1 Serial Algorithm of SCSRC Based on ADMM (SCSRC_S)

Input: Training sample set $\mathbf{A} \in \mathbf{R}^{L \times J}$, test sample set $\mathbf{X} \in \mathbf{R}^{L \times I}$

Initialization: Set $k = 0$, choose regularization parameters $\mu > 0, \lambda_1 > 0, \lambda_2 > 0, \varepsilon > 0$, initialize $S^{(0)}, V^{(0)}, D^{(0)}$

Do:

   Step 1. Calculate $S^{(k+1)} \leftarrow \arg\min_S l(S, V^{(k)}, D^{(k)})$

   $$S_\Lambda^{(k-1)} = (A^T A + 2\mu I)^{-1}(A^T X$$
   $$+ \mu(V_1^{(k)} + D_1^{(k)} + V_2^{(k)} + D_2^{(k)}))_\Lambda$$
   $$S_\Lambda^{(k+1)} = I$$

   there into, $\overline{\Lambda}$ denotes the complementary set of Λ.

   Step 2. Calculate

   $$V_1^{(k+1)} \leftarrow \arg\min_{V_1} \lambda_1 \| V_1 \|_{1,1} + \frac{\mu}{2} \left\| S^{(k+1)} - V_1 - D_1^{(k)} \right\|_F^2$$
   $$V_1^{(k+1)} = \text{soft}\left(D_1^{(k)} - S^{(k+1)}, \lambda_1 / \mu \right)$$

   there into, $\text{soft}(z, u) = \text{sign}(z) \max\{| z | -u, 0\}$

   Step 3. Calculate

   $$V_2^{(k+1)} \leftarrow \arg\min_{V_2} \frac{\mu}{2} \left\| S^{(k+1)} - V_2 - D_2^{(k)} \right\|_F^2$$
   $$+ \frac{\mu}{2} \left\| H V_2 - V_3^{(k)} - D_3^{(k)} \right\|_F^2$$

   Step 4. Calculate

   $$V_3^{(k+1)} \leftarrow \arg\min_{V_3} \lambda_2 \| V_3 \|_F^2 + \frac{\mu}{2} \left\| H V_2^{(k+1)} - V_3 - D_3^{(k)} \right\|_F^2$$

   Step 5. Calculate

   $$D^{(k+1)} = D^{(k)} - G S^{(k+1)} - B V^{(k+1)}$$

   Step 6. Calculate $k = k + 1$

While $\dfrac{\| S^{(k+1)} - S^{(k)} \|_F^2}{| S^{(k)} \|_F^2} > \varepsilon^2$

Step 7. $Class(x) = \min_{k=1,\ldots,K} \| x - A^k \hat{s}^k \|_2,$

   $\forall x$, x is a column of $\mathbf{X}$.

Output: Class labels of X.

End

In Steps 3 and 4 of SCSRC_S, the H operator independently transforms the matrix for each row; thus, the optimization problems of Steps 3 and 4 can be decomposed into several optimization subproblems, which are solved by the Gauss-Seidel [36,37] method.

It can be seen from Figure 4.4 that the data involved in Algorithm 4.1, including the hyperspectral data matrix $\mathbf{X}_{L \times 1}$, the coefficient matrix $\mathbf{S}_{J \times I}^{(k)}$, the training sample matrix $\mathbf{A}_{L \times J}$, and V_i and $D_i^{(k)}$, are all high-dimensional and have large data volume. Furthermore, the procedure of Algorithm 4.1 contains several loop executions, in which there are many complex calculations of matrix multiplication, matrix addition, matrix subtraction, matrix transposition, matrix inversion, sum of matrix elements, and so on. Therefore, the computation of Algorithm 4.1 is a typical computation- and data-intensive task, which is highly iterative and time and memory consuming. It is necessary to optimize the algorithm on our proposed distributed parallel framework, thus utilizing cloud computing technologies to efficiently accelerate hyperspectral data computations of SCSRC.

In the following, we describe the parallel and distributed implementation of the different phases of SCSRC algorithm, and further describe the architecture-related optimizations carried out in the development of the distributed parallel implementation.

4.4.2 STORAGE OPTIMIZATION

When applying the SCSRC method to classify a hyperspectral image, a main challenge is the capability of big data processing. In the case of a large-scale matrix of hyperspectral data, the processing algorithm involves a massive significant requirement of computing resources (e.g., especially memory), leading to extremely high computational complexity. Two optimization strategies coping with this problem are proposed. One solution is to reasonably partition the dataset. To be specific, the iteration variables in the algorithm should be distributed to the memories of individual computing nodes according to their spatial correlations. The other solution is to create a joint distributed storage matrix during algorithm execution for the purpose of minimizing data transmissions among nodes. Each node is responsible for processing a local hyperspectral image block, and all computing nodes work in parallel and independently; therefore, the data communication among nodes would be reduced to a great extent.

The resilient distributed dataset in Spark is capable of dividing data into partitions for distributed storage on multiple nodes while preserving the structures of RDD partitions for reducing data transmission [38]. Considering this advantage, we present an RDD partition and matrix joint storage strategy that not only achieves load balancing, but also reduces the amount of data transmissions across nodes, for improving computational efficiency.

4.4.2.1 Partitioning Strategy for Hyperspectral Data

The purpose of data partitioning is to create pipelines by taking full advantage of narrow dependencies and thus avoid unnecessary interpartition shuffle operations. A narrow dependency in Spark indicates that each partition of the generated RDD is dependent solely on a fixed partition of its parent RDD(s). In contrast, a wide dependency indicates that each partition of the generated RDD is dependent on all partitions of its parent RDD(s), as Figure 4.5 shows [39,40]. From an algorithmic point of view, narrow dependencies require that all partition data of parent RDD(s) have to be precomputed. Then, all computation results by parent RDD(s) are merged by performing interpartition shuffle operations. In the case where the partitions belong to different nodes, a data transmission across nodes, that is, a shuffle operation, will be triggered. Narrow dependencies support pipeline computation on a single node and therefore are more efficient. For this reason, it is preferred to use narrow dependencies, for example, map and mapPartition operations, during RDD programming.

In the implementation of a distributed parallel SCSRC algorithm, considering that adjacent pixel vectors are mutually dependent during computation, the spatial correlation information needs to be taken into account. Therefore, when performing hyperspectral data partitioning, it is necessary to organize the adjacent pixels of a certain local space into the same partition to avoid data requests

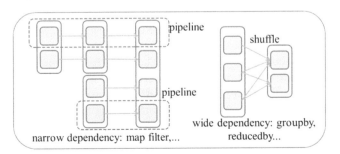

FIGURE 4.5   Narrow and wide dependencies in Spark.

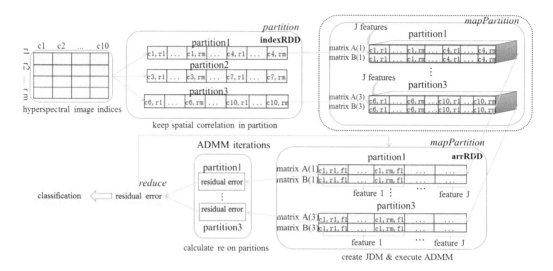

FIGURE 4.6 Instance of data partitions and parallel execution of SCSRC.

among different partitions. In addition, it is also important to split a hyperspectral image vertically by column in order to avoid data skew issues. By means of these procedures, it is possible to implement the parallel computation of spatial correlation information within a single partition. To further preserve the boundary information and improve classification accuracy, a redundant column-splitting scheme is employed. For instance, a hyperspectral image of size 10×10 is split into three partitions that store columns 1~4, columns 3~7, and columns 6~10, respectively, as illustrated in Figure 4.6.

4.4.2.2 Joint Storage for Iterative Matrices

Taking into consideration that Algorithm 4.1 is highly iterative, it is important to efficiently store the matrix variables involved in the loop executions. There are two straightforward solutions to cope with this problem. One is to create matrices at drivers before distributing them to workers. The other is to create an RDD for each iterative matrix variable. In the case of large-scale matrices, apparently the former solution not only consumes substantial network bandwidth of the cluster, but also places a heavy workload on the driver. This issue can be effectively addressed by the latter solution of creating individual RDDs for distributed storage. However, in this manner, the matrix computation may lead to shuffle operation across nodes, degrading computation efficiency.

To solve the above-mentioned problem, we have proposed a joint storage strategy for iterative matrices. By lowering the amount of data transmissions across nodes, this strategy can effectively improve computation efficiency and save network bandwidth. The proposed joint storage strategy is detailed as follows. Let us assume A and B are two $m \times n$ two-dimensional matrices. We first reshape A and B by column to vectors of size $(m \times n)$. We then merge A and B to form a new two-dimensional matrix, of which the first and second rows consist of A's elements and B's elements, respectively.

In the implementation of distributed parallel SCSRC algorithm, an observation is that most of the matrix variables used in the classification algorithm are $J \times I$ two-dimensional matrices that satisfy the requirement of joint distributed storage. Therefore, for each partition obtained according to the previously described partitioning strategy, we create a $(J \times I_i) \times k$ joint storage matrix, where J stands for the feature dimensionality of training matrix, I_i represents the one-dimensional index of matrix partition reshaped from the original hyperspectral matrix, and k denotes the number of matrix variables required during the computation procedure. Figure 4.6 provides an illustrative example of creating a joint storage matrix (JSM) for each partition [39]. The main advantage of this strategy is that all matrix elements have been aligned by column. Compared with the solution that creates matrices at drivers and distributes them to workers, this strategy alleviates the computation workload

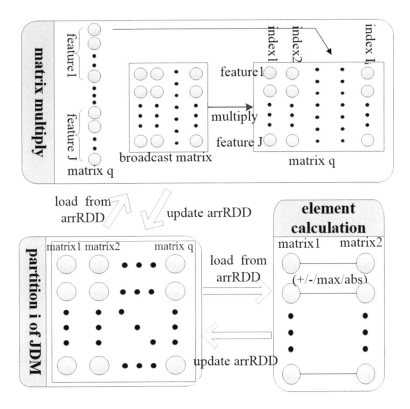

FIGURE 4.7 Matrix calculation in one partition of joint storage matrix.

at drivers and reduces the amount of data transmission. Compared with the solution that creates an RDD for each iterative matrix, this strategy eliminates the requirement of data transmission across nodes when matrix data of the same index are stored in different partitions. Since the spatial correlation information and iteration variables are included in a same partition, as long as the joint storage matrix has been created for a particular partition, the iterative procedure of the classification algorithm can be computed within this partition independently.

4.4.3 ALGORITHM OPTIMIZATION

Since the elements of all the joint storage matrix have been aligned by row, matrix addition/subtraction and other elementary operations can be easily done by the iterator in a row-by-row manner. However, matrix multiplication is relatively more complicated. We propose the following method to enable an efficient computation of matrix multiplication [39]. As shown in Figure 4.7, **arr_comb** is the joint storage matrix, **A** denotes the ith column of **arr_comb**, and **B** is a small-size matrix that will not be updated during iteration. We broadcast **B** to all nodes and perform multiplication at each node. The algorithmic flow of matrix multiplication is described in detail by the following pseudocode.

Algorithm 4.2 Joint Distributed Storage Matrix Multiplication

---

Matrix B = sc.broadcast(B)
arr_comb.mapPartition(
    iterator=>{
      iter_arr = iterator.toArray

Matrix A = iter_arr(:,i) //extract the elements of
        matrix's ith column
A = new DenseMatrix(row,col,A) //encapsulate the
        matrix as a Breeze matrix for fast
        computation
iter_arr(:,j) = (B*A).values //perform matrix
        multiplication and write back the
        result to j-th column of ***arr_comb***

   }

)

Based upon the above analysis, we developed a Spark-based parallel distributed SCSRC algorithm in cloud computing environments, as Algorithm 4.3 shows. The key contribution of this algorithm is that as long as the partitions are initialized, all partitions are mutually independent for computation. Except for the calculation of accumulated errors and collection of classification results in the final stage, all procedures can be computed independently within each partition. In this manner, the heavy iterative computation has been converted into a pipeline procedure, significantly improving the computation efficiency.

Algorithm 4.3   Parallel and Distributed SCSRC on Cloud (SCSRConCloud)

Input: Training samples loaded from HDFS $A = [A^1, A^2, \ldots A^c] \in R^{L \times J}$, test image $X = [x_1, x_2, \ldots x_I] \in R^{L \times I}$.

Initialization: Set $k = 0$, choose regularization parameters $\mu > 0, \lambda_1 > 0, \lambda_2 > 0, \varepsilon > 0$, initialize *tol*, *maxiter*, $S^{(0)}, V^{(0)}, D^{(0)}$.

Step 1. Compute multiplier matrix at driver end: $F = (A^T A + 2\mu I)^{-1}$, $B = A^T X$, and broadcast B to all nodes.

Step 2. Perform partitioning on X according to its column index I; each partition obtains I_i indices to form IndexRDD.

Step 3. Initialize the $(J \times I_i) \times k$ joint storage matrices for all partitions of IndexRDD, denoted by arrRDD.

Do:

Step 4.

$$\text{for } each\ partition\ i : \begin{cases} S_{I_i}^{k+1} = F \times (B_{I_i} + \mu \times subd_{I_i}) \\ S_{I_i \Lambda}^{k+1} = I \\ s_{I_i} = Gauss - Seidel(s_{I_i}, S_{I_i} - b_{I_i}) \end{cases}.$$

Step 5.

$$\text{for } each\ partition\ i, \atop e \in I_i \begin{cases} u_e^{k+1} = soft(b_e^k - S_e^k, \lambda_1 / \mu) \\ d_e^{k+1} = d_e^k - (S_e^k - s_e^k) \\ b^{k+1} = b_e^k - (S_e^k - u_e^k) \\ subd_e^{k+1} = \mu * (s_e^k + u_e^k + b_e^k + d_e^k) \end{cases}.$$

Step 6. for *each partition i : sum(tol$_{I_i}$)*.
Step 7. $k = k + 1$.

until k>maxiter or $\sum\limits_{i=1,2\cdots I} (\|S_{I_i}^k - SO_{I_i}^k\|_F^2) < $ tol.

Step 8. Collect S at driver end.

Step 9. $Class(\mathbf{x}) = \underset{k=1,...,K}{\arg\min} \| \mathbf{x} - \mathbf{A}^k \hat{\mathbf{s}}^k \|_2$,

$\forall x$, x is a column of X.

Output: Class labels of X.

4.4.4 EXPERIMENTAL PLATFORMS AND DATASETS

The efficiency of our distributed parallel implementation of the SCSRC algorithm was evaluated in terms of classification accuracy, parallel execution performance, and scalability for big data processing.

Two cloud computing platforms used for experimental evaluation in this work were built on clusters that include nine and seven nodes, respectively, as Figure 4.8 shows. The hardware and software specifications of the considered platforms are listed in Tables 4.2 and 4.3. Specifically, for Platform1, the master node is the virtual machine, built on a host (IBM X3650M3) equipped with two Intel Xeon E5630 CPUs (eight cores in total) at 2.53 GHz, with 16 GB RAM and a 292-GB SAS hard disk. The eight slave nodes are virtual machines built on four IBM BladeCenter HX5 blade computers. Each blade computer is equipped with two Intel Xeon E7-4807 CPUs (12 cores in total) at 1.86 GHz and is connected to a 12-TB disk array by SAS bus. After virtualization (using VMware* vSphere 5.1.0), every virtual machine (slave node) is allocated with six cores (logic processors) and 15 GB memory. During the Apache Spark execution, every slave node launches two worker instances, and each of them is allocated with two cores. For Platform2, the master node and each slave node are equipped with two Intel Xeon E5-2680 CPUs (24 cores in total) at 2.50 GHz, with 242 GB RAM. During the Apache Spark execution, every slave node launches four worker instances, and each of them is allocated with five cores. All nodes have Java 1.8.0_131, Apache Spark 2.1.1, Hadoop-2.73, and Ubuntu 16.04.2 LTS as operating system installed. The distributed parallel version on Spark is realized by Java and Scala hybrid programming, in which Java is used for the computation part coding and Scala is used for the logic control part coding. The Spark platform is running in Standalone mode.

In terms of the experimental dataset, we first chose a widely used real hyperspectral image of a typical vegetation scene, the Indian Pines image, for which ground truth information is available (Figure 4.9). It was collected by the Airborne Visible/Infrared Imaging Spectrometer sensor over NW Indiana in June 1992. The AVIRIS sensor collects 220 spectral bands in the range from 0.4 to 2.5 μm. In our experiments, the number of bands is reduced to 200 by removing 20 water absorption bands. This scene has a spectral resolution of 10 nm and a spatial resolution of 20 m by pixel, and the size in pixels is 145 × 145. The ground truth contains 16 land cover classes (including alfalfa, corn, pasture, trees, grass, oats, soybeans, wheat, woods, etc.) and a total of 10,366 labeled pixels, as Figure 4.10 shows [41]. The number of pixels in the smallest class is 20, while the number of pixels in the largest class is 2468.

It is worth noting that, in practical applications of supervised image classification, the number of training samples is usually limited. Therefore, a small number of training samples, that is, nearly 10% of the labeled pixels of each class, were randomly chosen as training samples, and the remaining labeled pixels were used as test samples, as shown in Table 4.4. The data size of original Indian Pines dataset (denoted as Dataset1) is 145*145*200, and the size of iterative data during execution is 1043*21025*8, nearly 1338 MB. In order to test the performance and scalability of the proposed

---

* http://www.vmware.com/

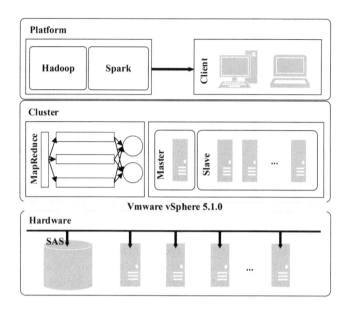

FIGURE 4.8   The architecture of the experimental platform.

TABLE 4.2

Hardware Specifications and Computing Capabilities of the Considered Platforms

	Specification	Spark Cloud Computing Platform1	Spark Cloud Computing Platform2
Master	Processor Number	Intel Xeon E5630	Intel Xeon E5-2680
	Processor Base Frequency	2.53 GHz	2.50 GHz
	Number of Cores	8	24
	Main Memory	16 GB	242 GB
Slaves	Processor	Intel Xeon E7-4807	Intel Xeon E5-2680
	Number of Slaves	8	6
	Processor Base Frequency	1.86 GHz	2.50 GHz
	Number of Cores	6	24
	Main Memory	15 GB	242 GB
	Number of Workers in Each Slave	2	4
	Number of Each Worker's Cores	2	5

TABLE 4.3

Software Specifications of the Considered Platforms

Specification	Version
Hadoop version	2.73
Spark version	2.1.1
Java version	1.8.0_131
Scala version	2.11.6
Compiler	IntelliJ IDEA 2017

(a)

(b)

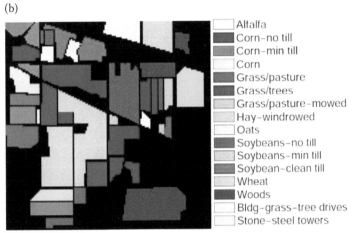

☐ Alfalfa
■ Corn–no till
▨ Corn–min till
☐ Corn
▨ Grass/pasture
■ Grass/trees
▨ Grass/pasture-mowed
▨ Hay–windrowed
☐ Oats
■ Soybeans–no till
▨ Soybeans–min till
▨ Soybean–clean till
▨ Wheat
■ Woods
☐ Bldg–grass–tree drives
☐ Stone–steel towers

FIGURE 4.9   AVIRIS Indian Pines hyperspectral image. (a) Data cube, (b) Ground truth as a collection of mutually exclusive classes.

method on big datasets, four larger datasets have been artificially generated by simply mosaicking the original dataset as follows: Dataset2, with 145*290 pixels and JSM size of 2677 M (1043*42050*8); Dataset3, with 145*580 pixels and JSM size of 5354 M (1043*84100*8); Dataset4, with 145*870 pixels and JSM size of 8031 M (1043*126150*8); and Dataset5, with 145*1160 pixels and JSM size of 10708 M (1043*168200*8).

Furthermore, another hyperspectral dataset of the AVIRIS North-South flightline (denoted as Dataset6, as Figure 4.10 shows) is used to evaluate the efficiency and computational capability of the proposed method on a real big hyperspectral dataset. It was collected by the AVIRIS sensor over a 25×6-mile portion of northwest Tippecanoe County, Indiana, in June 1992. The size in pixels is 2678×614. The ground truth contains 58 land cover classes (including corn, grass, oats, soybeans, wheat, etc.), and a total of 334,245 labeled pixels. In the experiments, the number of bands is also reduced to 200 by removing 20 water absorption bands. Nearly 10% of the labeled pixels of each class were randomly chosen as training samples, and the remaining labeled pixels were used as test samples.

FIGURE 4.10   Data cube of AVIRIS north-south flightline hyperspectral image.

The classification accuracies are measured by the overall accuracy (OA), average accuracy (AA), and kappa statistic (k) [42]. The OA is computed as the ratio between the correctly classified test samples and the total number of test samples.

$$OA = \sum_{i=1}^{K} M_{ii} / N.$$

The AA is the mean of the accuracies across the different classes.

$$AA = \frac{1}{K} \sum_{i=1}^{K} (M_{ii} / N_i).$$

TABLE 4.4

Training and Test Samples of AVIRIS Indian Pines Dataset

Class	Training Samples	Test Samples
Alfalfa	6	48
Corn-notill	144	1290
Corn-min	84	750
Corn	24	210
Grass/pasture	50	447
Grass/trees	75	672
Grass/pasture-mowed	3	23
Hay-windrowed	49	440
Oats	2	18
Soybeans-notill	97	871
Soybeans-min	247	2221
Soybean-clean	62	552
Wheat	22	190
Woods	130	1164
Building-grass-trees-drives	38	342
Stone-steel towers	10	85

The k statistic is computed by weighting the measured accuracies [12]. It incorporates both the diagonal and off-diagonal entries of the confusion matrix and is a robust measure of the degree of agreement.

$$k = \left(N \left(\sum_{i=1}^{K} M_{ii} \right) - \sum_{i=1}^{K} \left(\sum_{j=1}^{K} M_{ij} \sum_{j=1}^{K} M_{ji} \right) \right) \bigg/ \left(N^2 - \sum_{i=1}^{K} \left(\sum_{j=1}^{K} M_{ij} \sum_{j=1}^{K} M_{ji} \right) \right).$$

4.4.5 Experimental Results and Evaluation

The computational performance and classification accuracy have been assessed in our experiments, in which *maxiter* has been set to 40 and *tol* has been set to 1E-5. All of the measurements reported in the following experiments were achieved after 10 Monte Carlo runs. When submitting tasks, each partition of RDD is processed as an individual task. This indicates that the degree of parallelism depends on the number of partitions of RDD.

We first evaluated the accuracy and computational performance of our parallel and distributed implementation on Dataset1, as compared with the serial version of SCSRC (i.e., the number of partitions was 1). Table 4.5 reports a quantitative comparison of the accuracies and accelerate factors. From Table 4.5, we can see that the proposed serial and distributed parallel versions of SCSRC produce very high accuracies. It is worth mentioning that we performed spatial-domain partitioning; thus, every pixel vector (i.e., spectral signature) is stored in the same node.

As can be seen from Table 4.5, the most time-consuming part of the SCSRConCloud algorithm is iteration computation. Since the operations of read and broadcast are executed on the driver, the processing times of these parts are stable. Thus, the accelerate factors mainly depend on the iteration computation part. We can conclude that the proposed parallel and distributed implementation significantly accelerates the SCSRC computation on the tested dataset. The distributed parallel version achieved a high speedup of more than 10x on the considered platform (when the number of partitions was larger than 16).

TABLE 4.5

Computational Performance and Accuracies of SCSRConCloud on Dataset1

Number of Partitions	Read (Time/s)	Broadcast (Time/s)	Iteration Time (Time/s)	Total Time (Time/s)	Speedup of Iteration Computation	Total Speedup	OA	AA	K
1	8	11.4	3651.2	3670.6	1	1	98.3	95.9	98.1
4	8	11.3	779.7	799	4.7	4.39	98.3	95.9	98.1
8	8.7	11.9	479.4	500	7.6	7.34	98.3	95.9	98.1
16	8	11.9	287.1	319.3	12.71	11.50	98.2	95.7	97.9
32	8.9	11.4	229.5	249.8	15.9	14.74	97.6	95.2	97.3

TABLE 4.6

Computational Performance of SCSRConCloud on Larger Datasets

	Initialization (s)	Read (s)	Broadcast (s)	Iteration Time (s)	Write (s)	Percentage of I/O (%)	Total Time (s)
Dataset1	15.4	8.7	11.9	287.1	12.3	10.30	319.3
Dataset2	27.1	14.8	19.8	502.1	24.7	10.08	588.5
Dataset3	39.5	24.2	29.3	876.6	34.8	8.79	1004.4
Dataset4	65.7	31.8	43.2	1268.0	70.3	9.82	1479.0
Dataset5	92.6	43.8	53.8	1574.0	93.4	10.28	1857.6

To evaluate the efficiency of our proposed distributed parallel implementation of SCSRC algorithm on larger datasets, we also conducted experiments with larger datasets (i.e., Dataset2, Dataset3, Dataset4, Dataset5). In these cases, the serial version cannot run on one node due to the limitation of its memory resources. The results obtained using one master node and eight slaves with 16 partitions are given in Table 4.6 and Figure 4.11. From this plot, we can conclude that our proposed distributed parallel implementation scales almost linearly with the size of the dataset. Moreover, the parallel implementation was compute-bound for large datasets (e.g., nearly 10% of the time is consumed by I/O transfer for the execution on each dataset). This is an important consideration, as it indicates that the proposed implementation scales better as the data size becomes larger. This opens new avenues to exploit cloud computing technology in hyperspectral imaging problems.

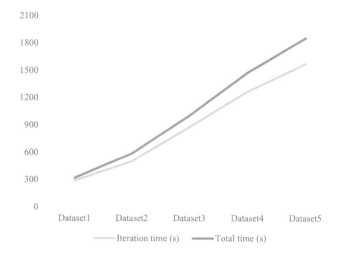

FIGURE 4.11　Execution time of SCSRConCloud with different datasets.

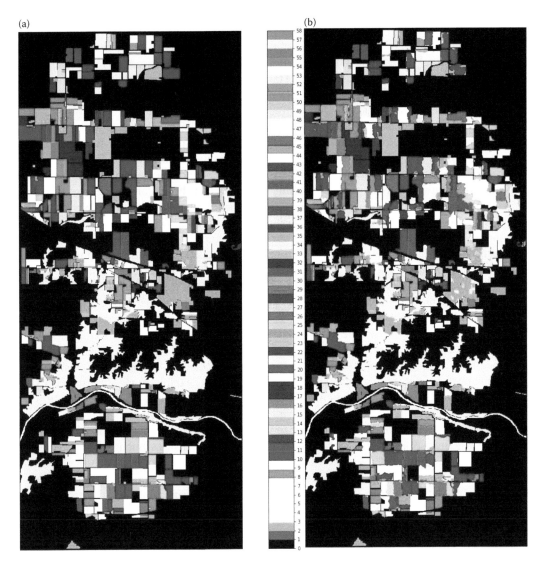

FIGURE 4.12   Ground truth and classification map of the AVIRIS north-south flightline hyperspectral image. (a) Ground truth as a collection of mutually exclusive classes, (b) The classification map.

Furthermore, we performed tests on Dataset6 by utilizing Platform2 to evaluate the efficiency and computational capability of the proposed method on a real big hyperspectral dataset. The classification results are graphically illustrated in Figure 4.12. The total computational time was 54,766 s, and the classification accuracy of OA was 93.01%, AA was 90.02%, and K was 0.9245. It can be seen from the results that for the significantly big hyperspectral dataset of size 78 times larger than Dataset1, our approach needed only 15.21 hours to classify it with high accuracy.

Experimental results show that the proposed parallel and distributed implementation SCSRConCloud greatly benefits from the dynamic storage of HDFS and in-memory cluster computing of Spark, which is specifically designed to optimize the distributed processing and management of very large files. As a result, the SCSRConCloud is compute-bound , even for large hyperspectral datasets. We have found that the use of HDFS and Spark greatly simplifies the implementation, leading to very good performance in our context.

4.5   DISCUSSION AND FUTURE DEVELOPMENTS

This section provides an in-depth discussion of the distributed and parallel computing framework, investigating several design and implementation issues: (1) compromised performance with an increasing number of computing resources, (2) task schedule strategy for exploiting data processing parallelism, (3) load-balancing mechanism for improving cloud system performance, (4) optimization techniques for minimizing cloud service usage cost in different cloud environments.

First, the distributed computation of the data partitions may promisingly accelerate the processing of large-scale remote sensing data. However, it is worth emphasizing the tradeoff between computation efficiency and communication overhead. Due to the overhead of data communication between the master and slave nodes, the speedup achieved by employing the MapReduce mechanism does not always grow linearly with the increasing number of distributed computing resources. To be specific, we perform a series of experiments to observe the runtime speedups achieved by executing the training process of a pan-sharpening algorithm with regard to various data sizes and numbers of workers. We recorded the execution time for each configuration and observed the speedup over serial execution time. For different sizes of training data, the speedup exhibits approximately a linear increase if the number of workers is below eight. However, if we keep on increasing the number of workers, the communication overhead becomes comparable to the execution time, leading to a degradation in terms of speedup. For example, the speedup is compromised to 9.39X for a dataset of size 156 MB. As a result, during the distributed processing of remote sensing big data, it would be of great importance to determine the optimal number of partitions into which divide the original dataset, and accordingly the number of worker nodes to be deployed on Apache Spark.

In addition to the parallel and distributed computing mechanism, a majority of existing cloud computing platforms rely on task-scheduling strategies to further exploit the parallelism in workflow processing and promote the utilization of cloud computing resources [43–45]. Similar to other general-purpose applications, a remote-sensing data-processing application can be divided into a set of subtasks that allow for the distribution of the application's computational load across cloud computing resources [9,46–50]. In general, there are certain dependencies among the tasks belonging to a specific application that impose an order of precedence on their execution. Adhering to this order is essential for the correct execution of the application. For instance, in many remotely sensed image super-resolution and pan-sharpening applications, a necessary preprocessing procedure has to be accomplished prior to the procedure of image reconstruction. When processing large-scale remote sensing data in cloud computing environments, a crucial step is the assignment of tasks to processing elements and the order of their execution. This step, which is referred to as Scheduling [51,52], fundamentally determines the remote sensing application's parallelism and eventually the efficiency of big data processing. Effectiveness and efficiency are two specifications used to evaluate scheduling algorithms. Generally, the scheduling problem is known to be computationally intractable in many cases [53]. Feasible solutions to solving the scheduling problem in a reasonable amount of time are heuristics and meta-heuristics [54,55]. The key of a heuristic is a deterministic heuristic rule that guides the searching procedure to find a good solution in the solution space both effectively and efficiently. Different from heuristics, meta-heuristics involve randomized factors to avoid being trapped into a local optimum. In the literature, a considerable number of meta-heuristics have already been proposed, for example, genetic algorithms (GAs) [56,57], ant colony optimization (ACO) [58–60], particle swarm optimization (PSO) [61,62], and quantum-inspired evolutionary algorithms (QEAs) [63,64].

Usually combined with the task-scheduling approach, load balancing is another common approach used to improve cloud system performance and capacity by means of balancing the application's workload across the computing resources [65,66]. In the parallel and distributed computing framework, the load-balancing mechanism can be incorporated to perform a range of specialized runtime workload distribution functions [67], including: (1) asymmetric distribution— considering the imbalanced workload from the task graph of a remote sensing big data application, larger workloads are expected to be issued to computing resources with higher processing capacities; and (2) workload prioritization—under certain circumstances, we may assign priority levels to the subtasks of remote sensing applications, and workloads are scheduled and distributed according to their priority levels. The load balancer is typically located on the communication path between the computing resources generating the workload and the resources performing the workload processing. Similar to task scheduling, this mechanism works as a system-level management technique that abstracts the computing resources performing their workloads.

The last important issue is that the aforementioned distributed parallel framework for remote sensing big data was limited in a private cloud environment. Accordingly, only resource constraints are considered in the optimization procedure [68,69]. However, due to the limits of the resource capacity of the private cloud, some tasks of emergent remote sensing big data processing applications can be outsourced to public clouds in order to finish those applications as fast as possible. However, this outsourcing method will generate additional costs. In future work, we expect to extend the current framework to be applicable for this hybrid cloud environment by taking into account additional constraints [70,71] (e.g., the budget of using public clouds' resources).

4.6 CONCLUSIONS

Recent technological advances in spaceborne and airborne remote sensing have exponentially increased the volume of remote sensing image data. Cloud computing offers the possibility to store and process massive amounts of remotely sensed hyperspectral data in a distributed way. After introducing the fundamentals of cloud computing technologies and giving a detailed case study of our proposed cloud computing implementation of the SCSRC algorithm on the Spark platform, we can conclude that it is important to exploit cloud computing architectures for parallel and distributed processing of remotely sensed hyperspectral big data.

For a certain task of processing a significantly large hyperspectral dataset, an efficient way is to utilize an HDFS to realize distributed storage, use Apache Spark as the computing engine, and develop a parallel algorithm based on the Map-Reduce parallel model. By taking full advantage of the high throughput access and high-performance distributed computing capabilities of cloud computing environments, it is possible to achieve high analysis accuracy with high computational performance and scale very well as the data size becomes larger and larger. It has the potential for further extensive application. Future work could be focused on the task-scheduling strategy for exploiting data processing parallelism, the load balancing mechanism for improving cloud system performance, and optimization techniques for minimizing cloud service usage cost in different cloud computing environments.

REFERENCES

1. National Institute of Standards and Technology. The NIST Definition of Cloud Computing. NIST Special Publication 800-145. Available online, http://dx.doi.org/10.6028/NIST.SP.800-145, September 2011.
2. S. Bhardwaj, L. Jain, and S. Jain, "Cloud computing: A study of infrastructure as a service (IaaS)," *International Journal of Engineering and Information Technology*, vol. 2, pp. 60–63, 2010.
3. R.O. Green, M.L. Eastwood, C.M. Sarture, T.G. Chrine, M. Aronsson, B.J. Chippendale, J.A. Faust, B.E. Pavri, C.J. Chovit, and M. Solis, "Imaging spectroscopy and the airborne visible/infrared imaging spectrometer (AVIRIS)," *Remote Sensing of Environment*, vol. 65, no. 3, pp. 227–248, 1998.
4. J.G. Lee and M. Kang, "Geospatial big data: Challenges and opportunities," *Big Data Research*, vol. 2, no. 2, pp. 74–81, 2015.
5. X. Wu, X. Zhu, G.Q. Wu, and W. Ding, "Data mining with big data," *IEEE Transactions on Knowledge and Data Engineering*, vol. 26, no. 1, pp. 97–107, 2014.
6. F. Lin, L. Chung, C. Wang, W. Ku, and T. Chou, "Storage and processing of massive remote sensing images using a novel cloud computing platform," *Giscience and Remote Sensing*, vol. 50, no. 3, pp. 322–336, 2013.

7. P. Cappelaere, S. Sanchez, S. Bernabe, A. Scuri, D. Mandl, and A. Plaza, "Cloud implementation of a full hyperspectral unmixing chain within the NASA web coverage processing service for EO-1," *IEEE Journal of Selected Topics in Applied Earth Observations and Remote Sensing*, vol. 6, no. 2, pp. 408–418, 2013.

8. Z. Wu, Y. Li, A. Plaza, J. Li, F. Xiao, and Z. Wei, "Parallel and distributed dimensionality reduction of hyperspectral data on cloud computing architectures," *IEEE Journal of Selected Topics in Applied Earth Observations and Remote Sensing*, vol. 9, no. 6, pp. 2270–2278, 2016.

9. V.A.A. Quirita, G.A.O.P. da Costa, P.N. Happ, R.Q. Feitosa, R.D.S. Ferreira, D.A.B. Oliveira, and A. Plaza, "A new cloud computing architecture for the classification of remote sensing data," *IEEE Journal of Selected Topics in Applied Earth Observations and Remote Sensing*, vol. 10, no. 2, pp. 409–416, 2017.

10. P. Wang, J. Wang, Y. Chen, and G. Ni, "Rapid processing of remote sensing images based on cloud computing," *Future Generation Computer Systems*, vol. 29, no. 8, pp. 1963–1968, 2013.

11. J. Dean and S. Ghemawat, "MapReduce: Simplified data processing on large clusters," *Symposium on Operating Systems Design & Implementation*. USENIX Association, pp. 137–147, 2004.

12. Z. Wu, Q. Wang, A. Plaza, J. Li, J. Liu, and Z. Wei, "Parallel implementation of sparse representation classifiers for hyperspectral imagery on GPUs," *IEEE Journal of Selected Topics in Applied Earth Observations and Remote Sensing*, vol. 8, no. 6, pp. 2912–2925, 2015.

13. A. Plaza, J.A. Benediktsson, J.W. Boardman, J. Brazile, L. Bruzzone, G. Camps-Valls, J. Chanussot, M. Fauvel, P. Gamba, A. Gualtieri, M. Marconcini, J.C. Tilton, and G. Trianni, "Recent advances in techniques for hyperspectral image processing," *Remote Sens. Environ.*, vol. 113, no. Supplement 1, pp. S110–S122, September 2009.

14. S. Li, B. Zhang, A. Li, X. Jia, L. Gao, and M. Peng, "Hyperspectral imagery clustering with neighborhood constraints," *IEEE Geoscience and Remote Sensing Letters*, vol. 10, no. 3, pp. 588–592, March 2013.

15. M. Fauvel, Y. Tarabalka, J.A. Benediktsson, J. Chanussot, and J.C. Tilton, "Advances in spectral-spatial classification of hyperspectral images," *Procedings of the IEEE*, vol. 101, no. 3, pp. 652–675, March 2013.

16. Y. Gu, C. Wang, D. You, Y. Zhang, S. Wang, and Y. Zhang, "Representative multiple kernel learning for classification in hyperspectral imagery," *IEEE Transactions on Geoscience and Remote Sensing*, vol. 50, no. 7, pp. 2852–2865, July 2012.

17. L. Gomez-Chova, G. Camps-Valls, L. Bruzzone, and J. Calpe-Maravilla, "Mean map kernel methods for semisupervised cloud classification," *IEEE Transactions on Geoscience and Remote Sensing*, vol. 48, no. 1, pp. 207–220, January 2010.

18. Y. Zhong and L. Zhang, "An adaptive artificial immune network for supervised classification of multi-/ hyperspectral remote sensing imagery," *IEEE Transactions on Geoscience and Remote Sensing*, vol. 50, no. 3, pp. 894–909, March 2012.

19. J. Li, J.M. Bioucas-Dias, and A. Plaza, "Spectral–spatial hyperspectral image segmentation using subspace multinomial logistic regression and Markov random fields," *IEEE Transactions on Geoscience and Remote Sensing*, vol. 50, no. 3, pp. 809–823, March 2012.

20. Y. Chen, N.M. Nasrabadi, and T.D. Tran, "Hyperspectral image classification via kernel sparse representation," *IEEE Transactions on Geoscience and Remote Sensing*, vol. 51, no. 1, pp. 217–231, January 2013.

21. J. Li, P. Reddy Marpu, A. Plaza, J. Bioucas-Dias, and J. Atli Benediktsson, "Generalized composite kernel framework for hyperspectral image classification," *IEEE Transactions on Geoscience and Remote Sensing*, vol. 51, no. 9, pp. 4816–4829, 2013.

22. G. Camps-Valls, L. Gomez-Chova, J. Munoz-Mari, J. Vila-Frances, and J. Calpe-Maravilla, "Composite kernels for hyperspectral image classification," *IEEE Geoscience and Remote Sensing Letters*, vol. 3, no. 1, pp. 93–97, January 2006.

23. M. Marconcini, G. Camps-Valls, and L. Bruzzone, "A composite semisupervised SVM for classification of hyperspectral images," *IEEE Geoscience and Remote Sensing Letters*, vol. 6, no. 2, pp. 234–238, February 2009.

24. Y. Chen, N.M. Nasrabadi, and T.D. Tran, "Hyperspectral image classification using dictionary-based sparse representation," *IEEE Transactions on Geoscience and Remote Sensing*, vol. 49, no. 10, pp. 3973–3985, October 2011.

25. X. Sun, Q. Qu, N.M. Nasrabadi, and T.D. Tran, "Structured priors for sparse-representation-based hyperspectral image classification," *IEEE Geoscience and Remote Sensing Letters*, vol. 11, no. 7, pp. 1235–1239, July 2014.

26. J. Liu, Z. Wu, Z. Wei, L. Xiao, and L. Sun, "Spatial correlation constrained sparse representation for hyperspectral image classification," *Journal of Electronics and Information Technology*, vol. 34, no. 11, pp. 2666–2671, October 2012.

27. Y. Tarabalka, M. Fauvel, J. Chanussot, and J.A. Benediktsson, "SVM-and MRF-based method for accurate classification of hyperspectral images," *IEEE Geoscience and Remote Sensing Letters*, vol. 7, no. 4, pp. 736–740, April 2010.

28. G. Camps-Valls, N. Shervashidze, and K.M. Borgwardt, "Spatio-spectral remote sensing image classification with graph kernels," *IEEE Geoscience and Remote Sensing Letters*, vol. 7, no. 4, pp. 741–745, April 2010.

29. X. Kang, S. Li, and J. Benediktsson, "Spectral-spatial hyperspectral image classification with edge-preserving filtering," *IEEE Transactions on Geoscience and Remote Sensing*, vol. 52, no. 5, pp. 2666–2677, May 2014.

30. Y. Tarabalka, J. Chanussot, and J.A. Benediktsson, "Segmentation and classification of hyperspectral images using watershed transformation," *Pattern Recognition*, vol. 43, no. 7, pp. 2367–2379, July 2010.

31. R. Ji, Y. Gao, R. Hong, Q. Liu, D. Tao, and X. Li, "Spectral-spatial constraint hyperspectral image classification," *IEEE Transactions on Geoscience and Remote Sensing*, vol. 52, no. 3, pp. 1811–1824, March 2014.

32. Y. Tarabalka, J.A. Benediktsson, and J. Chanussot, "Spectra-spatial classification of hyperspectral imagery based on partitional clustering techniques," *IEEE Transactions on Geoscience and Remote Sensing*, vol. 47, no. 8, pp. 2973–2987, August 2009.

33. L. Zhang, W.D. Zhou, P.C. Chang, J. Liu, Z. Yan, T. Wang, and F.Z. Li, "Kernel sparse representation based classifier," *IEEE Transactions on Signal Processing*, vol. 60, no. 4, pp. 1684–1695, April 2012.

34. J.M. Bioucas-Dias and M.A.T. Figueiredo, "Alternating direction algorithms for constrained sparse regression: Application to hyperspectral unmixing," *2nd Workshop WHISPERS*, pp. 1–4, 2010.

35. W. Deng, W. Yin, and Y. Zhang, "Group sparse optimization by alternating direction method," Department of Computational and Applied Mathematics, Rice University, TR11-06, 2011.

36. M.D. Iordache, J.M. Bioucas-Dias, and A. Plaza, "Total variation spatial regularization for sparse hyperspectral unmixing," *IEEE Transactions on Geoscience and Remote Sensing*, vol. 50, no. 11, pp. 4484–4502, October 2012.

37. T. Goldstein and S. Osher, "The split Bregman method for L1-regularized problems," *SIAM Journal of Imaging Sciences*, vol. 2, no. 2, pp. 323–343, February 2009.

38. Zaharia M, Chowdhury M, Franklin M J et al., "Spark: Cluster computing with working sets," *Hot Cloud*, vol. 10, p. 10, 2010.

39. J. Shen, Z. Kang, Z. Wu, Z. Wei, and Y. Zhu, "Distributed parallel optimization of hyperspectral image classification based on spacial correlation regularized sparse representation," *2017 IEEE International Geoscience and Remote Sensing Symposium*, Fort Worth, Texas, USA, July 23–July 28, 2017, pp. 2198–3201.

40. M. Zaharia, M. Chowdhury, T. Das et al. "Resilient distributed datasets: A fault-tolerant abstraction for in-memory cluster computing," *Proceedings of the 9th USENIX Conference on Networked Systems Design and Implementation*. USENIX Association, 2012, pp. 2–2.

41. Z. Wu, Q. Wang, A. Plaza, J. Li, L. Sun, and Z. Wei, "Parallel spatial-spectral hyperspectral image classification with sparse representation and Markov random fields on GPUs," *IEEE Journal of Selected Topics in Applied Earth Observations and Remote Sensing*, 2015, vol. 8, no. 6, pp. 2926–2937.

42. G.M. Foody, "Classification accuracy comparison: Hypothesis tests and the use of confidence intervals in evaluations of difference, equivalence and non-inferiority," *Remote Sensing of Environment*, vol. 113, no. 8, pp. 1658–1663, August 2009.

43. M.N. Kavyasri and B. Ramesh, "Comparative study of scheduling algorithms to enhance the performance of virtual machines in cloud computing," *Proceedings of the International Conference on Emerging Trends in Engineering, Technology and Science*, 2016, pp. 1–5.

44. N. Patil and D. Aeloor, "A review—Different scheduling algorithms in cloud computing environment," *Proceedings of the International Conference on Intelligent Systems and Control*, 2017, pp. 182–185.

45. C.W. Tsai and J.J.P.C. Rodrigues, "Metaheuristic scheduling for cloud: A survey," *IEEE Systems Journal*, vol. 8, no. 1, pp. 279–291, 2014.

46. W. Huang, L. Xiao, Z. Wei, H. Liu, and S. Tang, "A new pan-sharpening method with deep neural networks," *IEEE Geoscience and Remote Sensing Letters*, vol. 12, no. 5, pp. 1037–1041, 2015.

47. W. Huang, L. Xiao, H. Liu, and Z. Wei, "Hyperspectral imagery super-resolution by compressive sensing inspired dictionary learning and spatial-spectral regularization." *Sensors*, vol. 15, no. 1, pp. 2041–2058, 2015.

48. W. Li, H. Fu, Y. You, L. Yu, and J. Fang, "Parallel multiclass support vector machine for remote sensing data classification on multicore and many-core architectures," *IEEE Journal of Selected Topics in Applied Earth Observations & Remote Sensing*, vol. 10, no. 10, pp. 4387–4398, 2017.

49. C. Jiang, H. Zhang, H. Shen, and L. Zhang, "Two-step sparse coding for the pan-sharpening of remote sensing images," *IEEE Journal of Selected Topics in Applied Earth Observations & Remote Sensing*, vol. 7, no. 5, pp. 1792–1805, 2014.

50. W. Li and Q. Guo, "A new accuracy assessment method for one-class remote sensing classification," *IEEE Transactions on Geoscience and Remote Sensing*, vol. 52, no. 8, pp. 4621–4632, 2014.

51. F. Magoules, J. Pan, and F. Teng, *Cloud Computing: Data-Intensive Computing and Scheduling*. London, UK: Chapman & Hall, 2012.

52. H. El-Rewini, T.G. Lewis, and H.H. Ali, *Task Scheduling in Parallel and Distributed Systems*. New York, NY, USA: Prentice Hall, 2007.

53. G.L.N. Laurence and A. Wolsey, *Integer and Combinatorial Optimization*. Hoboken, NJ, USA: John Wiley & Sons, 1999.

54. D. Simon. *Evolutionary Optimization Algorithms*. Hoboken, NJ, USA: John Wiley & Sons, 2013.

55. X.-S. Yang. *Engineering Optimization: An Introduction with Metaheuristic Applications*. Hoboken, NJ, USA: John Wiley & Sons, 2011.

56. J.S.F. Barker, "Simulation of genetic systems by automatic digital computers," *Australian Journal of Biological Sciences*, vol. 11, no. 4, pp. 603–612, 1958.

57. J.H. Holland, *Adaptation in Natural and Artificial Systems*. Cambridge, MA, USA: MIT Press, 1992.

58. M. Dorigo and C. Blum, "Ant colony optimization theory: A survey," *Theoretical Computer Science*, vol. 344, no. 2–3, pp. 243–278, 2005.

59. C. Blum, "Ant colony optimization: Introduction and recent trends," *Physics of Life Reviews*, vol. 2, no. 4, pp. 353–373, 2005.

60. M. Dorigo, M. Birattari, and T. Stutzle, "Ant colony optimization," *IEEE Computational Intelligence Magazine*, vol. 1, no. 4, pp. 28–39, 2003.

61. R. Poli, J. Kennedy, and T. Blackwell, "Particle swarm optimization," *Swarm Intelligence*, vol. 1, no. 1, pp. 33–57, 2007.

62. Y. Shi and R.C. Eberhart, *Parameter Selection in Particle Swarm Optimization*. Berlin, Germany: Springer, 1998.

63. K.-H. Han and J.-H. Kim, "Quantum-inspired evolutionary algorithm for a class of combinatorial optimization," *IEEE Transactions on Evolutionary Computation*, vol. 6, no. 6, pp. 580–593, 2002.

64. G. Zhang, "Quantum-inspired evolutionary algorithms: A survey and empirical study," *Journal of Heuristics*, vol. 17, no. 3, pp. 303–351, 2011.

65. J. Baliga, R.W.A. Ayre, K. Hinton, and R.S. Tucker, "Green cloud computing: Balancing energy in processing, storage, and transport," *Proceedings of the IEEE*, vol. 99, no. 1, pp. 149–167, 2010.

66. F. Ramezani, J. Lu, and F.K. Hussain, "Task-based system load balancing in cloud computing using particle swarm optimization," *International Journal of Parallel Programming*, vol. 42, no. 5, pp. 739–754, 2014.

67. T. Erl, R. Puttini, and Z. Mahmood: *Cloud Computing: Concepts, Technology & Architecture*. New York, NY, USA: Prentice Hall, 2013.

68. L.F. Bittencourt, E.R.M. Madeira, and N.L.S. Da Fonseca, "Scheduling in hybrid clouds," *IEEE Communications Magazine*, vol. 50, no. 9, 2012.

69. H. Yuan, J. Bi, W. Tan, and B.H. Li, "Temporal task scheduling with constrained service delay for profit maximization in hybrid clouds," *IEEE Transactions on Automation Science & Engineering*, vol. 14, no. 1, pp. 337–348, 2017.

70. X. Zuo, G. Zhang, and W. Tan, "Self-adaptive learning PSO-based deadline constrained task scheduling for hybrid IaaS cloud," *IEEE Transactions on Automation Science & Engineering*, vol. 11, no. 2, pp. 564–573, 2014.

71. Y. Zhang, J. Sun, and J. Zhu, "An effective heuristic for due-date-constrained bag-of-tasks scheduling problem for total cost minimization on hybrid clouds." *International Conference on Progress in Informatics and Computing*, pp. 479–486, 2016.

Section III

Hyperspectral Vegetation Index Applications to Agriculture and Vegetation

5 Noninvasive Quantification of Foliar Pigments
Principles and Implementation

Anatoly Gitelson and Alexei Solovchenko

CONTENTS

5.1 INTRODUCTION

Pigments are central to the functioning of photosynthetic apparatus and hence for all vital functions of plants. Green chlorophylls (Chl), represented by Chl *a* and *b*, the primary photosynthetic pigments, absorb light energy and eventually convert it into chemical energy in the form of electron flow [1–3]. Yellow-to-orange carotenoids (Car) are the accessory pigments that augment Chl in light absorption and serve the indispensable function of protection of the photosynthetic apparatus from photooxidative damage, mostly via elimination of reactive oxygen species and thermal dissipation of excessively absorbed light energy via operation of the xanthophyll cycle [1,4,5]. Foliar Car are usually represented by carotenes, mostly β-carotene, and xanthophylls—lutein, zeaxanthin, violaxanthin, antheraxanthin, and neoxanthin [6]. The retention of carotenoids in the progress of chlorophyll breakdown has been suggested as a mechanism of photoprotection during leaf senescence [7,8]. Changes in leaf carotenoid content and its proportion to chlorophyll are widely used for diagnosing the physiological state of plants during development, senescence, acclimation, and adaptation to different environments and stresses [9].

Another widespread pigment group, flavonoids include red-colored anthocyanins (AnCs) and pale-yellow flavonols (Flv) important for optical shielding of plant tissues in the green and UV-to-blue regions of the spectrum, respectively [10–12]. In leaves, they localize in vacuoles of epidermal cells or those just below the adaxial epidermis, but occasionally also in the cells of abaxial epidermis, palisade, and spongy mesophyll [13]. The induction of AnC biosynthesis occurs as a result of deficiencies in nitrogen and phosphorus, wounding, pathogen infection, desiccation, low temperature, UV irradiation, and so on, so it is generally accepted that AnCs fulfill important physiological functions by being involved in adaptation to numerous stresses and environmental strain reduction [14–16]. Some lines of evidence suggest that the protective effects of anthocyanins are related to their ability, via screening and/or internal light trapping, to reduce the amount of excessive solar radiation reaching photosynthetic apparati [11,15].

There are also more exotic pigment groups like betalains [17,18] and secondary keto-carotenoids possessing optical properties very similar to those of AnC [7]. Apart from their photosynthetic and photoprotective functions, the pigments serve a plethora of other important functions like attraction of pollinators.

Since pigments are important for the function of plant organisms, their biosynthesis and catabolism are tightly regulated, making them informative markers of plant physiological condition and, ultimately, plant productivity. There are several reasons leaf pigmentation is important from an applied perspective to both land managers and ecophysiologists [19]. First, the amount of solar radiation absorbed by a leaf is largely a function of the foliar contents of photosynthetic pigments; therefore, Chl content in many situations determines the photosynthetic potential and hence primary production [20–22]. Second, much of leaf nitrogen is represented by and correlates with Chl, so quantifying Chl gives an indirect but precise measure of nitrogen nutrition status [22,23]. Third, plant stresses are manifested by directed and specific changes in pigment composition. Thus, the content of Car generally increases and that of Chl decreases under stress and during senescence [9]. The relative contents of photosynthetic pigments reflect the effects of abiotic factors such as light; for example, sun leaves have a higher Chl a/Chl b ratio [24] and so quantifying these proportions can provide important information about relationships between plants and their environment.

Traditional methods of "wet" pigment analysis (extraction with subsequent spectrophotometry or high-pressure liquid chromatography (HPLC) are destructive and do not permit repeated measurement on the same samples, so it is impossible to follow the changes of vegetation condition in time. These techniques are time consuming and expensive, thus making assessment of the vegetation state on the landscape and ecosystem scales impractical. An alternative solution for leaf pigment analysis is represented by nondestructive optical methods. Monitoring plant physiological status via measuring leaf optical properties such as absorbance and/or reflectance possesses a number of distinct advantages over traditional destructive approaches. The most important ones are simplicity, sensitivity, reliability, and a high throughput, translating into their applicability on a large spatial scale and saving a lot of labor [25].

Developing and implementing methods for quantification of pigment content and composition via nondestructive measurement of optical properties would provide a deeper insight into the physiology of photosynthetic apparatus (regulation of light harvesting and photochemical utilization, balance of photoprotection and photodamage) under favorable conditions and under stress [20,25,26].

The absorption of light by plant pigments allows tracking their content-affecting spectra of optical properties, absorbance, transmittance, and reflectance. Accurate estimation of pigment content using absorbance (α) and reflectance (ρ) spectra requires close linear relationships with the content of the specific pigment of interest. Importantly, these relationships should have a minimal effect of other pigments. Since the absorption bands of the pigments often overlap, it is a challenging problem; hence, development of quantitative measures of α and ρ response to specific pigment content is a prerequisite for assessment of the potential for nondestructive technologies based either on α or ρ spectroscopy.

Kubelka-Munk theory [27,28] laid a basis for reflectance spectroscopy, suggesting that the relationship between remission function, which in reality is reciprocal reflectance, ρ^{-1} [29], is related to the ratio of absorption to scattering coefficients. However, this assumption was not tested for leaves containing widely variable pigment content and composition, which makes the limits of reflectance spectroscopy, as well as whether the requirements for spectral regions could be used for estimating contents of four types of foliar pigments—chlorophylls, Chl; carotenoids, Car; anthocyanins AnC; and flavonoids, Flv—uncertain. Thus, development and implementation of reflectance-based techniques requires answering three pivotal questions. First, is it possible to describe the leaf as a medium with a close, linear ρ^{-1} vs. α relationship throughout the visible and near-infrared (NIR) ranges of spectrum, as required by Kubelka-Munk theory? Second, what are the spectral ranges where the abovementioned requirement is fulfilled? Third, what would be an objective criterion for discerning the ranges where it is not fulfilled? Answering these questions, requiring a thoughtful study of the ρ^{-1} vs. α relationship in leaves with widely variable pigment (Chl, Car, AnC, and Flv) contents, shall reveal the possibilities and limitations of reflectance-based techniques. It may lay a background for informed selection of spectral bands for devising new and improving existing models for reflectance-based estimation of pigments as well as for developing absorbance-based techniques applicable in cases where reflectance-based approaches fail.

Attempts to apply nondestructive methods based on optical spectroscopy for assessment of plant physiological state via measuring pigment content have been undertaken for several decades [30–34]. The situation has changed drastically during the last few decades, when a significant amount of research was dedicated to the development of techniques for nondestructive evaluation of leaf pigments. Here, we provide an overview of foliar absorbance and reflectance spectral features with an emphasis on *in situ* specific optical properties of the pigments. We also give the rationale for quantitative responses of absorbance and reflectance to each pigment content. Based on this, we demonstrate possibilities and limitations of reflectance- and absorbance-based approaches for estimating foliar pigment content. We attempt to find the spectral regions where assumption of a close linear ρ^{-1} vs. α relationship holds so reflectance spectroscopy provides accurate estimation of foliar pigment content. Furthermore, we tackle quantification of the pigments that absorb in spectral regions where reflectance is insensitive to the pigment content. In these regions, absorbance-based techniques are the only way to nondestructively assess pigment content. In view of these limitations, we present models for accurately estimating Chl, Car, AnC, and Flv, as well as generic algorithms for estimating Chl and AnC. Finally, we compare the efficiency of informative spectral band selection for pigment estimation by established methods (neural networks, partial least-square [PLS] regression, vegetation indices, uninformative variable elimination PLS) with that of the specific spectral responses introduced in this chapter.

5.2 SPECTRAL CHARACTERISTICS OF LEAVES

Three data sets were used in this study: Virginia creeper [*Parthenocissus quinquefolia* (L.) Planch.] with a widely varying content of pigments, especially of AnC and Flv [35]*; ANGERS recorded in 2003 at INRA in Angers (France), including 308 leaves of more than 40 plant species [36], as well as transmittance and reflectance spectra†; and the data set composed of 90 leaves (beech, chestnut, and maple) described in [29,37].‡

The leaves of the Virginia creeper illustrate how variable the inherent (absorbance and transmittance) and apparent (reflectance) optical properties are (Figure 5.1). In the blue range (400–500 nm), absorbance was very high, varying widely between 1 and 3 and increasing toward shorter wavelengths in all leaves studied. It was affected by Chl, Car, and Flv. Accordingly, transmittance of the leaves was below 0.1 and varied about 10-fold. In contrast, reflectance of the leaves converged to a narrow range around 0.1, showing small variability.

In the green range (500–600 nm), the absorbance and transmittance varied widely, with two ranges of convergence—around 500 and 600 nm, where the change in pigment content only slightly affected both traits (Figure 5.1a and b). Optical properties around 500 nm were governed by Chl, Car, and AnC and beyond 530 nm—by Chl and especially AnC, whose absorption *in situ* peaks around 550 nm [40]. Reflectance in the green was variable, but much less than absorbance and transmittance.

In the red range (600–690 nm), optical properties were affected by Chl absorption that peaked *in situ* around 670 nm (e.g., [38,39]); the magnitude of the absorbance peak increased gradually

---

* https://www.researchgate.net/publication/319213426_Foliar_reflectance_and_biochemistry_5_data_sets

† ANGERS Leaf Optical Properties Database (ecosis.org)

‡ https://www.researchgate.net/publication/319619724_Dataset_of_foliar_reflectance_spectra_and_corresponding_pigment_contents_for_Aesculus_hippocastanum_Fagus_Silvatica_Acer_platanoides_published_widely

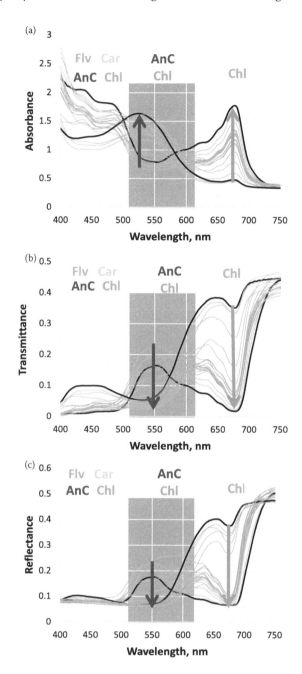

FIGURE 5.1 (a) Absorbance, (b) transmittance, and (c) reflectance spectra of 24 Virginia creeper leaves with widely varying pigment contents and composition. The spectral range shaded with green is solely governed by chlorophylls, Chl; the range shaded with red is governed predominantly by anthocyanins, AnC (when they are present) and Chl; the range shaded in gray is governed jointly by Chl, Car, and flavonoids, Flv, as well as AnC in the long-wave part of blue region. Arrows show direction of increasing leaf pigment content.

with the increase in [Chl] (Figure 5.1a). In the transmittance spectra, Chl absorption manifested itself as a trough whose depth steadily grew with an increase in [Chl] and transformed into a deep minimum in the spectra of leaves with a high [Chl] (Figure 5.1b). In the reflectance spectra, Chl absorption revealed itself as a trough (in the case of low-to-moderate [Chl]), although in the case of leaves with moderate-to-high [Chl], reflectance in the red converged to a narrow range

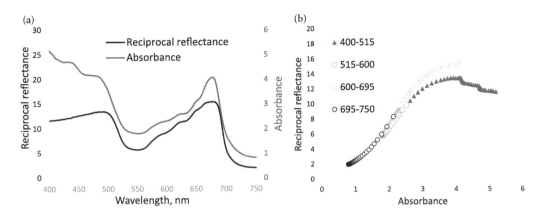

FIGURE 5.2 (a) Absorbance and reciprocal reflectance spectra of Virginia creeper leaf with a moderate [Chl] = 22 μg cm⁻² and [Car] = 5 μg cm⁻², a very high [Flv] = 165 μg cm⁻², and a small [AnC] = 0.07 μg cm⁻²; (b) reciprocal reflectance vs. absorbance of the same Virginia creeper leaf.

around 0.1 (Figure 5.1c). Thus, the behavior of reflectance in the red range differed from that of absorbance or transmittance.

The differences in the spectra depicting absorbance and reflectance were further studied in leaves with contrasting pigment content and composition (Figures 5.2 and 5.3). We compared absorbance and reciprocal reflectance of representative leaves from these data sets. In the Virginia creeper leaf with a moderate [Chl], α and ρ^{-1} were closely related in a linear manner in spectral ranges between 690–750 nm and 515–600 nm (determination coefficient $R^2 = 0.98$)—Figure 5.2. However, at wavelengths shorter than 515 nm and between 600 and 695 nm, the slope of the α vs. ρ^{-1} relationship decreased (Figure 5.2b). A strong hysteresis appeared in the range 600–695 nm: for the same absorbance, reciprocal reflectance was significantly higher than in the blue range 400–515 nm. Moreover, in the shortwave blue range with $\alpha > 2$, ρ^{-1} decreased with an increase in absorbance (Figure 5.2). Thus, in Virginia creeper leaves with $\alpha \geq 1$, reciprocal reflectance cannot be considered as a proxy for absorbance.

The ANGERS data set contained leaves with a wide [Chl] and [Car] variation ([31], Figure 5.3). In the spectral ranges 520–560 nm and 695–750 nm, the ρ^{-1} vs. α relationship was linear, with $R^2 = 0.99$ (Figure 5.3b). In the range 560–695 nm, the relationship was essentially nonlinear

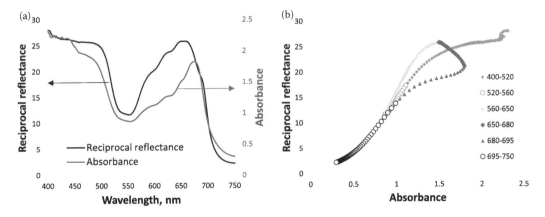

FIGURE 5.3 (a) Absorbance and reciprocal reflectance spectra of a leaf from the ANGERS data set with a high [Chl] = 54 μg cm⁻² and [Car] = 12 μg cm⁻², and a small [AnC] = 2.2 μg cm⁻²; (b) reciprocal reflectance vs. absorbance of the same leaf.

with a strong hysteresis and negative slope between 650–680 nm. At shorter wavelengths (beyond 520 nm), the slope of the relationship decreased drastically and was close to zero at wavelengths $\lambda < 500$ nm; that is, ρ^{-1} remained virtually invariant while absorbance varied widely (Figure 5.3a).

Thus, in both leaves of nonrelated species (Figures 5.2 and 5.3), the ρ^{-1} vs. α relationships were *close and linear in the green* (520–560 nm) *and red edge* (695–750 nm) ranges and *essentially nonlinear in the blue and red*. This circumstance imposes a very strict limitation on the possibility of foliar pigment content retrieval via reflectance spectroscopy.

5.3 *IN SITU*-SPECIFIC OPTICAL PROPERTIES OF FOLIAR PIGMENTS

ANGERS is a data set with probably the widest [Chl] variation among existing data [36]. However, it does not represent leaves with moderate-to-high [AnC]: only in 12 of 308 leaves [AnC] exceeded $5 \, \mu g \, cm^{-2}$, with a maximal value of $17 \, \mu g \, cm^{-2}$. Thus, we used this data set to study optical properties of leaves with high variability of [Chl] and [Car] contents against a slightly variable background of [AnC]. The Virginia creeper leaves had widely variable AnC content [35]; in addition, it is the only data set we know where [Flv] and optical properties are presented. This data set was used to study *in situ* optical properties of Flv and AnC.

To quantify the effect of each pigment's content, [p], on absorbance and reciprocal reflectance, α vs. [p] and ρ^{-1} vs. [p] relationships were established at each wavelength (λ) and for each pigment. We calculated the determination coefficient (R^2) for linear relationships α vs. [p] and ρ^{-1} vs. [p] and the slopes of these relationships at each λ. R^2 is a quantitative measure of how well the best-fit function performs as a predictor of α or ρ^{-1}, specifically, how much of their variability can be explained by the variation in the corresponding pigment content. Slopes of α vs. [p] and ρ^{-1} vs. [p] relationships represent sensitivity of absorbance and reciprocal reflectance to the pigment content. However, none of the parameters, either R^2 or the slope, is an accurate quantitative measure of each pigment's effect on α and ρ^{-1}. The spectra of the slope per se do not impart the strength of the corresponding relationships, as the R^2 spectra bear no information about the sensitivity of α or ρ^{-1} to each pigment's content. The quantitative measure of the effect of each pigment on absorbance combining these two parameters is a slope/NRMSE ratio for the α vs. [p] relationship [35], which can be calculated at each wavelength:

$$R\alpha = (d\alpha/d[p])/NRMSE \tag{5.1}$$

where $R\alpha$ is the response of α to a pigment content [*p*], and $d\alpha/d[p]$ and NRMSE are the first derivative and normalized root mean-square error of the α vs. [p] relationship, respectively.

In the same way, the quantitative measure of the effect of each pigment on reciprocal reflectance is the slope/NRMSE ratio of the ρ^{-1} vs. [p] relationship at the corresponding wavelength:

$$R\rho^{-1} = (d\rho^{-1}/d[p])/NRMSE \tag{5.2}$$

where $R\rho^{-1}$ is the response of ρ^{-1} to a pigment content [*p*], and $d\rho^{-1}/d[p]$ and NRMSE are the first derivative and normalized root mean-square error of the ρ^{-1} vs. [p] relationship.

Both measures, $R\alpha$ and $R\rho^{-1}$, represent the spectral response of absorbance and reciprocal reflectance to the content of a specific pigment.

The first question that needed an answer was how α and ρ^{-1} responded to [Chl] and how close the α vs. [Chl] and ρ^{-1} vs. [Chl] relationships are. We carried it out for the ANGERS data set [31] with the widest [Chl] variation. The spectra of R^2, α response ($R\alpha$), and ρ^{-1} response ($R\rho^{-1}$) to [Chl] are presented in Figure 5.4. The main feature of both traits was the disparate spectral behavior of $R\alpha$ and $R\rho^{-1}$. In the blue range (400–500 nm), R^2 for the α vs. [Chl] relationship was above 0.7, but

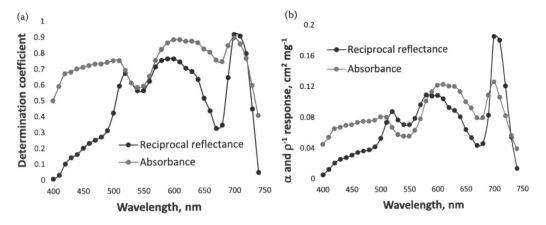

FIGURE 5.4   Characteristics of α vs. [Chl] and ρ^{-1} vs. [Chl] relationships calculated for 308 leaves constituting the ANGERS data set: (a) spectra of determination coefficient, and (b) spectra of α response, Rα, and ρ^{-1} response, Rρ^{-1}, to Chl content.

it was below 0.3 for the ρ^{-1} vs. [Chl] relationship (Figure 5.4a). The same was the case in the range 600–680 nm. Only *in the green and red edge ranges* the R^2 of both the α vs. [Chl] and ρ^{-1} vs. [Chl] relationships comparable, reaching R^2 around 0.9 in the red edge range, 700–710 nm. Importantly, (i) in the ranges of highest Chl absorption—the blue (400–500 nm) and the red (around 670 nm)—Rα response to [Chl] was twofold higher than Rρ^{-1} response (Figure 5.4b), and (ii) the green edge and red edge (700–710 nm) were the only spectral ranges where Rρ^{-1} > Rα and Chl was the main factor governing ρ^{-1} (Figure 5.4b).

The next step was to compare the spectral response of reciprocal reflectance to the contents of all three pigments identified in the ANGERS data set (Figure 5.5). The Rρ^{-1} spectra for Chl, Rρ^{-1}(Chl), and Car, Rρ^{-1}(Car) were almost identical. It is not surprising because in the Angers data set, [Car] correlated very closely (R^2 > 0.9) with [Chl]. Thus, [Chl] and [Car] were not really independent variables in this data set. Chl and Car contents often vary synchronously during ontogeny and senescence [7,9]. Such a conservative pigment composition is expected since the photosynthetic pigment apparatus is under tight regulation to achieve both maximum efficiency of carbon fixation by and mitigate the risk of photooxidative damage to the leaf [7].

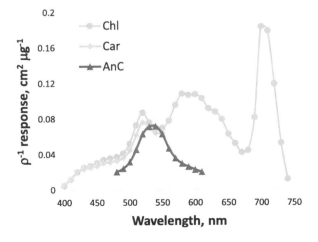

FIGURE 5.5   Spectral response of reciprocal reflectance, Rρ^{-1}, to Chl, Car, and AnC contents calculated for 308 leaves constituting the ANGERS data set.

The main spectral feature of the $R\rho^{-1}$ spectra was low values of this trait in the blue (absorption bands of Chl and Car) and red (absorption band of Chl) ranges (Figure 5.5). Two distinguishable peaks of $R\rho^{-1}(Chl)$ were around 600 and 700 nm. Importantly, both peaks were positioned in the ranges where absorption by Chl is much smaller than in the red absorption band of this pigment, where leaf reflectance is saturated at small $[Chl] < 20\ \mu g\ cm^{-2}$ [39].

$R\rho^{-1}(AnC)$ was high in the green range, around 550 nm (the main AnC absorption region *in situ*) [40,41]. However, $R\rho^{-1}(AnC) \cong R\alpha(Chl)$, so ρ^{-1} in this region was affected by Chl to the same degree as by AnC.

The responses $R\alpha$ and $R\rho^{-1}$ to [Chl], [AnC], and [Flv] were compared for Virginia creeper leaves with highly variable [AnC] and [Flv] and low-to-moderate [Chl]. As for the Angers data set containing leaves with much higher [Chl], $R\alpha(Chl)$ was higher than $R\rho^{-1}(Chl)$ in the ranges of highest Chl absorption, blue and red (Figure 5.6a). $R\rho^{-1}(Chl) > R\alpha(Chl)$ around 640 and 700 nm, located far from the red Chl absorption band. Notably, both α and ρ^{-1} responses to Chl were negative in the green range (Figure 5.6b) due to the negative slopes of α vs. [Chl] and ρ^{-1} vs. [Chl] relationships (with increases in [Chl], both responses, α and ρ^{-1}, decreased). A majority of leaves in this data set had high amounts of [AnC] and small amounts of [Chl], and leaves with moderate [Chl] contained small AnC. In leaves with small [Chl], absorbance in the green range was high, governed by [AnC], and with increasing [Chl], it decreased due to decreasing [AnC].

$R\alpha(AnC) > R\rho^{-1}(AnC)$ was recorded in the range 520–560 nm, where absorption of AnC peaks *in situ* (Figure 5.6b) due to saturation of reflectance at high [AnC] when α exceeded 2.5. In the range 400–460 nm, $R\alpha(Flv)$ was five- to sevenfold higher than $R\rho^{-1}(Flv)$ (Figure 5.6c). In this range Chl, Car, and Flv absorbance and reflectance saturated at a low [Chl] even in slightly green leaves; thus, ρ^{-1} became almost invariant with respect to pigment content (Figure 5.1c). This is illustrated well in Figure 5.7b, where $R\rho^{-1}(Chl)$ and $R\rho^{-1}(Flv)$ were indistinguishable and very small. In contrast, the responses of absorbance $R\alpha(Chl)$ and $R\alpha(Flv)$ were much higher, and in the narrow spectral range 400–430 nm, response $R\alpha(Flv)$ was higher than that of $R\alpha(Chl)$ (Figure 5.7a). This finding gives an important insight into identification of a spectral range suitable to [Flv] retrieval from absorbance spectra.

Above, we compared absorbance and reflectance vs. pigment content relationships in leaves using large data sets collected across plant species, developmental stages, and physiological states. The analysis made obvious certain limitations of reflectance-based quantification of the foliar pigments, especially in the blue and red manifesting itself as a failure of the close linear relationship between reciprocal reflectance and absorbance. These limitations can be understood in the frame of Kubelka-Munk theory, which was developed for the case of a relatively weak absorber evenly distributed in a thick layer of a highly reflective substance [27]. Considering the large extinction coefficients of Chl and other pigments [42], their high content in and structural complexity of the leaf and its photosynthetic apparatus [43], it becomes clear that in many cases, the foliar pigments violate these assumptions. Indeed, leaves with absorbance exceeding unity are rather "strong absorbers" in Kubelka-Munk terminology, and distribution of pigments in the cells is far from uniform [13,38]. Furthermore, superficial structures of plants such as leaf cuticle give rise to backscattering [44]. The contribution of light backscattered by weakly pigmented superficial structures of the leaf (cuticle and epidermis) to the total leaf reflectance bears no information about the leaf pigment composition and decreases the "information payload" of total reflected signal. This contribution increases dramatically toward shorter wavelengths of the visible part of the spectrum but bears scarce information on the biochemical composition of the leaf.

These limitations obviously affect the spectral ranges suitable for [Chl], [Car], [AnC], and [Flv] estimation. As a result, a reflectance-based approach can be implemented only in certain spectral ranges positioned outside the main absorption bands of the pigments, mainly in the long-wave part of the visible range, red edge, and NIR [33,34]. In view of these restrictions, it is important to have a quantitative criterion of the suitability of a certain spectral range for application of the reflectance-based techniques, which has not been defined so far. In this work, we try to close this gap by suggesting traits $R\alpha$ and $R\rho^{-1}$ as quantitative measures of the α and ρ^{-1} response to content of each pigment.

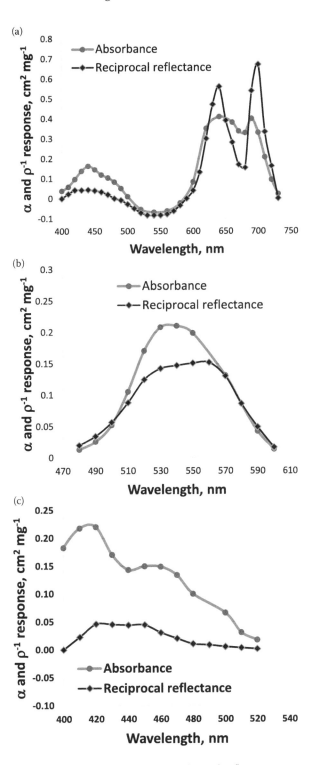

FIGURE 5.6 Spectra of absorbance response, Rα, and reciprocal reflectance response, Rρ^{-1}, to content of (a) chlorophyll [Chl], (b) anthocyanin [AnC], and (c) flavonoids [Flv] in 24 leaves of Virginia creeper with highly variable [AnC] and [Flv] and low to moderate [Chl].

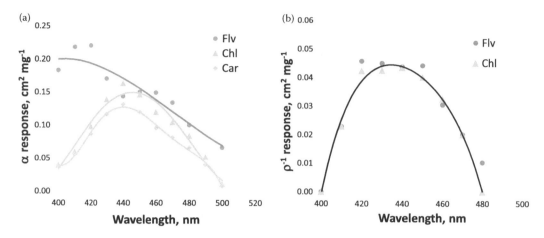

FIGURE 5.7   The responses of absorbance, Rα (a), and reciprocal reflectance, Rρ^{-1} (b), to pigment contents in the blue range of the spectrum in 24 Virginia creeper leaves. Note that ρ^{-1} responses to Flv and Chl are identical, showing that there is no way to distinguish between these pigments using reflectance, whereas the Rα response to Flv in the range 400–430 was higher than that to Chl. ((a): Modified from A. Gitelson et al. *Journal of Plant Physiology*, 218: 2017, 258–264. [35])

It was shown that responses Rα and Rρ^{-1} to Chl are very different across the spectral region and greatly depend on pigment content and composition. Thus, spectral responses Rα and Rρ^{-1} complement specific absorption coefficients, bringing the quantitative effect of each pigment *with background of other pigments* on α and ρ^{-1}. These findings using quantitative spectral responses to each pigment group are in accord with the results of previous studies that identified optimal spectral bands for retrieval of foliar pigment content [19,34,45,46].

5.4 PIGMENT CONTENT ESTIMATION

5.4.1 CHLOROPHYLLS

For individual leaves and two contrasting data sets, it was shown that responses Rα(Chl) and Rρ^{-1}(Chl) are very different across the whole spectral range and greatly depend on pigment content and composition. Essentially, in the *red edge region* (around 700 nm), Rρ^{-1} > Rα (Figures 5.4b and 5.6a) and the spectral shape and magnitude of the responses were almost identical despite great variation in pigment content in the data sets studied. This means that ρ^{-1} in the red edge region may be used as a term in algorithms for accurate and, probably, generic measure of [Chl] (Figures 5.8a and 5.9a). The only obstacle to achievement of a high accuracy of [Chl] estimation using ρ^{-1} is nonzero values of absorbance and reciprocal reflectance in the near-infrared spectral range where Chl does not absorb (Figure 5.1a). Merzlyak, Chivkunova, Melo, and Naqvi [47] have shown that this is apparent absorbance caused by uncertainties of absorbance and reflectance measurement. These uncertainties may affect accuracy of [Chl] estimation, especially for low-to-moderate [Chl] [29]. Thus, for accurate [Chl] estimation subtraction of α_{NIR} and ρ_{NIR}^{-1} (NIR beyond 760 nm) from α_{RE} and ρ_{RE}^{-1} (RE around 710 nm), is required:

$$Chl \propto \alpha_{RE} - \alpha_{NIR} \tag{5.3}$$

$$Chl \propto \rho_{RE}^{-1} - \rho_{NIR}^{-1} \tag{5.4}$$

Subtraction of α_{NIR} and ρ_{NIR}^{-1} makes ($\alpha_{RE} - \alpha_{NIR}$) and ($\rho_{RE}^{-1} - \rho_{NIR}^{-1}$) almost proportional to [Chl] (i.e., the relationships go to the origin) and it brings a significant increase in accuracy (Figures 5.8b

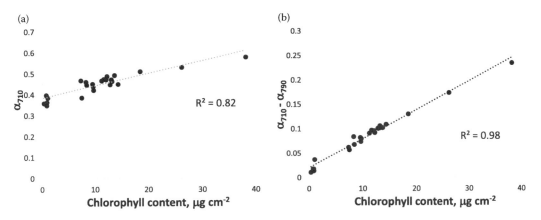

FIGURE 5.8 Chlorophyll content in the Virginia creeper leaves plotted versus (a) absorbance at 710 nm, α_{710}; (b) difference of absorbance at 710 nm and in the NIR at 790 nm, $\alpha_{710}-\alpha_{790}$.

and 5.9b). Both models yield very accurate [Chl] estimation using absorbance and reflectance, with the determination coefficient above 0.98 and NRMSE $< 2.4\%$.

A three-band model was suggested for estimating [Chl] in the form [29]:

$$\text{Chl} \propto (\rho_{RE}^{-1}-\rho_{NIR}^{-1}) \times \rho_{NIR} \qquad (5.5)$$

The third item, ρ_{NIR}, was introduced to take into account variability in leaf thickness and density that could affect [Chl] estimation. In leaves with the same [Chl] and different thickness/density, [Chl] retrieved from Equation 5.4 is smaller in thicker leaf than that in a thinner leaf. NIR reflectance of thicker leaves is higher than in thinner leaves, and the use of ρ_{NIR} allows decreasing uncertainty caused by variation in leaf thickness (Figures 5.9c and 5.10c). Reciprocal reflectance ρ_{710}^{-1} alone was also a very accurate measure of [Chl] in the ANGERS data set with a wide [Chl] variability (Figure 5.10a). To increase accuracy, we applied Equation 5.4 with a red edge band at 710 nm and NIR at 770–800 nm, estimating [Chl] in this data set (Figure 5.10b). The accuracy was very high ($R^2 = 0.94$), confirming the robustness of the approach. $CI_{red\ edge}$ taking into account different leaf thickness/density was the most accurate (Figure 5.10c).

Different techniques for foliar [Chl] estimation, neural network (NN), partial least-squares regression, and vegetation indices (VIs), for three unrelated plant species (maple, chestnut, and beech) were tested in [32]. Descriptive statistics of the relationships between pigment content estimated by all three techniques are presented in Table 5.1. All three techniques were found to estimate [Chl] accurately (Figure 5.11). Among the VIs tested, $CI_{red\ edge}$ was the most accurate. Compared to NN and PLS, $CI_{red\ edge}$ was also superior, with almost zero bias and a coefficient of variation (CV) below 12.1%. NN and PLS were very accurate, with CV below 11.8% for NN, with 8.9% positive mean normalized bias (MNB), and CV below 12.5% for PLS, with 9% negative MNB.

5.4.2 Carotenoids

Among other pigments, carotenoid estimation is probably the most challenging due to a strong overlap of Car absorption with those of Chl, AnC, and Flv, as well as the vast chemical (and hence spectral) diversity of Car, which is also easily changed by environmental stimuli. Another obstacle is the quite close relationship between [Chl] and [Car], so these variables are far from independent [48]. Thus, the precise estimation of [Car] with nondestructive spectral measurements has so far not reached accuracies comparable to the results obtained for [Chl] estimation.

It was found that maximal sensitivity of reflectance to [Car] is in the so-called green edge around 510–515 nm (Gitelson et al., 2002). However, reflectance in this range is greatly affected by Chl

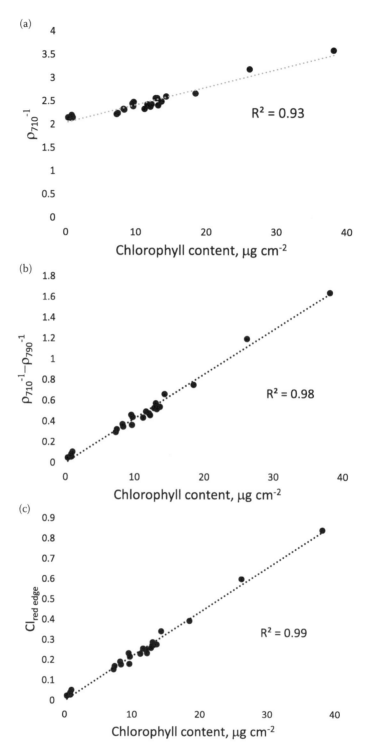

FIGURE 5.9 Chlorophyll content in the Virginia creeper leaves plotted versus (a) reciprocal reflectance at 710 nm, ρ_{710}^{-1}; (b) difference of reciprocal reflectances at 710 nm and in the NIR at 790 nm, $\rho_{710}^{-1} - \rho_{790}^{-1}$; (c) red edge chlorophyll index $CI_{red\ edge} = (\rho_{790}/\rho_{710})^{-1}$.

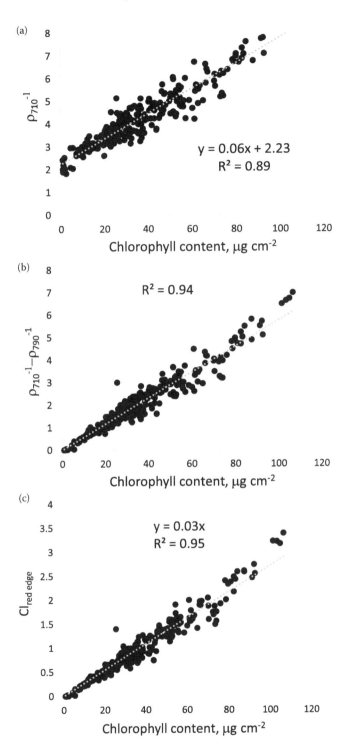

FIGURE 5.10 Chl content in the leaves of the ANGERS data set plotted versus (a) reciprocal reflectance at 710 nm, (b) ρ_{710}^{-1}, difference $\rho_{710}^{-1} - \rho_{790}^{-1}$, and (c) red edge chlorophyll index $CI_{red\ edge} = (\rho_{790}/\rho_{710}) - 1$.((c): Adopted from A. Gitelson, A. Solovchenko, *Geophysical Research Letters*, 2017. [66])

TABLE 5.1

Descriptive Statistics of the Relationships between Leaf Chlorophyll Content Measured and Estimated by Three Models, NN, PLS, and CI$_{\text{red edge}}$

	CV	R^2	MNB	NMB
CI$_{\text{red edge}}$	12.1	0.97	−0.8	−0.1
NN	11.8	0.97	8.9	−0.3
PLS	12.5	0.97	−6.2	0.6

Source: Modified from O. Kira et al. *International Journal of Applied Earth Observation and Geoinformation*, 38: 2015, 251–260. [37]

Note: CV = coefficient of variation, MNB = mean normalized bias, NMB = normalized mean bias; all three measures are in percent.

absorption. Thus, it was suggested to subtract the Chl effect from ρ_{510}^{-1} using reciprocal reflectance in the green and red edge that are quite accurate measures of [Chl]. Carotenoid reflectance indices were suggested in the forms:

$$\text{CRI}_{\text{green}} \propto (\rho_{510}^{-1} - \rho_{\text{green}}^{-1}) \tag{5.6}$$

$$\text{CRI}_{\text{RE}} \propto (\rho_{510}^{-1} - \rho_{\text{RE}}^{-1}) \tag{5.7}$$

To take into account likely differences in leaf thickness/density, CRI was modified and presented as:

$$\text{mCRI}_{\text{green}} \propto (\rho_{510}^{-1} - \rho_{\text{green}}^{-1}) \times \rho_{\text{NIR}} \tag{5.8}$$

$$\text{mCRI}_{\text{RE}} \propto (\rho_{510}^{-1} - \rho_{\text{RE}}^{-1}) \times \rho_{\text{NIR}} \tag{5.9}$$

However, [Car] was never estimated in AnC-containing leaves. As can be seen in Figure 5.5, ρ^{-1} response to Car and AnC in the range around 515 nm is almost even; thus. subtraction of this effect is necessary. Subtraction of α_{550} from α_{510} allowed quite accurate estimation of [Car] in AnC-containing Virginia creeper leaves (Figure 5.12a).

Kira, Linker, and Gitelson [37] compared accuracy of estimating [Car] by neural network, partial least-squares regression, and VIs in maple, chestnut, and beech leaves. In Table 5.2, descriptive statistics of the relationships between Car content estimated by all three techniques are presented. Relationships between NN and PLS models, red edge carotenoid reflectance index (CRI$_{\text{red edge}}$), and Car content were very close for each species. However, while CRI$_{\text{red edge}}$ vs. Car relationships for maple and chestnut had very similar slopes, the slope of the relationship for beech was much lower than for maple and chestnut; thus, the CRI$_{\text{red edge}}$ vs. Car relationship for all three species taken together was essentially not linear, with CV = 23%. For all three species taken together, NN vs. Car and PLS vs. Car relationships were much closer (with R^2 above 0.91) than that between CRI$_{\text{red edge}}$ and Car (Figure 5.13).

Fassnacht, Stenzel, and Gitelson [49] addressed the issue of nonlinearity of CRI and mCRI vs. [Car] relationships (Figure 5.13a). Using the same data set as Kira, Linker, and Gitelson [37] they examined the potential of the angular vegetation index (AVI) [49–51] to estimate total foliar [Car] of maple, chestnut, and beech. Based on an iterative search of all possible band combinations, a best-candidate AVI$_{\text{car}}$ was identified. The identified index used reflectances at wavelengths 410, 530, and 550 nm and showed a quite close but essentially not linear relation with Car contents of the

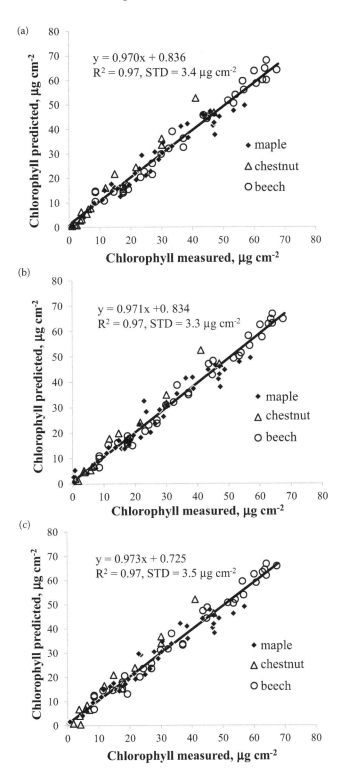

FIGURE 5.11 Chlorophyll content estimated by (a) red edge chlorophyll index, CI_red edge; (b) NN; and (c) PLS plotted versus measured chlorophyll content in maple, chestnut, and beech. (Modified from O. Kira et al. *International Journal of Applied Earth Observation and Geoinformation*, 38: 2015, 251–260. [37])

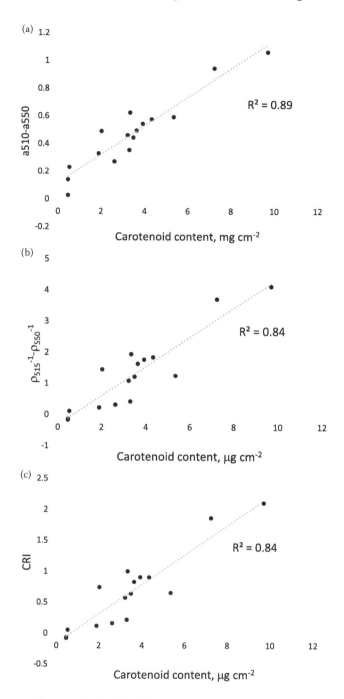

FIGURE 5.12   Carotenoid content in the Virginia creeper leaves plotted versus (a) difference of absorbance $\alpha_{510}-\alpha_{550}$; (b) difference of reciprocal reflectances at 510 and 550 nm, $\rho_{510}^{-1}-\rho_{550}^{-1}$; and (c) green carotenoid reflectance index $CRI_{green} = (\rho_{510}^{-1}-\rho_{550}^{-1}) \times \rho_{770-800}$.

examined species with increasing sensitivity to high [Car] and a lack of sensitivity to low [Car] for which both $mCRI_{RE}$ (Equation 5.9) and ρ_{760}/ρ_{500} [30] performed better. To make use of the advantages of both VI types, a simple merging procedure, which combined AVI_{car} with two earlier proposed carotenoid indices ($mCRI_{RE}$ and ρ_{760}/ρ_{500}), was developed. The merged indices had a close linear relationship with total Car content and outperformed all other examined indices. The merged indices

TABLE 5.2

Descriptive Statistics of the Relationships between Leaf Carotenoid Content Measured and Estimated by Three Models, NN, PLS, and CRI$_{red\ edge}$

	CV	R²	MNB	NMB
CRI$_{red\ edge}$	24.8	0.70	4.7	0.9
NN	15.6	0.88	1.7	1.4
PLS	16.7	0.86	−4.5	17.7

Source: Modified from O. Kira et al. *International Journal of Applied Earth Observation and Geoinformation*, 38: 2015, 251–260. [37]

Note: CV = coefficient of variation, MNB = mean normalized bias, NMB = normalized mean bias; all three measures are in percent.

were able to accurately estimate total [Car] with a normalized root mean square error (NRMSE) of 8.12% and a coefficient of determination of 0.88 (Figure 5.14). The findings were confirmed by simulations using the radiative transfer model PROSPECT-5. This strengthens the assumption that the proposed merged indices have a general ability to accurately estimate foliar [Car]. To prove the general applicability of the index for nondestructive estimation of Car from leaf reflectance data, further examination of the proposed merged indices for other plant species is desirable.

5.4.3 ANTHOCYANINS

In the Virginia creeper data set containing mainly red and dark red leaves, large differences in $R\alpha(AnC)$ and $R\rho^{-1}(AnC)$ responses in the range 520–550 nm were found (Figure 5.6b). This shows that the highest accuracy of [AnC] estimation may be achieved using absorbance at 550 nm, shown in Figure 5.15a. Reciprocal reflectance around 570 nm, where $R\alpha(AnC) = R\rho^{-1}(AnC)$, was also an accurate proxy of [AnC] (Figure 5.15c). However, Chl significantly affected α and ρ^{-1} in the range 550–570 nm. Thus, subtraction of this effect should be done using α and ρ^{-1} in the red edge range where they accurately represent [Chl]. The following models were capable of accurate estimation of [AnC] using absorbance and reflectance of leaves with widely variable pigment composition (Figure 5.15b and d):

$$[AnC] \propto \alpha_{550} - \alpha_{RE} \tag{5.10}$$

$$[AnC] \propto \rho^{-1}_{570} - \rho^{-1}_{RE} \tag{5.11}$$

5.4.4 FLAVONOIDS

In the blue range, 400–500 nm, where optical properties are affected by all three pigments, Chl, Car, and Flv, ρ^{-1} was either almost flat (Figure 5.3b) or even decreased with α increase (Figure 5.2b). $R\alpha$ and $R\rho^{-1}$ bring unique quantitative information on the responses of α and ρ^{-1} to [Chl] and [Flv], which is specific for each pigment. As can be seen from Figure 5.7b, the responses $R\rho^{-1}(Chl)$ and $R\rho^{-1}(Flv)$ were equal, showing that reflectance spectroscopy is unable to differentiate between these pigments. By contrast, in the range between 400 and 430 nm, the response $R\alpha(Flv)$ was higher than $R\alpha(Chl)$, suggesting that for the data used, it is the only spectral band where [Flv] may be estimated using absorbance spectroscopy (Figure 5.7a). Another important finding is that Chl's effect in the range 400–430 nm is still significant (Figure 5.7b), and its subtraction would

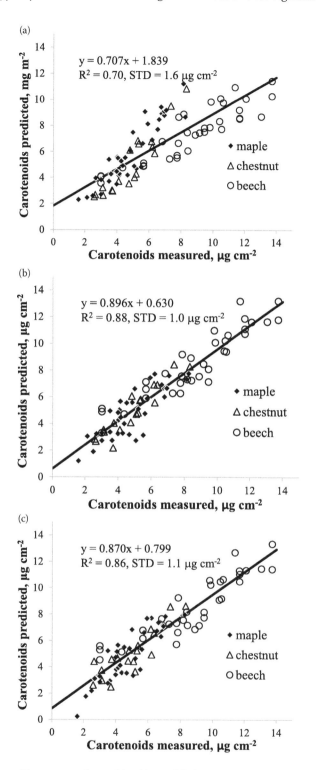

FIGURE 5.13 Carotenoid content estimated by (a) modified green carotenoid reflectance index, CRI$_{green}$; (b) NN; and (c) PLS plotted versus measured carotenoids content in maple, chestnut, and beech. (Reprinted from O. Kira et al. *International Journal of Applied Earth Observation and Geoinformation*, 38: 2015, 251–260. [37])

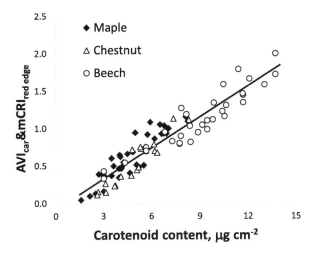

FIGURE 5.14 Merged AVI$_{car}$ and mCRI$_{red\ edge}$ plotted vs. total carotenoid content in maple, chestnut, and beech. The solid line is the linear best-fit function of the relationship between the index and total carotenoid content for all three species taken together. (Modified from F.E. Fassnacht et al. *Journal of Plant Physiology*, 176: 2015, 210–217 [49])

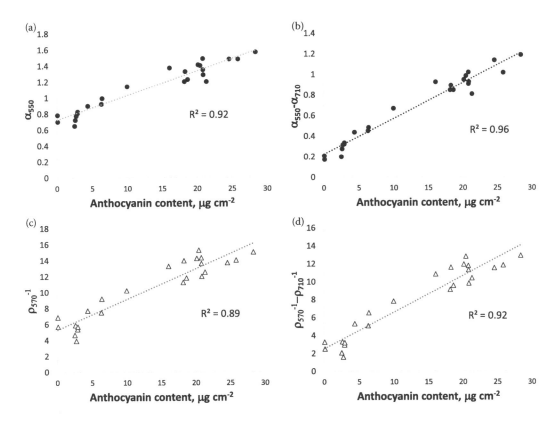

FIGURE 5.15 Anthocyanin content in 24 Virginia creeper leaves plotted versus (a) absorbance at 550 nm, α_{550}; (b) difference of absorbance at 550 and 710 nm, $\alpha_{550}-\alpha_{710}$; (c) reciprocal reflectance at 570 nm, ρ_{570}^{-1}; and (d) difference $\rho_{570}^{-1}-\rho_{710}^{-1}$.

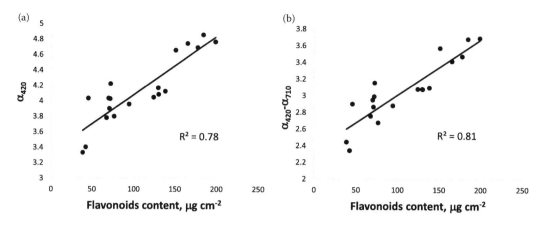

FIGURE 5.16 Flavonoid content in Virginia creeper leaves plotted versus (a) absorbance at 420 nm, α_{420}; (b) difference of absorbance at 420 and 710 nm, $\alpha_{420}-\alpha_{710}$. ((b): Modified from A. Gitelson et al. *Journal of Plant Physiology*, 218: 2017, 258–264. [35])

be necessary for accurate [Flv] estimation. Absorbance at 420 nm was quite an accurate proxy of [Flv] (Figure 5.16a). Subtraction of α_{710} allowed a decreasing Chl effect at 420 nm, and the following model is suggested for accurate nondestructive estimation of [Flv] in a wide range of their variation (Figure 5.16b):

$$[\text{Flv}] \propto \alpha_{420}-\alpha_{710} \tag{5.12}$$

5.5   KNOWLEDGE-BASED SELECTION OF SPECTRAL BANDS

The uninformative variable elimination PLS (UVE PLS) technique is not a band selection per se in the sense that one tries to find the best small subset of variables for fitting a model, but the elimination of those variables that are useless [52]. Kira, Linker, and Gitelson [37] investigated the magnitude of the reliability parameter *as an indicator of the information contained in the spectral bands* and compared the most informative bands with the spectral bands used by the other models (NN, PLS, and VIs—$CI_{red\,edge}$ and $CRI_{red\,edge}$). The results of UVE PLS for [Chl] and [Car] estimation are presented in Figure 5.17. The green spectral band around 560 nm retained in the NN and PLS models for Chl estimation (red areas in Figure 5.17a) coincided with highest values of the reliability parameter. In this spectral region, reflectance is governed by Chl hyperbolically decreasing with increase of [Chl] [53]. This spectral range was widely used for [Chl] estimation due to the high sensitivity of reflectance to Chl content in a wide range of its change in slightly green to dark green leaves [38,39,54–56].

Another maximum of the reliability parameter was found in the long wave end of the red edge region between 730 and 750 nm, where two factors govern reflectance. They include Chl absorption that is significant in leaves with Chl content above 400 mg m^{-2} (green to dark green leaves), and leaf structure and thickness affecting reflectance in the NIR range [43,55,57]. In previous studies, this band was used for accurately estimating foliar and total canopy Chl and nitrogen content using the red edge Chl index [23,29,58].

Despite the very low magnitude of the reliability parameter in the blue region (Figure 5.17a), the band around 480 nm was retained in the NN and PLS models for [Chl] estimation. In this region, absorption is saturated strongly and sensitivity of reflectance to [Chl] is minimal, as indicated by the magnitude of the reliability parameter. However, the blue range is suitable for reference reflectance in VIs for eliminating partially random variability of reflectance due to uncertainties of measurements as well as differences in leaf surface structure [59].

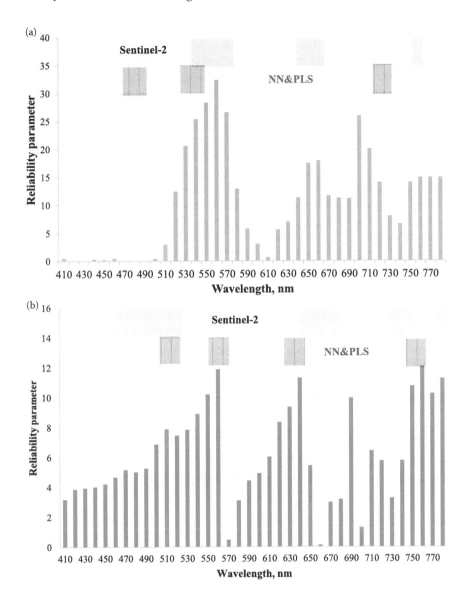

FIGURE 5.17 Reliability parameter calculated using uninformative variable elimination PLS (UVE PLS) plotted versus wavelength for (a) chlorophyll content and (b) carotenoid content estimation. The grey areas at the top correspond to positions of Sentinel-2 bands, which were found optimal for pigment estimation by NN and PLS models (Table 5.3). The red areas indicate the 20-nm-wide spectral bands that were found to be optimal for pigment content estimation by NN and PLS. Data containing maple, chestnut, and beech leaves (e.g., [32,38]) were used.

The reliability parameter for Chl estimation indeed provided a reliable indication of the usefulness of the spectral region: its maxima coincided with the positions of spectral bands used by the green and red edge Chl indices [29,34]. The other bands used in these indices (beyond 770 nm and either 540–560 nm in CI_{green} or 690–730 nm in $CI_{red\,edge}$) also correspond to regions in which the magnitude of the reliability parameter is substantial.

All four spectral bands retained in the NN and PLS models for [Car] estimation coincided with highest values of the reliability parameter (Figure 5.17b), as well as with spectral bands of the red edge CRI and green CRI [60]. The first band retained in the models was located around 510 nm, where maximal sensitivity of reflectance to [Car] was found [60]. In this region, both Chl and Car contents govern reflectance, and this band was used in both CRI_{green} and $CRI_{red\,edge}$ [34,60].

The second band retained in the models was located in the green range, where the magnitude of the reliability parameter is maximal. This band was used in CRI_{green} for subtraction of Chl's effect from reflectance around 510 nm [33]. The third and fourth spectral bands retained in the NN and PLS models were located around 630 and 740 nm. Both bands are located quite far from the main absorption bands of Chl and Car; at moderate to high Chl content, absorption around 630 and 740 nm is not saturated and thus reflectance in these regions is sensitive to [Chl]. The band centered at 630 nm has an additional advantage—absorption by both Chl a and Chl b at 630 nm are almost the same. Thus, in contrast to the 740-nm band where only Chl a absorbs, the 630-nm band accounts not only for Chl a but also for Chl b absorption. To the best of our knowledge, the 630-nm band has never been used in any estimation of either Chl or Car contents.

Importantly, the location of the spectral bands retained in the models did not correspond to the main absorption bands of pigments of interest Chl and Car, and this is not surprising. They coincided with the location of spectral bands where absorption by Chl remained strong enough to be sensitive to Chl content but far enough from the main Chl absorption bands to avoid saturation. Remarkably, for Car estimation, the NN and PLS models retained bands centered at 490–510 nm and 470–490 nm, respectively, which are the spectral regions where reflectance was found to be maximally sensitive to Car absorption while also being affected by Chl absorption [30,32,60]. This region is also close to the range where reflectance is sensitive to the xanthophyll cycle and used in the photochemical reflectance index [31].

The repeatability of the wavelength selection for the NN and PLS models is remarkable; they almost completely coincided (Figure 5.17). It is also worth noting that consistent results of Chl and Car estimation by NN and PLS were achieved using reflectance without any spectral transformation, for example, log(1/ρ), first derivative, second derivative [61]. The band selection was not dependent on the data used, and the bands retained in the NN and PLS models agreed well with those reported in other studies and known to explain the chemical variation in our data sets.

The expected accuracy of pigment content estimation by VIs, NN, and PLS with spectral bands of the multispectral instrument (MSI) on the Sentinel-2 satellite was assessed in [32] and is presented in Table 5.3. $CI_{red\,edge}$ with spectral bands centered at 705 and 775 nm (Table 5.4) was able to estimate

TABLE 5.3

Coefficient of Variation (in Percent) of Chlorophyll and Carotenoid Estimation by Neural Network (NN) and Partial Least-Squares (PLS) Regression with Simulated Expected Spectral Response of the Multispectral Instrument (MSI) aboard the Sentinel-2 Satellite

		CV					CV	
	Bands	NN	PLS			Bands	NN	PLS
Chlorophyll	B1-B7	13.6	16.0	Carotenoids		B1-B7	19.9	21.0
	B4-B7	13.6	13.6			B4-B7	21.1	21.1
	B5-B7	13.6	13.6			B5-B7	21.0	21.0
	B6-B7	14.1	14.1			B6-B7	20.9	20.9
	B5-B6	26.9	26.8			B5-B6	27.9	28.0
	B4-B6	20.9	20.9			B4-B6	25.9	25.9
	B2, B4, B6-B7	13.7	13.7			B2, B5, B6-B7	20.5	20.4
	B3, B6, B7	13.6	13.6			B2, B4, B6-B7	19.6	19.6
	B3, B5, B7	21.5	21.4					

Source: Reprinted from O. Kira et al. *International Journal of Applied Earth Observation and Geoinformation*, 38: 2015, 251–260. [37]

Note: The spectral bands of MSI are given in Table 5.4.

TABLE 5.4

Specifications of the Seven Spectral Bands (B1–B7) of the Multispectral Instrument (MSI) aboard the Sentinel-2 Satellite

Spectral Band	Center Wavelength (nm)	Band Width (nm)	Spatial Resolution (m)
B1	443	20	60
B2	490	65	10
B3	560	35	10
B4	665	30	10
B5	705	15	20
B6	740	15	20
B7	783	20	20

Source: Reprinted from O. Kira et al. *International Journal of Applied Earth Observation and Geoinformation*, 38: 2015, 251–260. [37]

[Chl] with CV = 13.1%, and using spectral bands at 740 and 775 nm CV was even lower − 12.6%. The minimal error of [Chl] estimation by both the NN and PLS models was higher than that of $CI_{red\,edge}$ and was achieved using all seven spectral bands (CV = 13.53%); using only three bands, 540–580, 732.5–747.5, and 770–780 nm, allowed for accurate estimation of Chl with CV = 13.63%.

$CRI_{red\,edge}$ employing bands centered at 490, 705, and 783 nm was able to estimate Car content with CV below 27%. CV was a little bit higher (28.5%) when the band at 705 nm was replaced by the band at 740 nm. Car estimation by NN and PLS was more accurate than that by $CRI_{red\,edge}$ (Table 5.3). *The highest accuracy (CV = 19.6%) was achieved using four spectral bands (B2, B4, B6, B7) in the blue, red, red edge, and NIR ranges of the spectrum.* The CV of [Car] estimation using MSI bands was about 3%–4% higher than that using 20-nm-wide optimal bands. This is likely due to the use of the band B2 positioned in the green edge region between 460 and 525 nm. The width of this band does not correspond to the required 10–15-nm width of the band positioned at 510 nm, where maximal sensitivity of reflectance to [Car] content was found [30,34,60].

While green and red edge CI were tested at close range at the canopy level [62] as well as using TM Landsat data [63], the Car and AnC models were not tested at the canopy level. It is also necessary to examine the presented techniques at other scales. For now, it remains unclear whether the found linear relationships between the models and pigment content also hold on a coarser spatial scale. It has to be examined if the proposed techniques are able to estimate Car content at the canopy scale using airborne and satellite data, which is typically influenced by atmosphere, bidirectional reflectance distribution function effects, canopy shadows, and soil background.

5.6 CONCLUDING REMARKS

The progress in the technology achieved over the last decade enabled precise and quick assessment of key plant pigments, including Chl, Car, and AnC *in situ*. Successful application of this approach, based mostly on reflectance spectroscopy, depends on the correct selection of informative spectral bands, which might not be trivial, especially in the case of pigments with strongly overlapping absorption bands. Furthermore, the comparison of the relationships between absorbance and reflectance vs. pigment content in leaves using large data sets collected across plant species, developmental stages, and physiological states obviated certain limitations of reflectance-based quantification of the foliar pigments, especially in the blue and red, manifesting itself as a failure of linear correlation between reciprocal reflectance and absorbance. In terms of Kubelka-Munk theory describing the behavior of relatively weak absorbers evenly distributed in a thick layer of a highly reflective substance

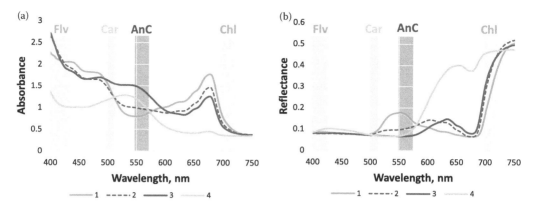

FIGURE 5.18 Absorbance and reflectance spectra of four Virginia creeper leaves with variable Chl, Car, AnC, and Flv contents (Table 5.5). Shaded areas represent spectral ranges found to be optimal for estimating pigment contents: [Chl]—green, model presented in the Equation 5.5; [AnC]—red, model presented in the Equation 5.11; [Car]—yellow, model presented in the Equation 5.8; and [Flv]—blue, model presented in the Equation 5.12. In models presented in the Equations 5.5 and 5.8, the 770–800 nm NIR range has been used.

[27], these limitations stem from (i) large extinction coefficients of Chl and other pigments [42], (ii) their high content in and (iii) structural complexity of the leaf and its photosynthetic apparatus [43]. Another plausible reason is backscattering from the superficial structures of plants such as leaf cuticle [44] contributing to the total leaf reflectance, especially in the blue, but bearing little or no information about the photosynthetic pigment composition of the leaf, hence decreasing the "information payload" of the total reflected signal.

In view of these difficulties, the reflectance-based approach is feasible only in certain spectral ranges positioned outside the main absorption bands of the pigments, mainly in the long-wave part of the visible range, red edge, and NIR [33,34]. To support the informed selection of suitable spectral bands for application of reflectance-based techniques, we proposed criteria, the traits $R\alpha$ and $R\rho^{-1}$, as quantitative measures of the α and ρ^{-1} responses to content of the pigment of interest. Importantly, spectral responses $R\alpha$ and $R\rho^{-1}$ complement specific optical properties, revealing the quantitative effect of each pigment on the background of other pigment absorption on α and ρ^{-1}.

The spectral bands selected based on these criteria are in line with the results of previous studies [19,34,45,46]. The task of spectral band selection can be further simplified by the elimination of useless spectral channels as outlined above, and the reduced set of spectral bands can be employed in different (NN-, PLS-, and VI-based) models for pigment content estimation. All three models were found to provide accurate estimations of foliar Chl content across three tree species. The Chl index using only two spectral bands in the red edge and NIR may be recommended for Chl estimation. NN and PLS with four spectral bands were the best for estimating carotenoid content; the NN model showed the highest accuracy. No techniques tested were species specific, allowing for estimating pigment content in different species without reparameterization of the model. All three

TABLE 5.5

Pigment Content (in $\mu g/cm^2$) in Four Leaves Presented in Figure 5.18

Leaf	Chl	Car	AnC	Flv*10⁻¹
1	38	9.7	0.14	27.1
2	18.5	5.3	14	30.2
3	13	6.3	46	24.7
4	0.4	0.7	40	10.4

techniques performed consistently well and yielded accurate estimations of pigment content when spectral bands were simulated in accord with the spectral response of the multispectral instrument on the Sentinel-2 satellite.

As a more generic approach capable of overcoming the limitations of reflectance-based models, we introduced the concept of specific absorbance response, objectively showing the contribution of each pigment group to light absorption, and deduced the *in situ* absorbance of foliar Chl, Car, AnC, and Flv, obviously free from the limitations typical of reflectance-based approaches. The absorbance-based algorithms demonstrated increased dynamic range and linear relationships with the leaf pigment content, especially pronounced in the shortwave part of the visible spectrum.

Based on the comparative account of advantages and drawbacks of reflectance- and absorbance-based pigment estimation, we argue that these approaches complement each other and can be used synergistically in advanced models for precision estimation of foliar pigments. We believe that the "response traits" are very instructive for understanding the *combined effect of pigments* on optical properties, which is at the foundation of knowledge-driven selection of spectral bands for creating new and improving existing models for noninvasive remote estimation of pigments.

The recently developed PROSPECT-D model includes, for the first time, the three main leaf pigments as independent constituents: Chl, Car, and AnC [43]. PROSPECT-D was tested on several data sets displaying many plant species with a large range of leaf traits and pigment composition, and showed very accurate estimation of Chl, Car, and AnC using hyperspectral transmittance and reflectance data. The accuracy of Chl estimation by the red edge Chl index ($CI_{red\ edge}$) was slightly better than that of Prospect D (4.5% vs. 5.5%), while the accuracy of AnC estimation by ARI (6.1%) and mARI (6.4%) was significantly better than that of Prospect D (9.5%). Significantly, Prospect D probably provides the only way to accurately estimate Car content in AnC-containing leaves [13]. Thus, the application of both data-driven and radiative transfer modeling are alternatives for developing generic algorithms estimating the content of all pigment groups. The combination of the two approaches brought desirable alternatives to extensive data collection (mandatory for the former) and high computational resources required by the latter [64,65].

Combining the approaches presented in this chapter, one can dramatically improve noninvasive estimation of the pigments absorbed in blue (flavonoids) and blue-green (carotenoids) on the background of strong overlapping absorption of other pigments. It is essential not only for those involved in remote estimation of pigments per se, but for plant biologists as well. Thus, our findings provide for a better understanding of light interaction with leaves, which translates into a deeper insight into the *in-situ* light absorption properties of all key plant pigment groups. This approach will constitute a handy tool for plant physiologists and photobiologists for dissecting the environmental stress effects in plants, especially for comparative analysis of interception of light by photosynthetic and photoprotective pigments as a function of physiological condition and developmental stage.

ACKNOWLEDGMENTS

This chapter is dedicated to late Mark N. Merzlyak, an extremely bright and productive scholar. He worked in many fields of biology and physiology. In each of them, his contributions were among the highest. Among others were free-radical oxidation of lipids, syndrome of lipid peroxidation in plants, phytoimmunology, and stress- and senescence-induced degradation of plant pigments. He contributed enormously in leaf optics and development techniques for foliar pigment retrieval. He was the best friend and we miss him tremendously.

We acknowledge the contributions of Drs. Veronica Ciganda, Robi Stark, Mark Steele, Andres Vina, and Yoav Zur. We are grateful to our colleagues Drs. Claus Buschmann, Olga B. Chivkunova, Harmut Lichtenthaler, and Donald C. Rundquist for helping us at different stages of this study. The support of Center for Advanced Land Management and Information Technologies at University of Nebraska, J. Blaustein Institute for Desert Research of Ben-Gurion University, Israel, and the Russian Science Foundation (to AS, grant 14–50-00029) is greatly appreciated.

REFERENCES

1. N. Liguori, X. Periole, S.J. Marrink, R. Croce, From light-harvesting to photoprotection: Structural basis of the dynamic switch of the major antenna complex of plants (LHCII), *Scientific Reports*, 5: 2015, 15661.
2. A.N. Tikhonov, A.V. Vershubskii, Computer modeling of electron and proton transport in chloroplasts, *Biosystems*, 121: 2014, 1–21.
3. R. Croce, H. van Amerongen, Natural strategies for photosynthetic light harvesting, *Nature Chemical Biology*, 10: 2014, 492–501.
4. B. Demmig-Adams, C. Cohu, O. Muller, W. Adams, III, Modulation of photosynthetic energy conversion efficiency in nature: From seconds to seasons, *Photosynthesis Research*, 113: 2012, 75–88.
5. P. Horton, Developments in research on non-photochemical fluorescence quenching: Emergence of key ideas, theories and experimental approaches, in: B. Demmig-Adams, G. Garab, W. Adams Iii, Govindjee (Eds.) *Non-Photochemical Quenching and Energy Dissipation in Plants, Algae and Cyanobacteria*, Springer, Netherlands, Dordrecht, 2014, pp. 73–95.
6. G. Giuliano, Plant carotenoids: Genomics meets multi-gene engineering, *Current Opinion in Plant Biology*, 19: 2014, 111–117.
7. A. Solovchenko, K. Neverov, Carotenogenic response in photosynthetic organisms: A colorful story, *Photosynthesis Research*, 2017.
8. M.N. Merzlyak, A.E. Solovchenko, Photostability of pigments in ripening apple fruit: A possible photoprotective role of carotenoids during plant senescence, *Plant Science*, 163: 2002, 881–888.
9. M. Merzlyak, A. Gitelson, O. Chivkunova, V. Rakitin, Non-destructive optical detection of pigment changes during leaf senescence and fruit ripening, *Plant Physiol*, 106: 1999, 135–141.
10. B. Winkel-Shirley, Flavonoid biosynthesis. A colorful model for genetics, biochemistry, cell biology, and biotechnology, *Am Soc Plant Biol*, 2001, 485–493.
11. N. Hughes, C. Morley, W. Smith, Coordination of anthocyanin decline and photosynthetic maturation in juvenile leaves of three deciduous tree species, *New Phytologist*, 175: 2007, 675–685.
12. G. Agati, C. Brunetti, M. Di Ferdinando, F. Ferrini, S. Pollastri, M. Tattini, Functional roles of flavonoids in photoprotection: New evidence, lessons from the past, *Plant Physiology and Biochemistry*, 72: 2013, 35–45.
13. M.N. Merzlyak, O.B. Chivkunova, A.E. Solovchenko, K.R. Naqvi, Light absorption by anthocyanins in juvenile, stressed, and senescing leaves, *Journal of Experimental Botany*, 59: 2008, 3903–3911.
14. B. Winkel-Shirley, Biosynthesis of flavonoids and effects of stress, *Current Opinion in Plant Biology*, 5: 2002, 218–223.
15. J.-H.B. Hatier, K.S. Gould, *Anthocyanin Function in Vegetative Organs, Anthocyanins*, Springer Science+Business Media, New York, NY, 2009, pp. 1–19.
16. K. Gould, Nature's Swiss army knife: The diverse protective roles of anthocyanins in leaves, *Journal of Biomedicine and Biotechnology*, 5: 2004, 314–320.
17. D. Strack, T. Vogt, W. Schliemann, Recent advances in betalain research, *Phytochemistry*, 62: 2003, 247–269.
18. Y. Tanaka, N. Sasaki, A. Ohmiya, Biosynthesis of plant pigments: Anthocyanins, betalains and carotenoids, *Plant Journal*, 54: 2008, 733.
19. A. Richardson, S. Duigan, G. Berlyn, An evaluation of noninvasive methods to estimate foliar chlorophyll content, *New Phytologist*, 2002, 185–194.
20. M.F. Garbulsky, J. Peñuelas, J. Gamon, Y. Inoue, I. Filella, The photochemical reflectance index (PRI) and the remote sensing of leaf, canopy and ecosystem radiation use efficiencies: A review and meta-analysis, *Remote Sensing of Environment*, 115: 2011, 281–297.
21. P.J. Curran, J.L. Dungan, H.L. Gholz, Exploring the relationship between reflectance red edge and chlorophyll content in slash pine, *Tree Physiology*, 7: 1990, 33–48.
22. I. Filella, L. Serrano, J. Serra, J. Penuelas, Evaluating wheat nitrogen status with canopy reflectance indices and discriminant analysis, *Crop Science*, 35: 1995, 1400–1405.
23. M. Schlemmer, A. Gitelson, J. Schepers, R. Ferguson, Y. Peng, J. Shanahan, D. Rundquist, Remote estimation of nitrogen and chlorophyll contents in maize at leaf and canopy levels, *International Journal of Applied Earth Observation and Geoinformation*, 25: 2013, 47–54.
24. H.K. Lichtenthaler, A. Ac, M.V. Marek, J. Kalina, O. Urban, Differences in pigment composition, photosynthetic rates and chlorophyll fluorescence images of sun and shade leaves of four tree species, *Plant Physiol Biochem*, 45: 2007, 577–588.
25. J. Cavender-Bares, J.A. Gamon, S.E. Hobbie, M.D. Madritch, J.E. Meireles, A.K. Schweiger, P.A. Townsend, Harnessing plant spectra to integrate the biodiversity sciences across biological and spatial scales, *Am J Bot*, 2017.

26. S.L. Ustin, J.A. Gamon, Remote sensing of plant functional types, *New Phytologist*, 186: 2010, 795–816.
27. G. Kortüm, *Reflectance Spectroscopy: Principles, Methods, Applications*, Springer, New York, 1969.
28. R. Kumar, L. Silva, Light ray tracing through a leaf cross section, *Applied Optics*, 12: 1973, 2950–2954.
29. A. Gitelson, Y. Gritz, M. Merzlyak, Non destructive chlorophyll assessment in higher plant leaves: Algorithms and accuracy, *Journal of Plant Physiology*, 160: 2003, 271–282.
30. E. Chappelle, M. Kim, J. McMurtrey III, Ratio analysis of reflectance spectra (RARS): An algorithm for the remote estimation of the concentrations of chlorophyll *a*, chlorophyll *b*, and carotenoids in soybean leaves, *Remote Sensing of Environment*, 39: 1992, 239–247.
31. J. Gamon, J. Penuelas, C. Field, A narrow-waveband spectral index that tracks diurnal changes in photosynthetic efficiency, *Remote Sensing of Environment*, 41: 1992, 35–44.
32. G. Blackburn, Quantifying chlorophylls and carotenoids at leaf and canopy scales: An evaluation of some hyperspectral approaches, *Remote Sensing of Environment*, 66: 1998, 273–285.
33. J. Gamon, J. Surfus, Assessing leaf pigment content and activity with a reflectometer, *The New Phytologist*, 143: 1999, 105–117.
34. A. Gitelson, G. Keydan, M. Merzlyak, Three-band model for noninvasive estimation of chlorophyll, carotenoids, and anthocyanin contents in higher plant leaves, *Geophysical Research Letters*, 33: 2006, L11402.
35. A. Gitelson, O. Chivkunova, T. Zhigalova, A. Solovchenko, *In situ* optical properties of foliar flavonoids: Implication for non-destructive estimation of flavonoid content, *Journal of Plant Physiology*, 218: 2017, 258–264.
36. S. Jacquemoud, S. Ustin, Application of radiative transfer models to moisture content estimation and burned land mapping, *4th International Workshop on Remote Sensing and GIS Applications to Forest Fire Management*, 2003.
37. O. Kira, R. Linker, A. Gitelson, Non-destructive estimation of foliar chlorophyll and carotenoid contents: Focus on informative spectral bands, *International Journal of Applied Earth Observation and Geoinformation*, 38: 2015, 251–260.
38. C. Buschmann, E. Nagel, *In vivo* spectroscopy and internal optics of leaves as basis for remote sensing of vegetation, *International Journal of Remote Sensing*, 14: 1993, 711–722.
39. A. Gitelson, M. Merzlyak, Quantitative estimation of chlorophyll-*a* using reflectance spectra: Experiments with autumn chestnut and maple leaves, *Journal of Photochemistry and Photobiology. B, Biology*, 22: 1994, 247–252.
40. A.A. Gitelson, M.N. Merzlyak, O.B. Chivkunova, Optical properties and nondestructive estimation of anthocyanin content in plant leaves, *Photochem Photobiol*, 74: 2001, 38–45.
41. M. Steele, A. Gitelson, D. Rundquist, M. Merzlyak, Nondestructive estimation of anthocyanin content in grapevine leaves, *American Journal of Enology and Viticulture*, 60: 2009, 87–92.
42. H. Lichtenthaler, Chlorophyll and carotenoids: Pigments of photosynthetic biomembranes, *Methods of Enzymology*, 1987, 331–382.
43. J.B. Féret, A.A. Gitelson, S.D. Noble, S. Jacquemoud, PROSPECT-D: Towards modeling leaf optical properties through a complete lifecycle, *Remote Sensing of Environment*, 193: 2017, 204–215.
44. P. Baur, K. Stulle, B. Uhlig, J. Schönherr, Absorption von strahlung im UV-B und blaulichtbereich von blattkutikeln ausgewählter nutzpflanzen, *Gartenbauwissenschaft*, 63: 1998, 145–152.
45. O. Kira, A.L. Nguy-Robertson, T.J. Arkebauer, R. Linker, A.A. Gitelson, Informative spectral bands for remote green LAI estimation in C3 and C4 crops, *Agricultural and Forest Meteorology*, 218: 2016, 243–249.
46. S.L. Ustin, A.A. Gitelson, S. Jacquemoud, M. Schaepman, G.P. Asner, J.A. Gamon, P. Zarco-Tejada, Retrieval of foliar information about plant pigment systems from high resolution spectroscopy, *Remote Sensing of Environment*, 113: 2009, S67–S77.
47. M.N. Merzlyak, O.B. Chivkunova, T.B. Melo, K.R. Naqvi, Does a leaf absorb radiation in the near infrared (780-900 nm) region? A new approach to quantifying optical reflection, absorption and transmission of leaves, *Photosynth Res*, 72: 2002, 263–270.
48. A.A. Gitelson, J.A. Gamon, A. Solovchenko, Multiple drivers of seasonal change in PRI: Implications for photosynthesis 1. *Leaf Level, Remote Sensing of Environment*, 191: 2017, 110–116.
49. F.E. Fassnacht, S. Stenzel, A.A. Gitelson, Non-destructive estimation of foliar carotenoid content of tree species using merged vegetation indices, *Journal of Plant Physiology*, 176: 2015, 210–217.
50. A. Palacios-Orueta, S. Khanna, J. Litago, M.L. Whiting, S.L. Ustin, Assessment of NDVI and NDWI spectral indices using MODIS time series analysis and development of a new spectral index based on MODIS shortwave infrared bands, *Proceedings of the 1st International Conference of Remote Sensing and Geoinformation Processing*, Trier, Germany. http://ubt.opus.hbz-nrw.de/volltexte/2006/362/pdf/03-rgldd-session2.pdf, 2006, pp. 207–209.

51. S. Khanna, A. Palacios-Orueta, M.L. Whiting, S.L. Ustin, D. Riaño, J. Litago, Development of angle indexes for soil moisture estimation, dry matter detection and land-cover discrimination, *Remote Sensing of Environment*, 109: 2007, 154–165.

52. V. Centner, D.-L. Massart, O.E. de Noord, S. de Jong, B.M. Vandeginste, C. Sterna, Elimination of uninformative variables for multivariate calibration, *Analytical Chemistry*, 68: 1996, 3851–3858.

53. A. Gitelson, M. Merzlyak, Signature analysis of leaf reflectance spectra: Algorithm development for remote sensing of chlorophyll, *Journal of Plant Physiology*, 148: 1996, 494–500.

54. H. Gausman, W. Allen, R. Cardenas, Reflectance of cotton leaves and their structure, *Remote Sensing of Environment*, 1: 1969, 19–22.

55. L. Fukshansky, A. Remisowsky, J. McClendon, A. Ritterbusch, T. Richter, H. Mohr, Absorption spectra of leaves corrected for scattering and distributional error: A radiative transfer and absorption statistics treatment, *Photochemistry and Photobiology*, 57: 1993, 538–555.

56. M. Merzlyak, A. Gitelson, Why and what for the leaves are yellow in autumn? On the interpretation of optical spectra of senescing leaves(*Acer platanoides* L.), *Journal of Plant Physiology*, 145: 1995, 315–320.

57. G. Le Maire, C. Francois, E. Dufrene, Towards universal broad leaf chlorophyll indices using PROSPECT simulated database and hyperspectral reflectance measurements, *Remote Sensing of Environment*, 89: 2004, 1–28.

58. J.G. Clevers, A.A. Gitelson, Remote estimation of crop and grass chlorophyll and nitrogen content using red-edge bands on Sentinel-2 and-3, *International Journal of Applied Earth Observation and Geoinformation*, 23: 2013, 344–351.

59. D. Sims, J. Gamon, Relationship between leaf pigment content and spectral reflectance across a wide range species, leaf structures and development stages, *Remote Sensing of Environment*, 81: 2002, 351–354.

60. A. Gitelson, Y. Zur, O. Chivkunova, M. Merzlyak, Assessing carotenoid content in plant leaves with reflectance spectroscopy, *Photochem Photobiol*, 75: 2002, 272–281.

61. Y. Grossman, S. Ustin, S. Jacquemoud, E. Sanderson, G. Schmuck, J. Verdebout, Critique of stepwise multiple linear regression for the extraction of leaf biochemistry information from leaf reflectance data, *Remote Sensing of Environment*, 56: 1996, 182–193.

62. A. Gitelson, S. Laorawat, G. Keydan, A. Vonshak, Optical properties of dense algal cultures outdoors and their application to remote estimation of biomass and pigment concentration in *Spirulina platensis* (Cyanobacteria), *Journal of Phycology*, 31: 1995, 828–834.

63. C. Wu, L. Wang, Z. Niu, S. Gao, M. Wu, Nondestructive estimation of canopy chlorophyll content using Hyperion and Landsat/TM images, *International Journal of Remote Sensing*, 31: 2010, 2159–2167.

64. J.-B. Féret, C. François, A. Gitelson, G.P. Asner, K.M. Barry, C. Panigada, A.D. Richardson, S. Jacquemoud, Optimizing spectral indices and chemometric analysis of leaf chemical properties using radiative transfer modeling, *Remote Sensing of Environment*, 115: 2011, 2742–2750.

65. J. Verrelst, G. Camps-Valls, J. Muñoz-Marí, J.P. Rivera, F. Veroustraete, J.G. Clevers, J. Moreno, Optical remote sensing and the retrieval of terrestrial vegetation bio-geophysical properties—A review, *ISPRS Journal of Photogrammetry and Remote Sensing*, 108: 2015, 273–290.

66. A. Gitelson, A. Solovchenko, Generic algorithms for estimating foliar pigment content, *Geophysical Research Letters*, 2017.

6 Hyperspectral Remote Sensing of Leaf Nitrogen Concentration in Cereal Crops

Tao Cheng, Yan Zhu, Dong Li, Xia Yao, and Kai Zhou

CONTENTS

6.1 INTRODUCTION

Cereal crops produce edible grains that are essential to feed the global population. The production of cereal crops, especially the three major cereals (maize, wheat, and rice), is heavily dependent on the use of nitrogen (N) fertilizer, which can affect in-season N use efficiency, the profitability of farming, and the sustainability of cropping systems (Peng et al., 2010; Ata-Ul-Karim et al., 2013). Production increases over 30% are still possible for cereals when nutrient overuse is eliminated (Mueller et al., 2012). The optimal management of N fertilizers would not only reduce the environmental impact of agricultural practices but also contribute to the stability or increase of cereal grain yield. It may be achieved by the accurate monitoring and diagnosis of crop N status. The monitoring of N status, particularly leaf nitrogen concentration (LNC), well accepted by agronomists, relies on the use of various analytical approaches such as chlorophyll meters (Piekkielek and Fox, 1992) and remote sensing tools (Hansen and Schjoerring, 2003; Xue et al., 2004). With the monitoring outcome as input, the diagnosis of N status is often based on various algorithms such as the N nutrition index (NNI) and critical N dilution curve (Lemaire et al., 2008).

Traditionally, the N status of cereals was assessed with a chlorophyll meter, from which the leaf chlorophyll content can be measured and used as a proxy of N status (Peng et al., 1993). This method is still in use due to its simplicity, but it is limited to the leaf level and not suitable for the monitoring of N status of the canopy. In recent decades, some active sensors such as Crop Circle and GreenSeeker have been used to monitor the N status at the canopy level (Cao et al., 2013, 2015). These portable sensors are easy to use in the field and can work under cloudy conditions, but they have only several spectral bands (less than 10) in the visible and near-infrared (VNIR) region. The multispectral band configuration limits their performance in the assessment of crop N status. In addition, multispectral bands may be used to estimate plant N accumulation (PNA) or plant N concentration (PNC) with moderate accuracy (Cao et al., 2015), but are barely used for the estimation of LNC in the literature. Even for PNA and PNC, the estimations with those multispectral bands would suffer from spectral saturation due to their close relationships with aboveground biomass. In fact, LNC is a fundamental indicator to understand the N status of cereal crops and the partitioning of N between organs. Alternatively, other researchers tend to use portable hyperspectral sensors that can provide spectral data in thousands of bands (Hansen and Schjoerring, 2003; Yao et al., 2013). As such, advanced techniques in hyperspectral remote sensing can be employed for more accurate monitoring of LNC.

Common measures for expressing the LNC of cereal crops are either area based (LNC_{area}, g/m^2) or mass based (LNC_{mass}, %). LNC_{area} is the nitrogen mass per unit leaf area, and LNC_{mass} is the ratio of nitrogen mass to leaf dry mass. These two measures can be converted between each other through leaf mass per area (LMA); that is, $LNC_{area} = LNC_{mass} \times LMA$ (Wright et al., 2004). Since destructive measurements of leaf nitrogen are often determined from dried and ground leaf samples with chemical procedures such as the micro-Kjedahl method, LNC_{mass} can be measured directly in the laboratory and LNC_{area} should be determined via the $LNC_{area} \sim LNC_{mass}$ relationship when the LMA is measured as well. Given its tight connection with photosynthetic capacity (Evans, 1989) and widespread use in fertilization management (Filella et al., 1995; Houlès et al., 2007), LNC_{mass} has received more attention and has been estimated from remotely sensed data more often than its counterpart LNC_{area} (Feng et al., 2008; Li et al., 2010; Ecarnot et al., 2013; Mutanga et al., 2015; Shi et al., 2015; Yao et al., 2015; He et al., 2016; Lepine et al., 2016). The term LNC hereafter refers to LNC_{mass} unless otherwise specified.

Given the important role of LNC in efficient and sustainable crop production, this chapter provides an overview of hyperspectral remote sensing of LNC with particular attention paid to cereal crops. The main objectives include:

1. Reviewing the principles and current status of the estimation of LNC with hyperspectral data
2. Discussing the theoretical and technical challenges and difficulties to be addressed with particular attention
3. Pointing out future directions for more effective use of remotely sensed data and spatial LNC mapping over large areas of cereal crop production

6.2 PRINCIPLE OF HYPERSPECTRAL REMOTE SENSING OF LEAF NITROGEN CONCENTRATION

The remote sensing of foliar chemistry lies in the characteristics of spectral variation associated with changes in the concentration of a specific chemical constituent being examined. The majority of nitrogen in the leaf exists in proteins. Leaf-level studies with reflectance spectra of dried and ground samples by Kokaly and Clark (1999) and Kokaly (2001) indicated that the physical basis for spectroscopic estimation of LNC was in the shape change of the broad 2100 nm absorption feature, which originates from two unique absorptions of N-H bonds at 2054 and 2172 nm. These two unique absorptions allow the estimation of LNC even though it is just a small fraction of the dry leaf (usually less than 5%) (Kokaly and Clark, 1999). The N absorption features are apparently visible in the reflectance spectra of dry leaves (Figure 6.1a) but not visible to naked eye in those of fresh leaves due to the influence of strong leaf water absorption (Figure 6.1b). It is important to

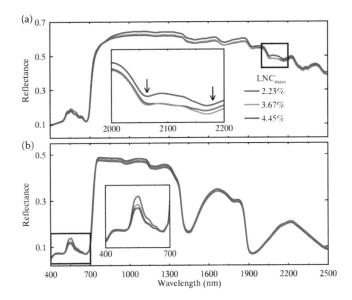

FIGURE 6.1   The reflectance spectra of three (a) fresh and (b) oven-dried wheat leaves corresponding to three levels of LNC. The downward arrows in (a) point to the wavelength positions of major nitrogen absorptions centered at 2060 and 2180 nm as documented by Curran (1989). Each spectrum in the figure represents the average of three samples for each leaf of wheat or rice plants.

note that LNC_{mass} is dependent on two variables, LNC_{area} and LMA. The leaf spectral responses to nitrogen change in Figure 6.1a also include noticeable variation in the absorption features of other dry matter constituents, such as the 2300 nm feature for lignin and cellulose.

At the canopy level, the masking effect of leaf water absorption on N-related absorption features is even stronger because of stronger water signals from crop canopies. Figure 6.2 illustrates the variation in canopy reflectance between plots with different N levels at the booting stage of wheat and rice crops. Apparently, higher LNC leads to lower reflectance in the red region due to the stronger chlorophyll absorption. The reflectance in the near-infrared (NIR) plateau increases with LNC. This variation is most likely caused by the vigorous leaf growth and the increase of leaf area index in response to N fertilization. The spectral difference in the shortwave-infrared (SWIR) region is less significant as water becomes the dominant factor controlling the reflectance amplitude and shape.

Since it is more practical to acquire reflectance measurements from fresh leaves and whole plants than from dried and ground leaves, the common way to estimate LNC with remote sensing is to use the reflectance spectra of fresh foliage at the leaf and canopy levels. As a consequence, the estimation of LNC is based on the characterization mostly of strong chlorophyll absorptions alone (Cho and Skidmore, 2006) and occasionally of chlorophylls and N absorptions in the full range (400–2500 nm) (Schlerf et al., 2010). Although chlorophylls are also N-containing compounds, their chemical structures and absorption characteristics are obviously different from those of proteins. The N atoms in chlorophylls are not attached to hydrogen atoms and do not exist in N-H bonds as in proteins (Kokaly, 2001). Nevertheless, N and chlorophyll concentrations are closely related for crops (Schlemmer et al., 2013) as for other plants (Croft et al., 2017), and their relationships could form the basis for the remote estimation of LNC with chlorophyll-sensitive spectral features, such as the red-edge position (Cho and Skidmore, 2006).

Although the impact of canopy structural variability on the physical interpretation of hyperspectral remote sensing of LNC was debated by the forest remote sensing community in the last decade (Ollinger et al., 2008; Knyazikhin et al., 2013; Ustin, 2013), numerous crop studies have reported the estimation of LNC for precision fertilization purposes with hyperspectral reflectance data, particularly the spectral information in the VNIR region (Lamb et al., 2002; Hansen and

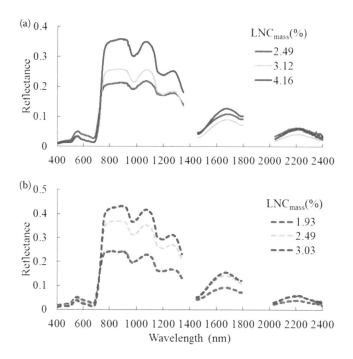

FIGURE 6.2   Canopy reflectance spectra for three N rates at the booting stage for (a) wheat and (b) rice. Each spectrum in the figure represents the average of three samples within a plot of wheat or rice plants.

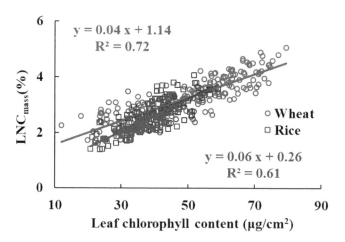

FIGURE 6.3   Relationships of leaf chlorophyll content with LNC_{mass} for wheat and rice crops. Red and blue lines represent best-fit regressions of the relationships for wheat and rice, respectively.

Schjoerring, 2003; Zhu et al., 2007a; Feng et al., 2008; Clevers and Kooistra, 2012; Tian et al., 2014; Yao et al., 2014). These estimations of LNC could be well justified by the close relationship between leaf chlorophyll content (LCC) and LNC, regardless of LNC_{mass} and LNC_{area} (Figures 6.3 and 6.4). Note that these relationships may vary with crop type, site condition, and growth stage as influenced by variation in physiological processes and solar irradiance. Even though the quantitative remote sensing community has devoted much effort to the hyperspectral estimation of LCC, the use of LCC in precision farming as a proxy of LNC may still suffer from uncertainty in LCC $\sim$ LNC relationships.

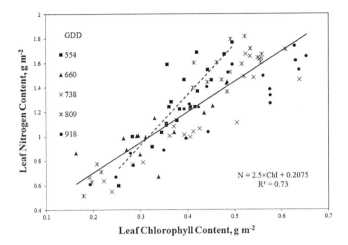

FIGURE 6.4 Relationship of leaf chlorophyll content with LNC_{area} for maize at five growth stages discriminated by thermal time (growing degree days, GDD). Solid and dashed lines are best-fit regressions for the whole season and the earliest stage with GDD = 554. (Reprinted from *International Journal of Applied Earth Observation and Geoinformation*, 25, Schlemmer, M. et al., Remote estimation of nitrogen and chlorophyll contents in maize at leaf and canopy levels, 47–54, Copyright 2013, with permission from Elsevier.)

6.3   CURRENT STATUS OF THE ESTIMATION OF LEAF NITROGEN CONCENTRATION

Based on the summary of relevant studies in the literature (Table 6.1), a review of the current status of hyperspectral remote sensing of LNC for cereal crops in terms of target crop types, coverage of growth stages, the level of observations, and the commonly used and emerging methods is examined here. The summary in Table 6.1 may not be an exhaustive list of all existing studies, but it is believed to be an appropriate representation of the advances and development trends on the topic. Note that this table also includes studies on the estimations of PNC and PNA that represent the N status of whole plants. The number of such studies is relatively small compared to that of LNC studies. For example, Inoue et al. (2012) concentrated on PNA, but it was the only airborne study on rice LNC monitoring. The inclusion here made a better representation of its category by level of observations in the summary.

6.3.1   CROP TYPE

It is clear that the previous work was conducted predominantly for rice and wheat, which are two major cereal crops cultivated in the world and provide more than 40% of the total dietary calories (Seck et al., 2012; Shiferaw et al., 2013). Compared to 12 studies for rice and 11 for wheat, only 4 were reported for maize and 1 for barley. No work has been seen for other cereal crops. The three major cereal crops accounting for 96% of the existing studies used 50% of the global N production, with 16% for maize, 16% for rice, and 18% for wheat (Ladha et al., 2016). Interestingly, most of the authors contributing to the rice and wheat studies are from China, where rice and wheat are the major staple crops to feed the largest population in the world. Apart from the demand of N fertilizer management and food supply, the widest attention paid to these two crops could also be explained by the appropriate plant height for operating hyperspectral instruments in practical applications (Milton et al., 2009). The height of rice and wheat crops is usually no more than 1.3 m, which makes it easy to acquire *in situ* canopy reflectance spectra on the ground with portable spectrometers throughout the entire growing season. For maize, the maximum canopy height is often approximately 3 m, and it would be difficult to directly locate portable spectrometers above the canopy after the midseason. This is supported by the frequent use of airborne platforms for maize research (Perry and Roberts, 2008; Cilia et al., 2014).

TABLE 6.1

A Summary of Studies on the Determination of N Status for Cereal Crops with Hyperspectral Data

Reference	Crop Type	Coverage of Growth Stages	Best VI or Analytical Method	Level of Observation	Variable of N Status
Xue et al. (2004)	Rice	Tillering, jointing, heading, and filling	R_{810}/R_{560}	Canopy	LNC_{mass}
Nguyen and Lee (2006)	Rice	Panicle initiation and booting	PLSR	Canopy	LNC_{mass}
Tang et al. (2007)	Rice	Booting, heading, milking, and maturing	SMLR	Canopy	LNC_{mass}
Zhu et al. (2007b)	Rice	Jointing, booting, heading, and filling	$(R_{1220} - R_{710})/(R_{1220} + R_{710})$	Canopy	LNC_{mass}
Zhu et al. (2007a)	Rice	Heading and filling	$(R_{1220} - R_{610})/(R_{1220} + R_{610})$	Canopy	LNC_{mass}
Stroppiana et al. (2009)	Rice	Tillering, stem elongation, booting, flowering	$(R_{503} - R_{483})/(R_{503} + R_{483})$	Canopy	PNC
Inoue et al. (2012)	Rice	Panicle formation stage	D_{740}/D_{522}	Canopy; aircraft	PNA
Wang et al. (2012)	Rice	Jointing, booting, heading, and filling	$(R_{924} - R_{703} + 2 \times R_{423})/(R_{924} + R_{703} - 2 \times R_{423})$	Canopy	LNC_{mass}
Yu et al. (2013)	Rice	Tillering, jointing, booting, heading, flowering, and filling	VIs and SMLR	Canopy	LNC_{mass}, PNC, PNA
Tian et al. (2014)	Rice	Tillering, jointing, booting, heading, filling, and milking	R_{553}/R_{537}	Canopy	LNC_{mass}
Cheng et al. (2016)	Rice	Booting, heading	CWA	Canopy	LNC_{mass}
Moharana and Dutta (2016)	Rice	Booting, heading, and filling	$R_{705}/(R_{717} + R_{491})$	Satellite	LNC_{mass}
Qin et al. (2016)	Rice	Jointing, heading, milking, ripening	R_{738}/R_{522}	Canopy	LNC_{mass}
Hansen and Schjoerring (2003)	Wheat	Early stem elongation to heading (BBHC-scale growth stages 30–51)	$(R_{750} - R_{734})/(R_{750} + R_{734})$	Canopy	LNC_{mass}
Zhu et al. (2007a)	Wheat	Initial heading, full heading, initial filling, full filling	$(R_{1220} - R_{610})/(R_{1220} + R_{610})$ & R_{1220}/R_{610}	Canopy	LNC_{mass}
Feng et al. (2008)	Wheat	Jointing, booting, anthesis, initial-filling, midfilling, late-filling	$R(\lambda) = Rs - (Rs - R_0) * exp(-(\lambda_0 - \lambda)^2/2\sigma^2)$	Canopy	LNC_{mass}
Li et al. (2010)	Wheat	Feekes 4–7, Feekes 8–10	$(R_{410} - R_{365})/(R_{410} + R_{365})$	Canopy	LNC_{mass}
Wang et al. (2012)	Wheat	Jointing, booting, heading, initial-filling, midfilling, late-filling	$(R_{924} - R_{703} + 2 \times R_{423})/(R_{924} + R_{703} - 2 \times R_{423})$	Canopy	LNC_{mass}

(Continued)

TABLE 6.1 (*Continued*)

A Summary of Studies on the Determination of N Status for Cereal Crops with Hyperspectral Data

Reference	Crop Type	Coverage of Growth Stages	Best VI or Analytical Method	Level of Observation	Variable of N Status
Li et al. (2014)	Wheat	Flowering	$(R_{694} - R_{302})/(R_{694} + R_{302})$	Canopy	LNC_{mass}
Yao et al. (2014)	Wheat	Reviving, jointing, booting, heading, anthesis, filling	$1.04 * (R_{514} - R_{469})/(R_{514} - R_{469} + 0.04)$	Canopy	LNC_{mass}
Feng et al. (2014)	Wheat	Jointing, booting, anthesis, grain filling	$(R_{755} + R_{680} - 2 \times R_{REP})/(R_{755} - R_{680})$	Canopy	LNC_{mass}
Yao et al. (2015)	Wheat	Jointing, booting, heading, anthesis	SVMs, ANNs, PLSR, SMLR, CR, $1.5 * (R_{1200} - R_{705})/(R_{1200} + R_{705} - 0.5)$	Canopy	LNC_{mass}
Feng et al. (2016)	Wheat	Reviving, jointing, booting, anthesis, initial-filling, midfilling, late-filling	$[(R_{735} - R_{720}) * R_{900}]/[R_{min}(930 - 980) * (R_{735} + R_{720})]$	Canopy	LNC_{mass}
He et al. (2016)	Wheat	Jointing, booting, heading, filling	$[R_{445} \times (R_{720} + R_{735}) - R_{573} \times (R_{720} - R_{735})]/[R_{720} \times (R_{573} - R_{445})]$	Canopy	LNC_{mass}
Perry and Roberts (2008)	Maize	V14, V15, R1	$TCARI = 3 \times [(R_{700} - R_{670}) - 0.2 \times (R_{700} - R_{550}) \times (R_{700}/R_{670})]$	Canopy; aircraft	LNC_{mass}
Chen et al. (2010)	Maize	V5~V9	$(R_{720} - R_{700})/(R_{700} - R_{670})/(R_{720} - R_{670} + 0.03)$	Canopy; aircraft	PNC
Schlemmer et al. (2013)	Maize	Vegetative to early reproductive	$(R_{755} + R_{680} - 2 \times R_{REP})/(R_{755} - R_{680})$ MCARI/MTVI2	Canopy	LNC_{area}
Cilia et al. (2014)	Maize	Stem elongation	$MCARI = (R_{700} - R_{670}) - 0.2 \times (R_{700} - R_{550}) \times (R_{700}/R_{670})$ $$MTVI2 = \frac{1.5 \times [1.2 \times (R_{800} - R_{550}) - 2.5 \times (R_{670} - R_{550})]}{\sqrt{(2 \times R_{800} + 1)^2 - [6 \times R_{800} - 5 \times \sqrt{R_{670}}] - 0.5}}$$	Aircraft	LNC_{area}
Xu et al. (2014)	Barley	Jointing to heading	OC	Canopy	LNC_{mass}

Note: References are categorized by crop type and ordered by year of publication within each crop type; R and λ denote reflectance and wavelength, respectively. Rs is the "shoulder" spectral reflectance, Ro is the minimum spectral reflectance at wavelength λo corresponding to the chlorophyll absorption well, and σ is the Gaussian function deviation parameter. R_{REP} is the reflectance at the red edge position. CWA = continuous wavelet analysis, OC = optimal combination, SMLR = stepwise multiple linear regression, PLSR = partial least-squares regression, PNA = plant nitrogen accumulation, PNC = plant nitrogen concentration.

6.3.2 COVERAGE OF GROWTH STAGES

Growth stage, which changes in days or weeks, is a crucial factor to consider for the timely monitoring of crop N status. As the growth stage progresses, dramatic changes can occur in the composition of canopy components and background materials (Figure 6.5). These changes pose a critical challenge for the use of concurrent spectral information to develop robust estimation models. To date, most of the existing studies for monitoring cereal crop LNC were reported for dense canopies from the reproductive period (from booting to filling for both rice and wheat). Only a few of them included the early growth stages (e.g., tillering for rice) when the canopy coverage was low, but the timing was also critical for fertilizer topdressing (Xue et al., 2004; Yu et al., 2013; Tian et al., 2014).

An early study by Xue et al. (2004) indicated that LNC was well related to vegetation indices (VIs) for each of the rice growth stages from tillering to filling, but the LNC ~ VI relationships varied by growth stage, and a generalizable relationship could not be established across all stages. One solution to this issue is to build models for stage groups instead of individual stages. For example, Yu et al. (2013) divided the entire dataset by growth stage into preheading and postheading groups and were able to build reasonable regression models per group. The LNC ~ VI relationships were found to be stronger for later stages than early stages of cereal crops, such as postheading versus preheading (Xue et al., 2004; Li et al., 2010; Yu et al., 2013).

6.3.3 LEVEL OF OBSERVATION

Hyperspectral data could be acquired on the ground and from aircraft or satellite platforms. Ground measurements are usually collected from approximately 1 m above the canopy of cereal crops with spectrometers. This represents the most common data source used for application-oriented nondestructive monitoring of LNC. It is also accepted most easily by users in agronomy and precision farming who may collect *in situ* hyperspectral measurements together with multispectral data from canopy sensors, such as Crop Circle (Cao et al., 2015) and GreenSeeker (Yao et al., 2013) for sensor comparison and intercalibration purposes. Compared to canopy-level reflectance measurements, only a small number of studies used airborne hyperspectral imagery for LNC monitoring and mapping, specifically from the Compact Airborne Spectrographic Imager (CASI) (Chen et al., 2010; Inoue et al.,

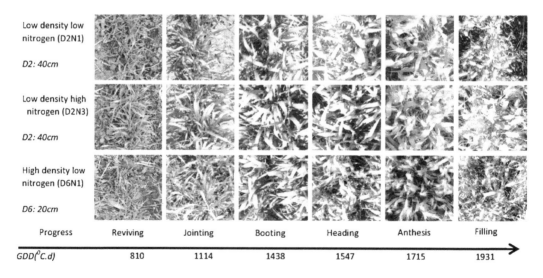

FIGURE 6.5 Snapshots of canopy components and background materials in a wheat field at different growth stages. (Reprinted from *International Journal of Applied Earth Observation and Geoinformation*, 32, Yao, X. et al., Detecting leaf nitrogen content in wheat with canopy hyperspectrum under different soil backgrounds, 114–124, Copyright 2014, with permission from Elsevier.)

2012), the Airborne Visible/Infrared Imaging Spectrometer (AVIRIS) (Perry and Roberts, 2008), and the AISA Eagle (Cilia et al., 2014) sensors. This reflects the limitation in the accessibility of aircraft hyperspectral imaging resources to the crop monitoring community. Given the high cost and logistic difficulties, the hyperspectral images in these studies were acquired only once in the growing season. Moharana and Dutta (2016) provided the single evaluation with spaceborne hyperspectral data from the EO-1 Hyperion instrument, also with only one acquisition in the season. Clearly, more frequent acquisitions are preferred for satisfying the requirements for in-season diagnosis of N status and management of fertilizer topdressings at several critical growth stages.

6.3.4 Methods for Estimating Leaf Nitrogen Concentration

Since hyperspectral data contain spectral information in hundreds or thousands of narrow bands contiguously distributed in the electromagnetic spectrum, various techniques used in vegetation analysis can be employed for estimating the LNC of cereal crops from hyperspectral data. The methods used for such a purpose mainly consist of vegetation indices, stepwise multiple linear regression (SMLR), partial least-squares regression (PLSR), and continuous wavelet analysis (CWA). Yao et al. (2015) provided a systematic comparison of six methods for estimating the LNC of wheat with a comprehensive dataset compiled from 10 years of *in situ* measurements over multiple sites. Three of them, that is, continuum removal (CR), artificial neural networks (ANNs), and support vector machines (SVMs), were exclusively evaluated in their study and therefore are not reviewed here. SMLR and PLSR are two multivariate regression methods that have been widely used in vegetation analysis over many decades. They are not covered here since the evaluation of their performance has been limited to a small number of studies (Nguyen and Lee, 2006; Tang et al., 2007; Yu et al., 2013). Instead, we focus on the status of VI development and the application of CWA, which separately represent the most common method in practical applications and an emerging technique for advanced spectroscopic analysis.

VIs derived from hyperspectral data are often calculated as ratios and normalized ratios of reflectance values or first derivative values at two or several narrow bands. They are easy to use and can be well understood by users from different disciplines. With multispectral data collected from a GreenSeeker or Crop Circle, one can develop new indices by choosing from a few band combinations. The advantage of using hyperspectral data is to increase the chances of creating new band combinations (Thenkabail et al., 2000), which can result in higher prediction accuracies and better consistencies between datasets (Yao et al., 2013).

Table 6.1 lists a set of 20 new VIs and 2 existing VIs found to date. Almost every publication concentrating on VI analysis developed a new VI for the purposes of adjusting to variations in growth stage (Zhu et al., 2007a,b), view angle (He et al., 2016), and geographic site (Inoue et al., 2012; Moharana and Dutta, 2016) and reducing the effects of optical saturation (Wang et al., 2012) and soil background (Yao et al., 2014). These adjustments or reductions were based on either the recalibration of band combinations in the same index form (Hansen and Schjoerring, 2003) or the extension from traditional two-band forms to the three-band (Chen et al., 2010) and four-band forms (He et al., 2016). The majority of VIs employed multiple bands in the 350–1000 nm range, with one band most often located in the red edge region (680–760 nm). Alternatively, Zhu et al. (2007a,b) used a band (1220 nm) in the longer wavelength region, and Li et al. (2014) used a band (302 nm) in the shorter wavelength region. The selected wavelengths in these VIs together with bandwidths could be used to guide the design of low-cost and easy-to-use portable two-band or three-band spectrometers (Wang et al., 2012). The VI-based models in these studies are ready to use for applications in real-time monitoring of LNC under similar conditions (Figure 6.6).

CWA is an emerging spectroscopic method that has recently been used to estimate LNC for cereal crops (Cheng et al., 2016). It makes use of full-spectrum information in the reflectance spectra and has the potential to improve the estimation of LNC by enhancing N absorption signals, as demonstrated in the estimations of leaf chlorophyll content for cereal crops (Li et al., 2017) and leaf

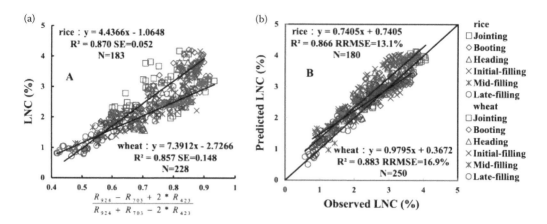

FIGURE 6.6 (a) Relationships between the newly developed three-band index and LNC for wheat (in red symbols) and rice (in blue symbols). (b) Comparison of observed and predicted LNC with the crop-specific models from (a) in the same color style. (Reprinted from *Field Crops Research*, 129, Wang, W. et al., Estimating leaf nitrogen concentration with three-band vegetation indices in rice and wheat, 90–98, Copyright 2012, with permission from Elsevier.)

water content (Cheng et al., 2011), dry matter content (Cheng et al., 2014b), and canopy water content (Cheng et al., 2014a) for other plant types.

After continuous wavelet transform (CWT), a reflectance spectrum is decomposed into a series of scale components, which have the same length as the original reflectance spectrum and are composed of wavelet coefficients as a function of scale and wavelength (Cheng et al., 2011). Wavelet coefficients have been proven to be superior to VIs in the characterization of absorption features of foliar chemicals present in reflectance spectra (Cheng et al., 2014a). Hence, a correlation scalogram was obtained by calculating coefficients of determination (R^2) between a foliar trait and a wavelet coefficient for all wavelengths and scales. The most significant wavelet coefficients representing the top ranked R^2 values (as shown in color in Figure 6.7) could be used as spectral features for estimating LNC (Cheng et al., 2014a). Cheng et al. (2016) derived the optimal wavelet coefficient (730 nm, scale 4) from rice samples and found it exhibited a strong relationship with LNC ($r^2 = 0.62$) but a weak relationship with leaf area index ($r^2 = 0.48$). This suggests the CWA method is promising for building LNC-sensitive models that are insensitive to canopy structural variation.

6.4 CHALLENGES FOR THE ESTIMATION OF LEAF NITROGEN CONCENTRATION IN CEREAL CROPS

Cereal crops have short life cycles when compared to other plant types that have also been extensively studied with remote sensing. Although significant advances in the hyperspectral remote sensing of LNC have been achieved for cereal crops in the past two decades, we have identified some challenges to be addressed for pushing foliar nitrogen monitoring to a higher level.

6.4.1 EFFECT OF CROP GROWTH STAGE ON MODEL ROBUSTNESS

These studies suggest that growth stage plays an important role in the rapid and accurate estimation of LNC and the transferability of estimation models to an independent dataset. As illustrated in Figure 6.8, four clusters of data points by growth stage can be observed when LNC is plotted against the widely used red edge chlorophyll index ($CI_{Red\text{-}edge}$). The LNC $\sim CI_{Red\text{-}edge}$ models are markedly different among the three early stages (early tillering, late tillering, and jointing) in the vegetative period. For the late stages (booting and filling) in the reproductive period, a single LNC $\sim CI_{Red\text{-}edge}$

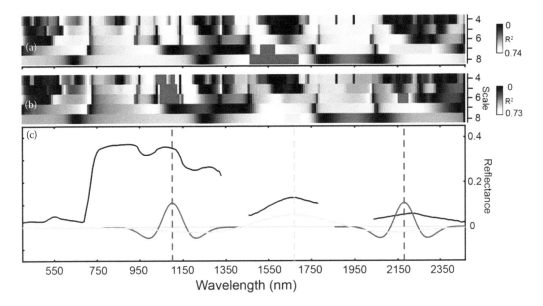

FIGURE 6.7   A correlation scalogram representing the squared correlation coefficient (R^2) for the relationship between (a) leaf area index and (b) canopy water content and wavelet coefficients. The color regions correspond to the top 5% R^2 values, and (c) is a visualization of the strongest wavelet coefficients from these color regions with corresponding wavelength locations and scales overlaid by a canopy reflectance spectrum. (Reprinted from *Remote Sensing of Environment*, 143, Cheng, T. et al., Detecting diurnal and seasonal variation in canopy water content of nut tree orchards from airborne imaging spectroscopy data using continuous wavelet analysis, 39–53, Copyright 2014, with permission from Elsevier.)

model could be fitted with a correlation coefficient higher than any of the early stages. This is mainly due to the decline of LNC from the start to the end of the entire growing season, more precisely, the N dilution effect from an agronomy perspective (Figure 6.9) (Ata-Ul-Karim et al., 2013). However, $CI_{Red-edge}$ or other VIs do not follow this simple trend.

The spectral signals from above the canopy at early stages are often confounded by background materials such as soil and water in the field (Yu et al., 2013; Gnyp et al., 2014; Sun et al., 2017).

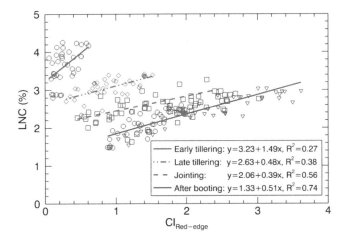

FIGURE 6.8   Best-fit linear relationships between LNC and $CI_{Red-edge}$ derived from canopy reflectance for different stages of paddy rice. Red lines represent individual stages before booting, and the blue line represents the group of booting, heading, and filling stages.

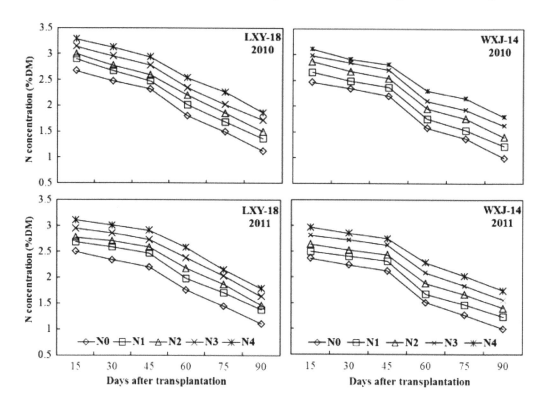

FIGURE 6.9 Changes of LNC with time (days after transplantation) for japonica rice under different N application rates in experiments conducted during 2010 and 2011. (Reprinted from *Field Crops Research*, 149, Ata-Ul-Karim, S.T. et al., Development of critical nitrogen dilution curve of Japonica rice in Yangtze River reaches, 149–158, Copyright 2013, with permission from Elsevier.)

The mixture of foliage and background in the field of view of observation instruments leads to less significant spectral contribution from the foliar chemicals and therefore weaker relationships between LNC and VIs. In the vegetative period of cereal crops, foliar biomass increases faster than nitrogen uptake and biomass contributes more to the variation in canopy reflectance (or VIs) (Mistele and Schmidhalter, 2008; Yu et al., 2013). In the reproductive period, the increase in foliar biomass stops, but nitrogen change continues as nitrogen is partitioned from leaves to panicles. As a result, the significant variation in LNC is more easily detected by remote sensors.

Moreover, the spectroscopic estimation of cereal crop LNC is often based on the relationship between LNC and LCC (Schlemmer et al., 2013). The stage-specific LNC ~ LCC relationships could also affect the universality of LNC ~ VI models across stages (Wang et al., 2014). This problem may be alleviated by further optimizing wavelengths for the selected VIs or adopting advanced machine learning algorithms (Yao et al., 2015). Although the use of those strategies may lead to higher estimation accuracies for the multiple stages, more effort should be devoted to understanding the influence of crop organ development on modifying the canopy spectral reflectance and the physical mechanism of reducing background effects (Yao et al., 2014). An accurate and simple-to-use approach is still highly desirable for applications to the entire growing season in the context of crop growth monitoring.

6.4.2   LACK OF PHYSICAL MODELS FOR LEAF NITROGEN CONCENTRATION INVERSION

The empirically based models for LNC estimation may lack of transferability to datasets from other species (Ferwerda et al., 2005; Kalacska et al., 2015), sites (Spaner et al., 2005), and growth stages

(Schlemmer et al., 2013). Instead, physically based models could be used to describe the general interaction of light with plant leaves and do not need parameterization for these conditions. The leaf optical properties model PROSPECT, developed by Jacquemoud and Baret (1990) more than two decades ago, is the most popular radiative transfer model for inversion of leaf biochemical parameters due to its ease of use, generalization capability, and free access (Jacquemoud et al., 2009). It takes a few leaf biochemical and structural parameters as input data and simulates the leaf hemispherical reflectance and transmittance spectra in the 400–2500 nm range. The model can be run in the forward mode to generate simulations and in the inverse mode to estimate leaf biochemical parameters. In the earliest version (PROSPECT-1), only leaf chlorophyll and water contents with strong absorptions were included. In the second version (PROSPECT-2), protein, cellulose, and lignin contents were added, but the inversion accuracy for protein as a proxy of LNC was poor (Jacquemoud et al., 1996). In PROSPECT-3, Baret and Fourty (1997) introduced LMA to assemble the major dry matter constituents, including protein, cellulose, hemicellulose, and lignin. Since then, leaf protein content has no longer been one of the PROSPECT input parameters (Féret et al., 2008, 2017), which has precluded the use of model inversion methods in recent studies on LNC estimation.

The search for the use of physical models in LNC estimation continues. The LIBERTY model proposed by Dawson et al. (1998) included LNC_{mass} as one of the input parameters, but it was designed for needle leaves and is hard to apply to the modeling of leaf optical properties for cereal crops. Recently, Wang et al. (2015a) used a strategy of combining empirical and physical models and estimated LNC_{area} at an R^2 of 0.58 through its linear relationship with equivalent water thickness (EWT) inverted from the PROSPECT model. However, they did not consider the estimation of LNC_{mass} due to its lower correlation of leaf traits that could be inverted from PROSPECT. The strongest correlation of LNC_{area} with EWT rather than chlorophyll and dry matter contents, found for broadleaf tree species, has not been reported for cereal crops.

Subsequently, Wang et al. (2015b) recalibrated the specific absorption coefficients specifically for protein and cellulose + lignin and estimated the protein content in fresh leaves at a moderate accuracy ($R^2 = 0.47$) with the newly calibrated PROSPECT model. This recalibration was evaluated on a common dataset composed of a large number of plant species, and how the model performs over cereal crops remains unknown. Compared with higher accuracies as measured by R^2 with empirical approaches, the accuracy at this level still needs to be improved. In addition, the area-based protein content in their study should be converted to mass-based protein content and then to LNC. There is still an important step from the moderately accurate estimation of area-based protein content to the highly accurate estimation of LNC for wider use in the monitoring of cereal crops.

6.4.3 CORRECTION FOR THE CONFOUNDING EFFECT OF CANOPY STRUCTURE

It is well known that the canopy spectral reflectance in the NIR region is mainly affected by canopy structural properties. However, Ollinger et al. (2008) reported a strong positive correlation between NIR reflectance and LNC for temperate forest species. We indeed observed the reflectance in the NIR region increases with LNC for cereal crops in previous studies (Yao et al., 2015; Zhu et al., 2007a), as well as in Figure 6.2, but could never generate accurate estimates of LNC with NIR reflectance. Knyazikhin et al. (2013) claimed that the counterintuitive positive correlation was an artifact and should be attributed to canopy structural variation rather than variation in the chemical concentration. To support their argument, they proposed a new spectral feature, called the canopy scattering coefficient (CSC), to estimate LNC by correcting for the effect of canopy structural variation. Their theory indicates that CSC should be derived from canopy reflectance spectra with negligible contribution of background materials, which is the reason it was applied to forest studies (Croft et al., 2017).

How canopy structure affects the estimation of LNC as argued by Knyazikhin et al. (2013) remains unclear for cereal crops. To examine the confounding effect of canopy structural variation on LNC estimation, we may apply the correction with CSC to cereal crops at reproductive stages but not at vegetative stages when the soil or water background can be clearly seen. Considering the dynamics of canopy cover during the entire season and the essence of growth monitoring in the early season, we need more sophisticated techniques to correct for this effect under low canopy cover conditions.

6.5   FURTHER CONSIDERATIONS AND OUTLOOK

Apart from the technical and theoretical challenges, we have further considerations about expansions from current studies to improved estimation and mapping of LNC, including the increased use of SWIR reflectance, observations in nadir and off-nadir angles, and the spatial mapping of LNC over cereal crop production areas. The attention to SWIR reflectance and multiangle observations is already exercised for other plant types, and the advantages can accrue for cereal crops (Hilker et al., 2008; Dechant et al., 2017). The regional mapping of LNC over cereal crop production areas is dependent on the availability of adequate hyperspectral data and may be a priority in the future.

6.5.1   INCREASED USE OF SHORTWAVE-INFRARED REFLECTANCE

In contrast to strong absorptions in the VNIR region, the absorption features in the SWIR reflectance spectra of fresh leaves are masked by strong leaf water absorption and hence not clearly visible as compared with dry leaves (Yoder and Pettigrew-Crosby, 1995; McClure et al., 2002; Schlerf et al., 2010; Pacheco-Labrador et al., 2014). Consequently, the reflectance spectra of fresh leaves in the SWIR region have seldom been used for LNC estimation. The use of SWIR reflectance could help us better understand the spectral responses to LNC variation through nitrogen absorptions (Curran, 1989; Kokaly and Clark, 1999). A major factor constraining the accurate estimation of LNC is the effect of leaf water absorption on reflectance spectra, which should be removed to within 10% as suggested by Kokaly and Clark (1999).

Recently, a few studies applied a water removal technique proposed by Gao and Goetz (1994) to leaf and canopy reflectance spectra in the full range so as to improve the estimation of LNC (Schlerf et al., 2010; Ramoelo et al., 2011). These studies indicate that more wavelengths selected from water removed spectra using simple multiple linear regression were related to known absorption features of nitrogen than those from the spectra without water removal. To further understand the physical mechanism underlying the estimation models, we should investigate the use of SWIR reflectance alone for the estimation of LNC as a supplementary way of using VNIR reflectance in current studies. Particular attention should be paid to the performance of water removal and the absorption features of nitrogen present in the residual spectra from water removal.

6.5.2   MULTIANGLE OBSERVATIONS

Cereal crops are often cultivated in rows along the north-south direction to facilitate the use of environmental resources and agricultural machinery. The composition of crops and background in the field of view is dependent on the solar position and viewing geometry. For crop monitoring with passive sensors, spectral reflectance measurements are often collected in the midday (10:00–14:00 local time) under sunny and clear conditions. The nadir-view observations for the highest solar elevation angles of the day would allow the maximal visibility of background materials such as soil and water in the field, which makes a significant contribution to the canopy-level reflectance signals. Numerous studies have demonstrated the non-negligible influence of background materials on the estimation of biochemical parameters (Darvishzadeh et al., 2008; Yao et al., 2014; Yu et al., 2014).

In addition, it is harder for us detect bottom-layer leaves of the canopy than middle- and upper-layer leaves through nadir observations, especially when crop canopy closure occurs (Huang et al., 2014). Since plants including cereal crops often exhibit a vertical gradient in LNC (Dreccer et al., 2000; Hikosaka, 2016), detection from nadir observations may not be representative of the LNC for the entire canopy. Recently, a few studies have shown that the use of off-nadir observations yielded more accurate estimations of LNC than the commonly used nadir observations (He et al., 2016; Jay et al., 2017). One of the reasons for such improvements might be due to the reduced confounding effect of soil background under off-nadir observations.

Note that specular reflection may occur on the water surface in paddy rice fields under certain viewing geometry conditions, and the water background for paddy rice could create more complex viewing problems than the soil background for other cereal crops. Future research can be directed toward the use of off-nadir observations in the solar principal plane for a specific solar angle or nadir observations for different solar angles. These observations could be feasible with hyperspectral sensors on the ground or onboard low-altitude unmanned aerial vehicles.

6.5.3 REGIONAL MAPPING

To date, the hyperspectral estimation of LNC for cereal crops has still been focused on the canopy level and applied to small field trial plots with portable spectrometers (Table 6.1). A few studies reported the spatial mapping of LNC for maize (Perry and Roberts, 2008; Cilia et al., 2014), wheat (Chen et al., 2010), and rice (Inoue et al., 2012) with airborne imaging spectroscopy data, but the image coverage areas were still limited. These studies laid the foundation for developing new spectral indices and algorithms for the hyperspectral remote sensing of LNC but were inadequate for examining the spatial pattern of LNC and understanding fertilization management over major crop production regions.

In this regard, spaceborne imaging spectrometers with much larger coverage areas can be a better choice for mapping LNC of cereal crops at the regional scale. Nevertheless, the amount and quality of spaceborne spectrometers remains relatively low compared to multispectral instruments. The gap in regional and global imaging spectroscopy data will be filled when hyperspectral instruments such as the Environmental Mapping and Analysis Program (EnMAP) are launched into space. In particular, the upcoming EnMAP instrument will cover a 30-km-wide area with a ground sampling distance of 30 m (Siegmann et al., 2015). EnMAP will also provide hyperspectral imagery data with a revisit time of approximately 4 days at the equator and even quicker at higher latitudes (Guanter et al., 2015). Mapping the LNC over major crop cultivation regions in the world will be an urgent task for the spaceborne imaging spectroscopy community. It has great potential in the guidance of the optimal fertilizer management and the reduction of the environmental impact from agricultural activities.

6.6 CONCLUSIONS

The N status of cereal crops is a crucial indicator for the precision management of N fertilizers in agronomic practices and needs to be monitored in a rapid and nondestructive way with advanced technologies. Driven by this application-oriented demand, hyperspectral remote sensing has been successfully applied to the estimation of the LNC with concentration on the three staple food crops (i.e., rice, wheat, and maize). The most common data source for such estimations was *in situ* canopy reflectance spectra that are easy to access and widely acceptable by agronomists. The hyperspectral data were then analyzed to develop a broad collection of VIs in different formulas and wavelength combinations for various site conditions and crop types. In contrast to the use of several bands in the VIs, the full-spectrum information could be used by wavelet analysis for improved estimations of LNC. Although the application of CWA for this purpose has just begun, we foresee that more interesting findings will come about from this emerging method as numerous studies have already reported its superior performance in the estimation of other chemicals at both leaf and canopy levels.

During the progression of these research activities, we have realized major challenges in the hyperspectral estimation of cereal crop LNC. The VI-based predictive models or the wavelength combinations in the optimized VIs were often unstable and largely affected by the coverage of growth stages in the input data. Since the major cereal crops of interest are annuals, their occupancy against the background materials in the field and their chemical composition can change rapidly between consecutive stages. However, these changes may not be reflected effectively by the current hyperspectral techniques. In addition, the estimation of LNC could be confounded by canopy structural variation that is also a significant controlling factor of canopy spectral variability. This structural effect has been well studied for forest canopies but not yet so for crop canopies. It may be corrected for by introducing a canopy scattering coefficient or decomposing the spectral responses of structural and chemical variables. The common way to analyze the structural effect on the estimation of foliar chemistry is to use a canopy-level radiative transfer model coupled with a leaf optical properties model. However, such coupled models do not have LNC as one of the input parameters. If LNC can be incorporated into the models, the direct inversion of LNC from canopy reflectance spectra will be feasible and the physical mechanism of LNC estimation would clearly be better understood.

Despite these challenges, our effort should also be devoted to new directions that could lead to improved estimation of LNC from the spectral domain and operational mapping of LNC from the spatial domain. Since we have made considerable achievements in the use of nadir-view canopy reflectance spectra in the VNIR region, how to use the spectral information from the SWIR region and off-nadir observations for obtaining improved LNC estimates would be the priority area in the next decade. These research activities will enable us to build N absorption–based estimation techniques and better suppress the influence of background materials. In contrast to the field-level monitoring in previous studies, the spatially explicit mapping of LNC over crop production regions would depend on the progress of hyperspectral satellite missions in preparation by the international community. It is important that scientists in hyperspectral remote sensing work closely with agronomists, farmers, and agricultural authorities at all levels to ensure up-to-date regional maps of LNC can be generated for the precision management of fertilizer application and productivity prediction over major regions of cereal crop production.

ACKNOWLEDGMENTS

The authors are supported by the National Key R&D Program (2016YFD0300601), the National Natural Science Foundation of China (31470084, 31671582, and 31725020), Jiangsu Distinguished Professor Program, and the Academic Program Development of Jiangsu Higher Education Institutions (PAPD). We are grateful to the faculty and staff members from the National Engineering and Technology Center (NETCIA) at Nanjing Agricultural University for their fruitful discussions and continuous support. Our thanks also go to the former and current graduate students who have been involved with this work. The reviewers are appreciated for their constructive comments, which improved the quality of this chapter.

REFERENCES

Ata-Ul-Karim, S.T., Yao, X., Liu, X., Cao, W., & Zhu, Y. 2013. Development of critical nitrogen dilution curve of Japonica rice in Yangtze River reaches. *Field Crops Research*, 149, 149–158.

Baret, F., & Fourty, T. 1997. Estimation of leaf water content and specific leaf weight from reflectance and transmittance measurements. *Agronomie*, 17, 455–464.

Cao, Q., Miao, Y., Feng, G., Gao, X., Li, F., Liu, B., Yue, S., Cheng, S., Ustin, S.L., & Khosla, R. 2015. Active canopy sensing of winter wheat nitrogen status: An evaluation of two sensor systems. *Computers and Electronics in Agriculture*, 112, 54–67.

Cao, Q., Miao, Y., Wang, H., Huang, S., Cheng, S., Khosla, R., & Jiang, R. 2013. Non-destructive estimation of rice plant nitrogen status with Crop Circle multispectral active canopy sensor. *Field Crops Research*, 154, 133–144.

Chen, P., Haboudane, D., Tremblay, N., Wang, J., Vigneault, P., & Li, B. 2010. New spectral indicator assessing the efficiency of crop nitrogen treatment in corn and wheat. *Remote Sensing of Environment*, 114, 1987–1997.

Cheng, T., Li, D., Zheng, H., Yao, X., Tian, Y., Zhu, Y., & Cao, W. 2016. Towards decomposing the effects of foliar nitrogen content and canopy structure on rice canopy spectral variability through multi-scale spectral analysis. *International Geoscience and Remote Sensing Symposium*, pp. 3508–3511.

Cheng, T., Riaño, D., & Ustin, S.L. 2014a. Detecting diurnal and seasonal variation in canopy water content of nut tree orchards from airborne imaging spectroscopy data using continuous wavelet analysis, *Remote Sensing of Environment*, 143, 39–53.

Cheng, T., Rivard, B., & Sánchez-Azofeifa, A. 2011. Spectroscopic determination of leaf water content using continuous wavelet analysis. *Remote Sensing of Environment*, 115, 659–670.

Cheng, T., Rivard, B., Sánchez-Azofeifa, A.G., Féret, J.-B., Jacquemoud, S., & Ustin, S.L. 2014b. Deriving leaf mass per area (LMA) from foliar reflectance across a variety of plant species using continuous wavelet analysis. *ISPRS Journal of Photogrammetry and Remote Sensing*, 87, 28–38.

Cho, M.A., & Skidmore, A.K. 2006. A new technique for extracting the red edge position from hyperspectral data: The linear extrapolation method. *Remote Sensing of Environment*, 101, 181–193.

Cilia, C., Panigada, C., Rossini, M., Meroni, M., Busetto, L., Amaducci, S., Boschetti, M., Picchi, V., & Colombo, R. Nitrogen status assessment for variable rate fertilization in maize through hyperspectral imagery. *Remote Sensing*, 2014, 6, 6549–6565.

Clevers, J.G.P.W., & Kooistra, L. 2012. Using hyperspectral remote sensing data for retrieving canopy chlorophyll and nitrogen content. *IEEE Journal of Selected Topics in Applied Earth Observations and Remote Sensing*, 5, 574–583.

Croft, H., Chen, J.M., Luo, X., Bartlett, P., Chen, B., & Staebler, R.M. 2017. Leaf chlorophyll content as a proxy for leaf photosynthetic capacity. *Global Change Biology*, 23, 3513–3524.

Curran, P.J. 1989. Remote sensing of foliar chemistry. *Remote Sensing of Environment*, 30, 271–278.

Darvishzadeh, R., Skidmore, A., Atzberger, C., & van Wieren, S. 2008. Estimation of vegetation LAI from hyperspectral reflectance data: Effects of soil type and plant architecture. *International Journal of Applied Earth Observation and Geoinformation*, 10, 358–373.

Dawson, T.P., Curran, P.J., & Plummer, S.E. 1998. Liberty: Modeling the effects of leaf biochemical concentration on reflectance spectra. *Remote Sensing of Environment*, 65, 50–60.

Dechant, B., Cuntz, M., Vohland, M., Schulz, E., & Doktor, D. 2017. Estimation of photosynthesis traits from leaf reflectance spectra: Correlation to nitrogen content as the dominant mechanism. *Remote Sensing of Environment*, 196, 279–292.

Dreccer, M.F., Oijen, M.V., Schapendonk, A.H.C.M., Pot, C.S., & Rabbinge, R. 2000. Dynamics of vertical leaf nitrogen distribution in a vegetative wheat canopy. Impact on canopy photosynthesis. *Annals of Botany*, 86, 821–831.

Ecarnot, M., Compan, F., & Roumet, P. 2013. Assessing leaf nitrogen content and leaf mass per unit area of wheat in the field throughout plant cycle with a portable spectrometer. *Field Crops Research*, 140, 44–50.

Evans, J.R. 1989. Photosynthesis and nitrogen relationships in leaves of C3 plants. *Oecologia*, 78, 9–19.

Feng, W., Guo, B.-B., Wang, Z.-J., He, L., Song, X., Wang, Y.-H., & Guo, T.-C. 2014. Measuring leaf nitrogen concentration in winter wheat using double-peak spectral reflection remote sensing data. *Field Crops Research*, 159, 43–52.

Feng, W., Yao, X., Zhu, Y., Tian, Y.C., & Cao, W. 2008. Monitoring leaf nitrogen status with hyperspectral reflectance in wheat. *European Journal of Agronomy*, 28, 394–404.

Feng, W., Zhang, H.-Y., Zhang, Y.-S., Qi, S.-L., Heng, Y.-R., Guo, B.-B., Ma, D.-Y., & Guo, T.-C. 2016. Remote detection of canopy leaf nitrogen concentration in winter wheat by using water resistance vegetation indices from *in-situ* hyperspectral data. *Field Crops Research*, 198, 238–246.

Féret, J.-B., François, C., Asner, G.P., Gitelson, A.A., Martin, R.E., Bidel, L.P.R., Ustin, S.L., le Maire, G., & Jacquemoud, S. 2008. PROSPECT-4 and 5: Advances in the leaf optical properties model separating photosynthetic pigments. *Remote Sensing of Environment*, 112, 3030–3043.

Féret, J.-B., Gitelson, A.A., Noble, S.D., & Jacquemoud, S. 2017. PROSPECT-D: Towards modeling leaf optical properties through a complete lifecycle. *Remote Sensing of Environment*, 193, 204–215.

Ferwerda, J.G., Skidmore, A.K., & Mutanga, O. 2005. Nitrogen detection with hyperspectral normalized ratio indices across multiple plant species. *International Journal of Remote Sensing*, 26, 4083–4095.

Filella, I., Serrano, L., Serra, J., & Penuelas, J. 1995. Evaluating wheat nitrogen status with canopy reflectance indices and discriminant analysis. *Crop Science*, 35, 1400–1405.

Gao, B.C., & Goetz, A.F.H. 1994. Extraction of dry leaf spectral features from reflectance spectra of green vegetation. *Remote Sensing of Environment*, 47, 369–374.

Gnyp, M.L., Miao, Y., Yuan, F., Ustin, S.L., Yu, K., Yao, Y., Huang, S., & Bareth, G. 2014. Hyperspectral canopy sensing of paddy rice aboveground biomass at different growth stages. *Field Crops Research*, 155, 42–55.

Guanter, L., Kaufmann, H., Segl, K., Foerster, S., Rogass, C., Chabrillat, S., Kuester, T., Hollstein, A., Rossner, G., & Chlebek, C. 2015. The EnMAP spaceborne imaging spectroscopy mission for earth observation. *Remote Sensing*, 7, 8830–8857.

Hansen, P.M., & Schjoerring, J.K. 2003. Reflectance measurement of canopy biomass and nitrogen status in wheat crops using normalized difference vegetation indices and partial least squares regression. *Remote Sensing of Environment*, 86, 542–553.

He, L., Song, X., Feng, W., Guo, B.-B., Zhang, Y.-S., Wang, Y.-H., Wang, C.-Y., & Guo, T.-C. 2016. Improved remote sensing of leaf nitrogen concentration in winter wheat using multi-angular hyperspectral data. *Remote Sensing of Environment*, 174, 122–133.

Hikosaka, K. 2016. Optimality of nitrogen distribution among leaves in plant canopies. *Journal of Plant Research*, 129, 299–311.

Hilker, T., Coops, N.C., Hall, F.G., Black, T.A., Wulder, M.A., Nesic, Z., & Krishnan, P. 2008. Separating physiologically and directionally induced changes in PRI using BRDF models. *Remote Sensing of Environment*, 112, 2777–2788

Houlès, V., Guérif, M., & Mary, B. 2007. Elaboration of a nitrogen nutrition indicator for winter wheat based on leaf area index and chlorophyll content for making nitrogen recommendations. *European Journal of Agronomy*, 27, 1–11.

Huang, W., Yang, Q., Pu, R., & Yang, S. 2014. Estimation of nitrogen vertical distribution by bi-directional canopy reflectance in winter wheat. *Sensors (Basel)*, 14, 20347–20359.

Inoue, Y., Sakaiya, E., Zhu, Y., & Takahashi, W. 2012. Diagnostic mapping of canopy nitrogen content in rice based on hyperspectral measurements. *Remote Sensing of Environment*, 126, 210–221.

Jacquemoud, S., & Baret, F. 1990. PROSPECT: A model of leaf optical properties spectra. *Remote Sensing of Environment*, 34, 75–91.

Jacquemoud, S., Ustin, S.L., Verdebout, J., Schmuck, G., Andreoli, G., & Hosgood, B. 1996. Estimating leaf biochemistry using the PROSPECT leaf optical properties model. *Remote Sensing of Environment*, 56, 194–202.

Jacquemoud, S., Verhoef, W., Baret, F., Bacour, C., Zarco-Tejada, P.J., Asner, G.P., François, C., & Ustin, S.L. 2009. PROSPECT+SAIL models: A review of use for vegetation characterization. *Remote Sensing of Environment*, 113, S56–S66.

Jay, S., Maupas, F., Bendoula, R., & Gorretta, N. 2017. Retrieving LAI, chlorophyll and nitrogen contents in sugar beet crops from multi-angular optical remote sensing: Comparison of vegetation indices and PROSAIL inversion for field phenotyping. *Field Crops Research*, 210, 33–46.

Kalacska, M., Lalonde, M., & Moore, T.R. 2015. Estimation of foliar chlorophyll and nitrogen content in an ombrotrophic bog from hyperspectral data: Scaling from leaf to image. *Remote Sensing of Environment*, 169, 270–279.

Knyazikhin, Y., Schull, M.A., Stenberg, P., Mottus, M., Rautiainen, M., Yang, Y., Marshak, A., Latorre Carmona, P., Kaufmann, R.K., Lewis, P., Disney, M.I., Vanderbilt, V., Davis, A.B., Baret, F., Jacquemoud, S., Lyapustin, A., & Myneni, R.B. 2013. Hyperspectral remote sensing of foliar nitrogen content. *Proceedings of the National Academy of Sciences of the United States of America*, 110, E185–E192.

Kokaly, R.F. 2001. Investigating a physical basis for spectroscopic estimates of leaf nitrogen concentration. *Remote Sensing of Environment*, 75, 153–161.

Kokaly, R.F., & Clark, R.N. 1999. Spectroscopic determination of leaf biochemistry using band-depth analysis of absorption features and stepwise multiple linear regression. *Remote Sensing of Environment*, 67, 267–287

Ladha, J.K., Tirol-Padre, A., Reddy, C.K., Cassman, K.G., Verma, S., Powlson, D.S., van Kessel, C., de, B.R.D., Chakraborty, D., & Pathak, H. 2016. Global nitrogen budgets in cereals: A 50-year assessment for maize, rice, and wheat production systems. *Scientific Reports*, 6, 19355.

Lamb, D.W., Steyn-Ross, M., Schaare, P., Hanna, M.M., Silvester, W., & Steyn-Ross, A. 2002. Estimating leaf nitrogen concentration in ryegrass (*Lolium* spp.) pasture using the chlorophyll red-edge: Theoretical modelling and experimental observations. *International Journal of Remote Sensing*, 23, 3619–3648.

Lemaire, G., Jeuffroy, M.-H., & Gastal, F. 2008. Diagnosis tool for plant and crop N status in vegetative stage. *European Journal of Agronomy*, 28, 614–624.

Lepine, L.C., Ollinger, S.V., Ouimette, A.P., & Martin, M.E. 2016. Examining spectral reflectance features related to foliar nitrogen in forests: Implications for broad-scale nitrogen mapping. *Remote Sensing of Environment*, 173, 174–186.

Li, D., Cheng, T., Zhou, K., Zheng, H., Yao, X., Tian, Y., Zhu, Y., & Cao, W. 2017. WREP: A wavelet-based technique for extracting the red edge position from reflectance spectra for estimating leaf and canopy chlorophyll contents of cereal crops. *ISPRS Journal of Photogrammetry and Remote Sensing*, 129, 103–117.

Li, F., Miao, Y., Hennig, S.D., Gnyp, M.L., Chen, X., Jia, L., & Bareth, G. 2010. Evaluating hyperspectral vegetation indices for estimating nitrogen concentration of winter wheat at different growth stages. *Precision Agriculture*, 11, 335–357.

Li, F., Mistele, B., Hu, Y., Chen, X., & Schmidhalter, U. 2014. Reflectance estimation of canopy nitrogen content in winter wheat using optimised hyperspectral spectral indices and partial least squares regression. *European Journal of Agronomy*, 52, 198–209.

McClure, W., Crowell, B., Stanfield, D., Mohapatra, S., Morimoto, S., & Batten, G. 2002. Near infrared technology for precision environmental measurements: Part 1. Determination of nitrogen in green- and dry-grass tissue. *Journal of Near Infrared Spectroscopy*, 10, 177–185.

Milton, E.J., Schaepman, M.E., Anderson, K., Kneubühler, M., & Fox, N. 2009. Progress in field spectroscopy. *Remote Sensing of Environment*, 113, S92–S109.

Mistele, B., & Schmidhalter, U. 2008. Estimating the nitrogen nutrition index using spectral canopy reflectance measurements. *European Journal of Agronomy*, 29, 184–190.

Moharana, S., & Dutta, S. 2016. Spatial variability of chlorophyll and nitrogen content of rice from hyperspectral imagery. *ISPRS Journal of Photogrammetry and Remote Sensing*, 122, 17–29.

Mueller, N.D., Gerber, J.S., Johnston, M., Ray, D.K., Ramankutty, N., & Foley, J.A. 2012. Closing yield gaps through nutrient and water management. *Nature*, 490, 254–257.

Mutanga, O., Adam, E., Adjorlolo, C., & Abdel-Rahman, E.M. 2015. Evaluating the robustness of models developed from field spectral data in predicting African grass foliar nitrogen concentration using WorldView-2 image as an independent test dataset. *International Journal of Applied Earth Observation and Geoinformation*, 34, 178–187.

Nguyen, H.T., & Lee, B.W. 2006. Assessment of rice leaf growth and nitrogen status by hyperspectral canopy reflectance and partial least square regression. *European Journal of Agronomy*, 24, 349–356.

Ollinger, S.V., Richardson, A.D., Martin, M.E., Hollinger, D.Y., Frolking, S.E., Reich, P.B., Plourde, L.C. et al. 2008. Canopy nitrogen, carbon assimilation, and albedo in temperate and boreal forests: Functional relations and potential climate feedbacks. *Proceedings of the National Academy of Sciences of the United States of America*, 105, 19336–19341.

Pacheco-Labrador, J., Gonzalez-Cascon, R., Pilar Martin, M., & Riano, D. 2014. Understanding the optical responses of leaf nitrogen in Mediterranean Holm oak (*Quercus ilex*) using field spectroscopy. *International Journal of Applied Earth Observation and Geoinformation*, 26, 105–118.

Peng, S., García, F. V., Laza, R.C., & Cassman, K.G. 1993. Adjustment for specific leaf weight improves chlorophyll meter's estimate of rice leaf nitrogen concentration. *Agronomy Journal*, 85, 987–990.

Peng, S., Buresh, R.J., Huang, J., Zhong, X., Zou, Y., Yang, J., Wang, G. et al. 2010. Improving nitrogen fertilization in rice by site-specific N management: A review. *Agrony for Sustainable Development*, 30, 649-656.

Perry, E.M., & Roberts, D.A. 2008. Sensitivity of narrow-band and broad-band indices for assessing nitrogen availability and water stress in an annual crop. *Agronomy Journal*, 100, 1211–1219.

Piekkielek, W.P., & Fox, R.H. 1992. Use of a chlorophyll meter to predict sidedress nitrogen. *Agronomy Journal*, 84, 59–65.

Qin, Z., Chang, Q., Xie, B., & Shen, J. 2016. Rice leaf nitrogen content estimation based on hysperspectral imagery of UAV in Yellow River diversion irrigation district. *Transactions of the Chinese Society of Agricultural Engineering*, 32, 77–85.

Ramoelo, A., Skidmore, A.K., Schlerf, M., Mathieu, R., & Heitkonig, I.M.A. 2011. Water-removed spectra increase the retrieval accuracy when estimating savanna grass nitrogen and phosphorus concentrations. *ISPRS Journal of Photogrammetry and Remote Sensing*, 66, 408–417.

Schlemmer, M., Gitelson, A., Schepers, J., Ferguson, R., Peng, Y., Shanahan, J., & Rundquist, D. 2013. Remote estimation of nitrogen and chlorophyll contents in maize at leaf and canopy levels. *International Journal of Applied Earth Observation and Geoinformation*, 25, 47–54.

Schlerf, M., Atzberger, C., Hill, J., Buddenbaum, H., Werner, W., & Schüler, G. 2010. Retrieval of chlorophyll and nitrogen in Norway spruce (*Picea abies* L. Karst.) using imaging spectroscopy. *International Journal of Applied Earth Observation & Geoinformation*, 12, 17–26.

Seck, P.A., Diagne, A., Mohanty, S., & Wopereis, M.C.S. 2012. Crops that feed the world 7: Rice. *Food Security*, 4, 7–24.

Shi, T., Wang, J., Liu, H., & Wu, G. 2015. Estimating leaf z crop plants from hyperspectral reflectance. *International Journal of Remote Sensing*, 36, 4652–4667.

Shiferaw, B., Smale, M., Braun, H.-J., Duveiller, E., Reynolds, M., & Muricho, G. 2013. Crops that feed the world 10. Past successes and future challenges to the role played by wheat in global food security. *Food Security*, 5, 291–317.

Siegmann, B., Jarmer, T., Beyer, F., & Ehlers, M. 2015. The potential of pan-sharpened EnMAP data for the assessment of wheat LAI. *Remote Sensing*, 7, 12737–12762.

Spaner, D., Todd, A.G., Navabi, A., Mckenzie, D.B., & Goonewardene, L.A. 2005. Can leaf chlorophyll measures at differing growth stages be used as an indicator of winter wheat and spring barley nitrogen requirements in eastern Canada? *Journal of Agronomy & Crop Science*, 191, 393–399.

Stroppiana, D., Boschetti, M., Brivio, P.A., & Bocchi, S. 2009. Plant nitrogen concentration in paddy rice from field canopy hyperspectral radiometry. *Field Crops Research*, 111, 119–129.

Sun, T., Fang, H., Liu, W., & Ye, Y. 2017. Impact of water background on canopy reflectance anisotropy of a paddy rice field from multi-angle measurements. *Agricultural and Forest Meteorology*, 233, 143–152.

Tang, Y., Huang, J., Cai, S., & Wang, R. 2007. Nitrogen contents of rice panicle and paddy by hyperspectral remote sensing. *Pakistan Journal of Biological Sciences: PJBS*, 10, 4420–4425.

Thenkabail, P.S., Smith, R.B., & Pauw, E.D. 2000. Hyperspectral vegetation indices and their relationships with agricultural crop characteristics. *Remote Sensing of Environment*, 71, 158–182.

Tian, Y.-C., Gu, K.-J., Chu, X., Yao, X., Cao, W.-X., & Zhu, Y. 2014. Comparison of different hyperspectral vegetation indices for canopy leaf nitrogen concentration estimation in rice. *Plant and Soil*, 376, 193–209.

Ustin, S.L. 2013. Remote sensing of canopy chemistry. *Proceedings of the National Academy of Sciences of the United States of America*, 110, 804–805.

Wang, W., Yao, X., Yao, X., Tian, Y., Liu, X., Ni, J., Cao, W., & Zhu, Y. 2012. Estimating leaf nitrogen concentration with three-band vegetation indices in rice and wheat. *Field Crops Research*, 129, 90–98.

Wang, Y., Wang, D., Shi, P., & Omasa, K. 2014. Estimating rice chlorophyll content and leaf nitrogen concentration with a digital still color camera under natural light. *Plant Methods*, 10, 36.

Wang, Z., Skidmore, A.K., Darvishzadeh, R., Heiden, U., Heurich, M., & Wang, T. 2015a. Leaf nitrogen content indirectly estimated by leaf traits derived from the PROSPECT model. *IEEE Journal of Selected Topics in Applied Earth Observations and Remote Sensing*, 8, 3172–3182.

Wang, Z., Skidmore, A.K., Wang, T., Darvishzadeh, R., & Hearne, J. 2015b. Applicability of the PROSPECT model for estimating protein and cellulose plus lignin in fresh leaves. *Remote Sensing of Environment*, 168, 205–218.

Wright, I.J., Reich, P.B., Westoby, M., Ackerly, D.D., Baruch, Z., Bongers, F., Cavender-Bares, J. et al. 2004. The worldwide leaf economics spectrum. *Nature*, 428, 821–827.

Xu, X.G., Zhao, C.J., Wang, J.H., Zhang, J.C., & Song, X.Y. 2014. Using optimal combination method and *in situ* hyperspectral measurements to estimate leaf nitrogen concentration in barley. *Precision Agriculture*, 15, 227–240.

Xue, L.H., Cao, W.X., Luo, W.H., Dai, T.B., & Zhu, Y. 2004. Monitoring leaf nitrogen status in rice with canopy spectral reflectance. *Agronomy Journal*, 96, 135–142.

Yao, X., Huang, Y., Shang, G., Zhou, C., Cheng, T., Tian, Y., Cao, W., & Zhu, Y. 2015. Evaluation of six algorithms to monitor wheat leaf nitrogen concentration. *Remote Sensing*, 7, 14939–14966.

Yao, X., Ren, H., Cao, Z., Tian, Y., Cao, W., Zhu, Y., & Cheng, T. 2014. Detecting leaf nitrogen content in wheat with canopy hyperspectrum under different soil backgrounds. *International Journal of Applied Earth Observation and Geoinformation*, 32, 114–124.

Yao, X., Yao, X., Jia, W., Tian, Y., Ni, J., Cao, W., & Zhu, Y. 2013. Comparison and intercalibration of vegetation indices from different sensors for monitoring above-ground plant nitrogen uptake in winter wheat. *Sensors*, 13, 3109–3130.

Yoder, B.J., & Pettigrew-Crosby, R.E. 1995. Predicting nitrogen and chlorophyll content and concentrations from reflectance spectra (400–2500 nm) at leaf and canopy scales. *Remote Sensing of Environment*, 53, 199–211.

Yu, K., Lenz-Wiedemann, V., Chen, X., & Bareth, G. 2014. Estimating leaf chlorophyll of barley at different growth stages using spectral indices to reduce soil background and canopy structure effects. *ISPRS Journal of Photogrammetry and Remote Sensing*, 97, 58–77.

Yu, K., Li, F., Gnyp, M.L., Miao, Y., Bareth, G., & Chen, X. 2013. Remotely detecting canopy nitrogen concentration and uptake of paddy rice in the Northeast China Plain. *ISPRS Journal of Photogrammetry and Remote Sensing*, 78, 102–115.

Zhu, Y., Tian, Y., Yao, X., Liu, X., & Cao, W. 2007a. Analysis of common canopy reflectance spectra for indicating leaf nitrogen concentrations in wheat and rice. *Plant Production Science*, 10, 400–411.

Zhu, Y., Zhou, D., Yao, X., Tian, Y., & Cao, W. 2007b. Quantitative relationships of leaf nitrogen status to canopy spectral reflectance in rice. *Crop and Pasture Science*, 8, 1077–1085.

7 Optical Remote Sensing of Vegetation Water Content

Colombo Roberto, Busetto Lorenzo, Meroni Michele, Rossini Micol, and Panigada Cinzia

CONTENTS

7.1 INTRODUCTION

The presence of water within the Earth's physical system sustains the environment and the life that depends on it. Plants, animals, and humans rely on the water in the atmosphere and environment, as well as what is in their bodies, to maintain the basic functions of life.

Water is one of the most important factors regulating plant growth and development in ecosystems [1], and it is required for the maintenance of leaf structure and shape, thermal regulation, and photosynthesis. The knowledge of the spatial and temporal variability of vegetation water content is very useful in fire risk assessment (e.g., [2–4]) and detecting vegetation physiological status (e.g., [5–8]) and provides important information for irrigation decisions in agriculture (e.g., [9–11]).

In the last 50 years, several studies have been conducted to provide reliable methods for the estimation of vegetation water content at different scales: leaf, field, airborne, or spaceborne observations. Currently, near real-time measurements of vegetation water content from space are also being developed [12]. Most of the studies exploit spectral data collected in the visible (VIS, 400–750 nm), near infrared (NIR, 750–1300 nm), and short-wave infrared (SWIR, 1300–2500 nm) spectral regions, while experiments using thermal and microwave data for the retrieval of vegetation water content are rare [13,14]. In this paper, different approaches to optical remote sensing (RS) for estimating leaf and canopy water content are presented and discussed.

7.2   LABORATORY AND FIELD MEASUREMENTS OF VEGETATION WATER CONTENT

The amount of water in vegetation (leaves or canopies) can be measured by remote sensing, laboratory analysis, and field surveys. Vegetation water content can be expressed by means of several variables that are functions of different leaf and canopy parameters.

The equivalent water thickness of a leaf (EWT_L) corresponds to the hypothetical thickness of a single layer of water averaged over the whole leaf area (e.g., [15]) and can be computed in laboratory by measuring fresh and dry weights (FW and DW, respectively, g) and the one-sided leaf area (A, cm^2):

$$EWT_L = \frac{FW - DW}{A} \, (g/cm^2 \text{ or cm})$$

In addition to the EWT_L, for RS applications, it is also important to consider the water content per unit of ground area and therefore to measure EWT at the canopy level. For grassland and some crops, canopy EWT (EWT_C) can be directly estimated by measuring the fresh and dry weight of vegetation harvested in a known area (kg/m^2) [16,17] or by scaling the plant water content at the field level using the areal stand density of crops [18]. The amount of water in forest canopies is instead generally obtained by scaling leaf EWT with leaf area index (LAI, m^2/m^2), which is usually estimated indirectly from canopy gap fraction measurements:

$$EWT_C = LAI \times EWT_L \, (kg/m^2)$$

Sometimes, vegetation fractional cover is also used to scale foliar content to canopy EWT [19].

In the framework of fire risk analysis, leaf water content is frequently expressed as the percentage of leaf fresh weight (range between 0 and 100%) or as a percentage of leaf dry weight, namely the fuel moisture content (FMC, %) [20], which may assume values above 100%:

$$FMC = \frac{FW - DW}{DW} = \frac{EWT}{LMA} * 100(\%)$$

FMC is related to two different leaf biochemical parameters that affect leaf optical properties: the EWT_L and the specific leaf dry mass area (LMA), which is the weight of dry matter per leaf unit area.

The water quantities described above are generally not considered driving variables in canopy functioning models. For plant status analysis, relative water content (RWC) and leaf water potential have frequently been investigated from optical RS. Leaf RWC compares the water content of a leaf with the maximum water content at full turgor and can be considered an indicator of vegetation status [21]. RWC can be obtained from laboratory measurements of leaf weight and leaf turgid weight (TW) according to the following expression [22]:

$$RWC = \frac{FW - DW}{TW - DW} * 100(\%)$$

Both FMC and RWC are not directly or physically linked to the absorption processes and do not scale with LAI from the leaf to canopy level. However, leaf RWC is sometimes averaged at the canopy level using information about vegetation cover and height to derive a canopy relative water content with arbitrary units [23].

7.3   EFFECTS OF WATER CONTENT ON SPECTRAL REFLECTANCE

Water absorption features result from the vibrational processes of OH bonds (i.e., small displacements of the atoms about their resting positions) and appear at the overtones of their fundamental frequencies.

Water absorption coefficients are extremely low in the visible part of the electromagnetic spectrum, whereas in the near- and short-wave infrared, four major absorption peaks are present (Figure 7.1). These peaks are located at approximately 975, 1175, 1450, and 1950 nm and increase in magnitude with wavelength.

Reflectance variations due to leaf water content differences can, for example, be appreciated by analyzing simulations conducted with the PROSPECT leaf radiative transfer model [24,25] (see Section 7.4.5). Figure 7.2 shows leaf reflectance and transmittance simulation results for a leaf characterized by six levels of EWT_L.

Visible and NIR reflectance are primarily influenced by chlorophyll content, leaf internal structure, and dry matter content, whereas leaf reflectance decreases with increasing EWT at longer wavelengths of NIR and SWIR spectral regions. Leaf EWT slightly affects reflectance values in the weaker absorption bands of the NIR and strongly influences reflectance in the SWIR at 1450 and 1950 nm [26–30]. Leaf internal structure and leaf dry matter content also affect reflectance values in the NIR and SWIR spectral regions so that their variations may originate differences in leaf reflectance that may be unrelated to water content variations (e.g., [15,21,31]).

In general, leaf reflectance is measured by coupling a spectroradiometer with an integrating sphere. This allows measurement of leaf bihemispherical reflectance and transmittance over a selected spectral region. Very often, leaf stacks rather than single leaves are measured (e.g., [8,32,33]) to determine the so-called infinite reflectance, which is defined as the maximum reflectance of an optically thick medium. This quantity approximates the reflectance of closed deciduous canopies characterized by high LAI in which the effect of soil background and understory on reflectance is very low. It should, however, be noted that infinite reflectance may show an excessive magnitude of NIR reflectance with respect to real canopy spectra and is not a direct representation of the whole-canopy spectrum [34,35].

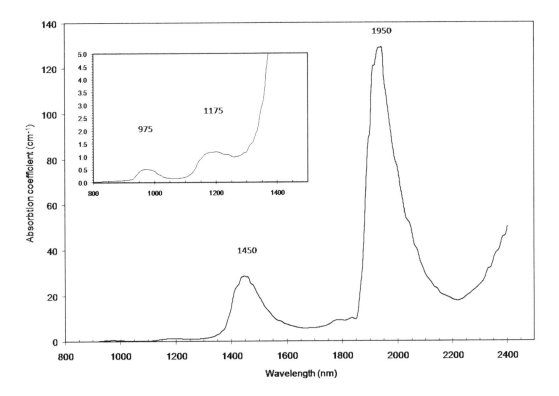

FIGURE 7.1   Specific absorption coefficients of water as used in the PROSPECT model. (Adapted from Feret J.B. et al., *Remote Sensing of Environment*, 112(6), 3030, 2008. [24])

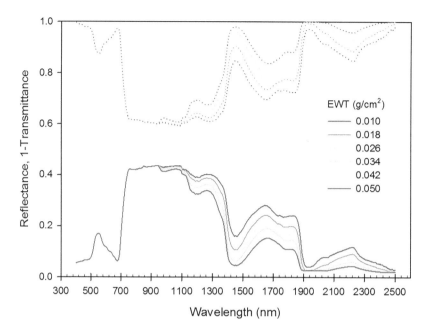

FIGURE 7.2   Example of leaf reflectance (R, continuous lines) and transmittance (plotted as 1-T, dotted lines) spectra, simulated with the PROSPECT model for different levels of EWT.

At the canopy level, spectral variations due to water content can be well appreciated by analyzing PROSAILH (PROSPECT & SAILH) radiative transfer model simulations [25,36,37] (see Section 7.4.5). Figure 7.3 shows the spectral reflectance simulated by PROSAILH for a canopy with different values of LAI and EWT_L.

The visible and the first part of the near-infrared plateau are affected only by LAI variations, whereas wavelengths above 900 nm are influenced by both LAI and EWT_L. A major difficulty when estimating canopy water content from reflectance data at the landscape level comes from the fact that the effect of LAI and background may vary spatially, temporally, and with sensor view

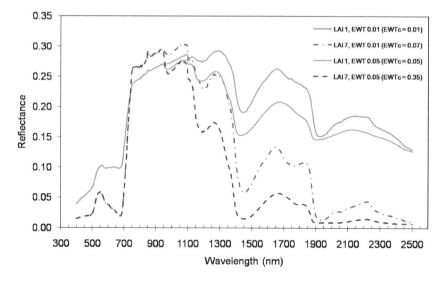

FIGURE 7.3   Canopy spectral reflectance simulated with PROSAILH for different LAI and EWT values.

angle, so these variations may cancel out water-related features in spectral reflectance. Additionally, vegetation water content retrieval from aerial and satellite data requires a very accurate estimation of the reflectance factor starting from the measured at-sensor radiance, so that fine geometric, radiometric, and atmospheric corrections have to be carefully pursued. Since the effect of absorption of water vapor in the atmosphere greatly reduces the energy reaching the ground surface around 1450 and 1950 nm, these water absorption bands are generally not exploited for landscape-level studies. Some signal remains around the weaker water absorption peaks located in the NIR spectral region, but any change in canopy reflectance at the 975 and 1175 nm wavelengths due to variations in water content should be decoupled from changes due to the variations of water vapor in the atmosphere. The depth of penetration for radiation within the canopy differs in the NIR and SWIR spectral regions [23,38,39]. Wavelengths that are weakly absorbed penetrate more deeply into the canopy and are expected to sense a larger portion of the total water content, whereas wavelengths that are strongly absorbed may be sensitive only to the water in the upper layers of the canopy and become quickly saturated.

The spectral regions most suitable for optical RS of vegetation water content at the landscape level are therefore located in the SWIR wavelengths where leaf water shows intermediate absorption (around 1520–1540, 1650, and 2130–2200 nm) and in the weak absorption bands in the NIR region (near 975 and 1150–1260 nm).

7.4 METHODS USED FOR ESTIMATING VEGETATION WATER CONTENT

Several empirical and physically based models have been developed in the last decades to retrieve vegetation water content from remotely sensed data. The possibility of using the different methods strongly depends on the spectral resolution of the available RS data. For example, studies based on the use of low spectral resolution satellite data (e.g., Landsat TM) are usually based on the use of broad-band spectral indices exploiting the wide water absorption features in the SWIR region. High-resolution field spectrometers and the recently developed hyperspectral aerial or satellite sensors allow instead the acquisition of spectral data in hundreds of contiguous spectral bands so that slight changes in reflectance at specific wavelengths affected by water absorption can be detected and analyzed [40]. This allows the use of narrow-band spectral indices also exploiting the weaker absorption regions in the NIR, and also more sophisticated methods such as continuum removal, derivative analysis, and curve-fitting techniques. Moreover, the inversion of physically based models has come to the fore as the most promising technique for retrieving biophysical and biochemical vegetation parameters at the leaf and canopy scale starting from high spectral resolution data.

7.4.1 HYPERSPECTRAL VEGETATION INDICES SENSITIVE TO WATER CONTENT

Spectral indices allow estimation of leaf and canopy water content by means of empirical approaches established using regression techniques with field-measured water content or related variables. Table 7.1 shows the main hyperspectral vegetation indices sensitive to leaf and canopy water content developed in RS studies.

Table 7.1 reports the band center wavelengths used to compute spectral indices sensitive to water content from hyperspectral data. The devices and bandwidths refer instead to the instrument used in the original study where the spectral index was first proposed and to its nominal bandwidth. In different studies focused on computing and comparing spectral indices, band centers and bandwidths may change because of the use of different devices.

In general, 2.5–5 nm bandwidths were used to develop narrow-band indices from spectral data collected from field spectroradiometers (such as GER or ASD FS) or airborne hyperspectral airborne sensors (such as CASI and AISA), and 10–50 nm bandwidths were used for indices based on Airborne Visible/Infrared Imaging Spectrometer (AVIRIS), HYMAP, Probe-1, DAIS, MIVIS, HYPERION,

TABLE 7.1

Summary of Reflectance-Based Indices Developed for Estimating Vegetation Water Content at Different Scales (Leaf L, Ground G, Airborne A, Spaceborne S)

Index	Full Name	Formulation	References	Estimated Variable	Level	Device and Bandwidth
		NIR (750–1300 nm)				
WI	Water Index	ρ_{900}/ρ_{970}	[6]	RWC	G	Spectron, 15 nm
NDWI	Normalized Difference Water Index	$(\rho_{858}-\rho_{1240})/(\rho_{858}+\rho_{1240})$	[41]	EWT$_C$	A	MODIS, 20–35 nm
WI/NDVI	Ratio WI Normalized Difference Vegetation Index	$\dfrac{(\rho_{900}/\rho_{970})}{(\rho_{800}-\rho_{680})/(\rho_{800}+\rho_{680})}$	[42]	FMC	G	Spectron, 15 nm
RDI	Relative Depth Index @1175	$(\rho_{max}-\rho_{min})/\rho_{max}$	[16]	EWT$_C$	G	GER IRIS MK IV, 2, 4 nm
SRWI	Simple Ratio Water Index	ρ_{858}/ρ_{1240}	[43]	EWT$_L$	S	MODIS, 20–35 nm
R975	Ratio @975	$2\overline{\rho_{960-990}}/(\overline{\rho_{920-940}}+\overline{\rho_{1090-1100}})$	[44]	FMC	L	ASD FieldSpec Pro Fr, 1.4, 2 nm
		SWIR (1300–2500 nm)				
NDII	Normalized Difference Infrared Index	$(\rho_{820}-\rho_{1650})/(\rho_{820}+\rho_{1650})$	[45]	EWT$_C$	S	TM, 140, 195 nm
R5/R7	Ratio of TM Band 5 to Band 7	ρ_{1650}/ρ_{2218}	[46]	EWT$_L$	A	ATM 195, 340 nm
LWCI	Leaf Water Content Index	$\dfrac{-\log(1-(\rho_{820}-\rho_{1600}))}{-\log(1-(\rho_{820,FT}-\rho_{1600,FT}))}$	[47]	RWC	L	MK I TM, 140, 195 nm
MSI	Moisture Stress Index	ρ_{1600}/ρ_{820}	[48,21]	RWC, EWT$_L$	L	GER VIRIS, 3, 12 nm
DRI	Datt Reflectance Index	$(\rho_{816}-\rho_{2218})/(\rho_{816}+\rho_{2218})$	[33]	EWT$_L$	L	GER IRIS MK IV, 2, 4 nm
GVMI	Global Vegetation Moisture Index	$\dfrac{(\rho_{820,rect}+0.1)-(\rho_{1600}+0.02)}{(\rho_{820,rect}+0.1)+(\rho_{1600}+0.02)}$	[28]	EWT$_C$	S	SPOT VGT 20, 170 nm
RDI1450	Relative Depth Index @1450	$(\rho_{max}-\rho_{min})/\rho_{max}$	[49]	EWT$_L$	G	ASD FieldSpec Pro Fr, 1.4, 2 nm
NDWI2130	Normalized Difference Water Index @2130	$(\rho_{858}-\rho_{2130})/(\rho_{858}+\rho_{2130})$	[50]	EWT$_C$	S	MODIS, 24, 50 nm
NMDI	Normalized Multi-band Drought Index	$\dfrac{\rho_{858}-(\rho_{1640}-\rho_{2130})}{\rho_{858}+(\rho_{1640}-\rho_{2130})}$	[51]	EWT$_L$	S	MODIS, 24, 50 nm
MSI/SR	Ratio MSI/Simple Ratio	$\dfrac{\rho_{1600}}{\rho_{820}}\Big/\dfrac{\rho_{895}}{\rho_{675}}$	[52]	EWT$_L$	A	MIVIS, 2, 9 nm

and MODIS sensors, while bandwidths larger than 100 nm were used to compute broad-band spectral indices starting from satellite data such as Landsat TM, SPOT VEGETATION, ASTER, and NOAA AVHRR. Broad-band indices can, in any case, be computed starting from data acquired by hyperspectral sensors, either by selecting the bands roughly corresponding to the band centers of the sensor used to develop the index or by resampling the data to a lower spectral resolution. On the contrary, the computation of spectral indices developed from high spectral resolution sensors on low-resolution data (for those indices where the band centers coincide) may lead to erroneous results, particularly when the spectral index considered exploits the use of the weak and narrow water absorption features in the NIR region. We believe that, more than bandwidth, the main limitations connected with low spectral resolution sensors are related to the reduced number of bands that may prevent the selection of the best spectral regions for index computation. It is difficult to identify the ideal instrument bandwidth that should be used for the retrieval of vegetation water content from spectral indices, since no studies have explicitly and systematically tested this issue. Some studies show that, in general, spectral indices computed with bandwidths ranging from 2.5 to 50 nm have similar performances for leaf water content and other biochemical compound estimation, while the use of broader bandwidths may hamper the accuracy of the estimates [32,53,54]. However, spectral resolution may not be critical when the investigated spectral feature is quite broad. Instead, the accuracy of leaf and canopy water content retrieval can be more sensitive to the signal-to-noise ratio; thus, resampling hyperspectral data to a lower spectral resolution (e.g., 20 nm) may be beneficial because it suppresses part of the instrument noise.

7.4.1.1 Near-Infrared-Based Spectral Indices

The ability of the water index (WI) to evaluate physiological status and water content is documented in several studies exploiting leaf, ground, and airborne measurements [6,55–59]. Peñuelas et al. [6,42] showed that WI closely tracks changes in RWC, especially when there are no important changes in LAI and architectural canopy parameters. To minimize structural effects and therefore maximize sensitivity to water content, Peñuelas et al. [42] and Piñol et al. [60] successfully tested the ratio of WI with the normalized difference vegetation index (NDVI) to estimate FMC in Mediterranean ecosystems from field spectroradiometer measurements. At the landscape level, by using AVIRIS imagery, Serrano et al. [23] identified canopy structure as the main source of variability in WI and other spectral indices sensing water content, although they also appeared to respond to canopy relative water content. Serrano et al. [23] showed that NIR-based spectral indices such as WI and normalized difference water index (NDWI) were more sensitive to changes in canopy RWC than those formulated using SWIR bands and curve-fitting techniques (Section 7.4.4). Sims and Gamon [61], using the GER 2600 instrument at ground level, also showed that water spectral indices formulated at 960 and 1180 mm have a strong correlation with EWT_C and provide better correlation than that obtained by using indices based on SWIR wavelengths.

NDWI was originally developed to retrieve vegetation water content from quasi-hyperspectral spaceborne measurements (e.g., MODIS) and successfully tested on imaging data acquired with AVIRIS [41]. The 860 and 1240 nm wavelengths, which compose NDWI, are located in the high reflectance plateau where the contribution of vegetation scattering to reflectance is similar. NDWI should also be insensitive to foliar dry matter biochemical compounds (lignin, cellulose, and protein), which typically absorb at 1500–2500 nm (e.g., [32]). Several studies have demonstrated the ability of this index to estimate leaf and canopy EWT using airborne hyperspectral sensors and MODIS data (e.g., [23,52,62]) and for estimating FMC at leaf, ground [63–65], and satellite data (e.g., [66,67]). To estimate EWT_C at ground level, Rollin and Milton [16] developed the relative depth index (RDI) at the water absorption feature centered around 1150 nm in a way that the employed reference wavelengths (at the shoulder and bottom of the absorption well) are so close that it is possible to assume that the contribution of other canopy factors is constant. RDI was found to be well correlated with canopy EWT as the first derivative (see Section 7.4.3) and insensitive to data smoothing.

7.4.1.2 Short-Wave Infrared-Based Spectral Indices

Spectral bands outside the main water vapor-absorbing bands in the SWIR have been investigated for a long time, starting from the pioneering studies of Tucker [68], who identified the 1550–1750 nm spectral range as the best suited for monitoring vegetation water content. The leaf water content index (LWCI), moisture stress index (MSI), normalized difference infrared index (NDII), and SWIR ratio were specifically designed for satellite application, mainly exploiting broad-bands 5 (1550–1750 nm) and 7 (2080–2350 nm) of the Landsat TM. Several studies exploited broad-band indices and narrow-band spectroscopy for plant status analysis at different scales. Results of these studies are somewhat controversial because in certain cases spectral indices were found to be strongly related to RWC and leaf water potential, while other studies found no significant statistical evidence of correlation (e.g., [8,11,69–72]). Hunt and Rock [48] and Hunt [73] have shown that the MSI computed at the leaf and landscape level by means of Daedalus Thematic Mapper Airborne Simulator data performed better for the estimation of leaf EWT, EWT_C, and LAI, rather than for the estimation of vegetation status indicators. Satellite RS studies showed that SWIR-based indices, such as NDII, allow successful tracking of seasonal variability of EWT_C in crops and forest ecosystems (e.g., [50,74–76]) and of FMC in grassland and shrubland landscapes [4,13,77]. In the context of soil moisture experiments, Anderson et al. [18] and Jackson et al. [78] tested SWIR-based indices (and greenness indices, such as NDVI) for monitoring seasonal EWT_C in agricultural areas by means of ground reflectance and multitemporal Landsat data, showing that spectral indices sensitive to water content are less susceptible to saturation than NDVI at high levels of LAI.

7.4.1.3 Greenness Indices

Indirect effects of water content were found using a visible spectral range at 400 nm, in the red-edge region at 680–780 nm [79,80], and on spectral vegetation indices expected to respond to canopy greenness such as NDVI [81]. Greenness indices derived from airborne measurements and coarse resolution satellite data have been successfully used to track seasonality of EWT_C and FMC in grassland and savanna ecosystems, while poor results were in general found for shrubs and forest environments [4,82–85].

In general, NIR- and SWIR-based spectral indices are more efficient than greenness indices in estimating vegetation water content. The correlation between greenness indices and EWT_C or FMC is spurious and due to the strong correlation between NDVI and LAI and to plant chlorophyll content, which are in turn affected by plant water status.

7.4.2 Absorption-Band-Depth Analysis and Continuum-Removed Spectral Indices

These methods have largely been exploited for estimating mineralogical composition in geology and soil science and only fairly recently employed in vegetation analysis. Kokaly and Clark [32] proposed a methodology to standardize laboratory-based NIR spectroscopy measurements to estimate vegetation biochemical parameters. This method benefits from continuum removal, calculation of band depth, normalization, and statistical methods to derive foliar concentration of various biochemical constituents. Briefly, the values associated with a wavelength are expressed as band depth normalized to the waveband at the center of the absorption feature or as the area of the absorption feature, then related to water content by means of multiple linear regression. This approach has been found accurate for estimating water content at the leaf level; it performs better than standard first derivative analysis and is promising for RS application at the canopy level [86].

Spectral indices computed from continuum-removed spectra have been used to estimate FMC at the leaf and ground level [44,87]. Colombo et al. [52] found that continuum-removed spectral indices based on absorption features around 1200 nm allow good estimates of EWT_C at the landscape level by means of airborne Multispectral Infrared and Visible Imaging Spectrometer (MIVIS) data.

Similar to relative depth indices or derivative techniques, the good performances of continuum-removed spectral indices based on water absorption features are probably due to their ability to minimize the effects of the background signal and albedo variations on reflectance spectra, thus allowing a better estimation of canopy water content.

7.4.3 Derivative Techniques

Derivative analysis of absorption spectra is a well-established technique in analytical chemistry, where it is used to suppress background signals and resolve overlapping spectral features [88,89]. These techniques are usually based on the analysis of the relationships between the slope of the reflectance curve in the proximity of the main water absorption bands and leaf or canopy water content. Derivative techniques have been less frequently used than narrow-band spectral indices for estimating vegetation water content, although they appear to be less affected by leaf and canopy structure variations and can be advantageous in standardizing for the overall level of reflectance. For example, Danson et al. [15] found that the first derivative of the reflectance spectrum at 1450 nm (measured at the leaf level by a GER IRIS Mark IV specroradiometer) is closely correlated with leaf water content and insensitive to leaf structure. More recently, Kumar [53] also found a statistically significant correlation between FMC and first derivative spectra at the leaf level. At ground level, Rollin and Milton [16] found fairly good correlations between the first derivative at 1156 nm and EWT_C in grasslands. Clevers et al. [90] recently showed that the spectral derivatives for wavelengths positioned on the right slope of the water absorption feature at 970 nm (over the 1015–1050 nm spectral interval) may be successfully used to estimate EWT_C of grassland ecosystems from ASD field measurements and airborne HYMAP reflectance data and provide better results than NIR-based spectral indices and continuum-removal techniques.

7.4.4 Fitting of the Water Absorption Spectra

This approach was originally developed to simultaneously retrieve the water vapor and liquid water from AVIRIS data [81,91]. Absorption due to the presence of canopy liquid water can be separated from water vapor absorption in the atmosphere owing to an approximately 40 nm shift toward longer wavelengths for liquid water (e.g., [92]). The equivalent water thickness can be estimated by combining the MODTRAN radiative transfer model and a simple vegetation reflectance model and using spectral fitting techniques (e.g., [81]). Twenty AVIRIS spectral bands between 865 and 1085 nm are generally used to fit measured radiance as a function of the equivalent transmittance spectrum of a water sheet modeled using a Beer-Lambert model modified to account for the scattering contribution [23,81,91–96]. Assuming that reflectance can be substituted for transmittance [26], the EWT is calculated as the slope of the plot of the natural log of reflectance as a function of known absorption coefficients for pure water in a selected spectral region (e.g., [19,61,97]). It is interesting to note that water thickness derived with such a technique has been proved to be sensitive to canopy structure and very useful for estimating LAI, with better performance than those found using traditional NDVI (e.g., [23,98]). Curve fitting has also been implemented for application with the Probe-1 airborne hyperspectral sensor over the 850–1200 nm range [99] and with multi-angular HYMAP measurements over the 860–1320 nm range [100].

In general, studies comparing the performance of curve-fitting techniques with spectral indices sensitive to water content and other techniques at the ground, aerial, and satellite levels have demonstrated good agreement between different methods and with field data (e.g., [19,92,95,96,99,101]). Roberts et al. [101] compared the estimates of EWT from HYPERION and AVIRIS reflectance across two wavelength regions (865–1088 and 1088–1200 nm) and found that the measures based on the 1200 nm band are more effective than that based on the 980 nm liquid water band to map canopy water content across a wide range of vegetation types.

7.4.5 Radiative Transfer Models

Radiative transfer models simulate reflectance and transmittance of leaves or canopies by describing the interactions (i.e., absorption and scattering) of solar radiation with the different vegetation elements. At the leaf level, reflectance is generally simulated as a function of the main biochemical and structural constituents of the leaf (e.g., chlorophyll, dry matter, and water content). Several leaf optical models exist, and they use different approaches for broadleaf and needleleaf species (e.g., [25,102,103]). At the canopy level, reflectance is instead simulated as a function of leaf reflectance and canopy structural and biophysical parameters (e.g., LAI, fractional cover), also taking into account factors such as background reflectance and viewing and illumination geometry. Canopy radiative transfer models differ in the description of the radiative regime within the canopy, from one-dimensional formulation where the canopy is described as a sparse homogeneous medium to three-dimensional descriptions of heterogeneous canopies (see [104]). In the framework of vegetation water content studies, leaf and canopy radiative transfer models have been primarily used in performing sensitivity analysis, optimizing/designing spectral vegetation indices, and retrieving leaf and canopy EWT from model inversion on RS data.

7.4.5.1 Sensitivity Analysis Studies on Reflectance and Spectral Indices

Sensitivity analysis is used to quantify the contribution of canopy biophysical and biochemical properties to canopy reflectance. This makes it possible to analyze the magnitude of reflectance changes as a function of vegetation water content, to understand its interactions with other leaf and canopy parameters, and to select appropriate wavelengths and spectral indices suitable for the retrieval of leaf and canopy water content. For example, the sensitivity studies performed using the PROSPECT model by Ceccato et al. [28] and Bowyer and Danson [29] showed that the stronger SWIR absorption bands around 1450, 1940, and 2480 nm are more sensitive to leaf water content variations than the weaker water absorption bands of the NIR region. Model simulations have underscored the important contribution of dry matter content to reflectance around the water absorption feature at 970 nm (which may hamper the estimation of EWT_L with the WI spectral index) and the rather high stability at 860 nm at changing of water content (this explains why this wavelength is often used as a reference for formulating spectral indices sensitive to water content). Asner [105], using the SAIL model [36,37], showed the deepening of the two water absorption features within the NIR as LAI increased and their stability at changes of mean leaf angle. The sensitivity study by Bacour et al. [106], performed using the PROSAIL model, showed that EWT_L explained 50% of the total variance in the SWIR spectral region but that this contribution is reduced in favor of LAI in the regions where the specific absorption of pure liquid water is higher (i.e., 1950 and 2500 nm). Bowyer and Danson [29] carried out a canopy sensitivity analysis using PROSAILH and PROGEOSAIL (PROSPECT & GEOSAIL [107]), showing that EWT_L has a strong influence only at the longer wavelengths of the NIR and in the SWIR spectral region, although the importance of EWT_L and LAI differs depending upon which model is used.

Canopy radiative transfer models have also been used to evaluate the performance of known or proposed spectral vegetation indices sensitive to water content. This process involves the use of physical models to simulate a set of reflectance spectra corresponding to canopies characterized by different water content and other leaf and canopy parameters. For example, Ceccato et al. [28] optimized a new water spectral index by exploiting the SWIR spectral region able to minimize various biophysical factors, geometry of observation, and atmospheric conditions, which was recently successfully applied in the estimation of EWT_C from SPOT-VEGETATION data [108]. Dawson et al. [19] showed the potential of LIBERTY [102] and FLIGHT [109] models in simulating leaf and canopy reflectance and found that NIR-based spectral indices were closely related to leaf and canopy EWT. The modeling analysis conducted by Zarco-Tejada et al. [62] also demonstrated that NIR-based spectral indices (MODIS NDWI and the simple ratio water index [SRWI]) efficiently respond to the change in EWT in the LAI range 4 and 10 m^2/m^2, while at lower and higher LAI values, background and saturation effects dominate the spectral indices. Colombo et al. [52] used PROSAILH to evaluate the sensitivity of double ratio indices to leaf EWT and their capability to minimize LAI effects at the landscape level with airborne hyperspectral data. Simulation results underlined that NIR- and SWIR-based double ratio indices are sensitive to leaf EWT but are strongly influenced by LAI and soil background in sparse canopy conditions. Finally, other studies used PROSAILH to develop spectral indices sensitive to water content that minimize disturbing influences of soil background in sparse coverage conditions [110,111].

7.4.5.2 Retrieval of Vegetation Water Content from Model Inversion

The inversion of a radiative transfer model consists of finding the model parameters (i.e., model input) that provide a "good match" between the modeled (i.e., model output) and observed reflectance (for a review of radiative transfer model inversion on RS data, see Kimes et al. [112]). Model inversion allows the estimation of both leaf and canopy properties from RS data and thus allows estimation of EWT_L and EWT_C (e.g., [52,62,95,108,113–120]).

At the leaf level, uncertainties in the estimation of leaf EWT have been demonstrated to be caused by the difficulties of radiative transfer models to completely decouple leaf EWT from leaf structure and dry matter parameters [25,121,122]. At the canopy level, the estimation of leaf water content and canopy structural parameters is also complicated by the coupling between the quantity of leaf-absorbing material and vegetation structure/architecture, so that some studies suggest that EWT_C may be more conveniently estimated from canopy reflectance data than from leaf EWT. Among the few studies that inverted radiative transfer models starting from reflectance computed from airborne hyperspectral sensors, Kötz et al. [117] found reasonable accuracy in both canopy structure and foliage water content estimates in heterogeneous coniferous forest canopies using the GEOSAIL and FLIGHT models on DAIS data. More recently, Colombo et al. [52] also showed that PROSAILH can be successfully inverted for the retrieval of both leaf and canopy EWT in poplar plantations starting from MIVIS data. New approaches that combine hyperspectral data and light detection and ranging (LiDAR) technologies or microwave information should make it possible to obtain detailed observations of vegetation and provide accurate estimation of biophysical/biochemical parameters, canopy height, and other 3D structural parameters useful in different applications. For example, LiDAR data can be used to estimate canopy structural parameters that may help to stabilize the radiative transfer model inversion to improve the retrieval of leaf EWT with imaging spectroscopy [123].

It is important to underline that for applications at the regional and global level, the inversion of radiative transfer models for vegetation water content estimation requires an appropriate parameterization of the structural parameters of the different ecosystems and the use of appropriate regularization techniques to solve the ill-posed problem in order to obtain stable and reliable solutions [96,124,125]. Detailed land cover classifications, which make it possible to account for mixed vegetation composition, are moreover necessary to accurately determine vegetation water content, especially when using medium to coarse spatial resolution satellite imageries.

7.4.6 OTHER TECHNIQUES

In addition to the methods described above, other techniques have been developed to estimate vegetation water content from remotely sensed data. For example, artificial neural networks have been successfully used to estimate EWT from leaf spectral data and AVIRIS and MODIS imageries in different ecosystems (e.g., [126–130]). Dawson et al. [130] used an artificial neural network to estimate the concentration of biochemical parameters in stacked fresh slash pine needles. Five wavelengths, located in correspondence to absorption features due to leaf water, lignin-cellulose, and nitrogen concentrations, were selected as input for training an artificial neural network, which was then used to successfully estimate FMC and the other biochemical compounds. Pinzon et al. [131] proposed the use of hierarchical foreground/background analysis to detect leaf water content.

This approach is based on a nonlinear spectral mixing technique that makes it possible to highlight sharp absorption features that may be directly related to the desired property (i.e., water content). More recently, Li et al. [132] proposed a new technique based on a genetic algorithm and partial least-squares regression to estimate vegetation water content from leaf spectral data and AVIRIS imageries.

7.5 SUMMARY AND CONCLUSIONS

Measuring vegetation water content remotely is an appealing prospect, but also a challenging one. Here, we have reviewed progress toward this goal and discussed the main approaches and techniques used to estimate vegetation water content from remotely sensed data at the leaf, ground, aircraft, and satellite level. The methods for retrieval of vegetation water content from optical RS have made much progress in the last decades. In particular, the increasing availability of high-resolution field spectrometers and hysperspectral imaging offers the possibility of exploiting several techniques to accurately estimate vegetation water content and evaluate its spatial and temporal variability for different ecosystems.

RS data are better suited for the estimation of leaf and canopy EWT, rather than FMC and RWC. At the leaf level, studies that explicitly tested the performances of different vegetation indices or employed radiative transfer models for performing sensitivity analysis have shown that the stronger absorption bands in the SWIR are more responsive to water content variations than the weaker water absorption bands of the NIR spectral region. At the canopy level, it appears that spectral indices based on reflectance around 1175 nm and other techniques exploiting contiguous spectral bands within this water absorption peak are better suited for the prediction of water content than that based on 975 nm and the longer wavelengths in the SWIR. In addition to other methods, canopy reflectance models represent an essential tool in the analysis of optical data, and their inversion has been demonstrated to be useful in producing spatial and temporal maps of both leaf and canopy water content.

Most studies indicate that the reflectance signal near the main water absorption bands is largely influenced by leaf structure and LAI variations and, over partially vegetated areas, also by soil and background reflectance. Therefore, prediction of vegetation water content remains a challenge when both leaf structure and canopy architecture vary within the image. Issues associated with scaling measurements from leaf to canopy levels and the existence of major covariances between vegetation structure and water content make the deconvolution of these vegetation properties quite challenging. Two developments show great potential for minimizing canopy effects, such as the availability of physically based reflectance models that contain explicit descriptions of canopy structure and leaf biochemical compounds and the possibility of simultaneous LiDAR measurements that could be used to measure and minimize the influence of canopy structure and architecture on the acquired spectra.

In summary, RS is a powerful tool that provides accurate estimations of vegetation water content and offers the only possibility of generating maps at different spatial and temporal scales that can be assimilated in ecological studies that include vegetation as a dynamic component.

REFERENCES

1. Kramer, P. J., & Boyer J. S., *Water Relations in Plants and Soils*, Academic Press, San Diego, CA, 1995.
2. Nelson, R. M., Water relations of forest fuels. In E. A. Johnson & K. Miyanishi (Eds.), *Forest Fires: Behavior and Ecological Effects*, Academic Press, San Diego, CA, pp. 79–149, 2001.
3. Carlson, J. D., & Burgan, R. E., Review of user needs in operational fire danger estimation: The Oklahoma example. *International Journal of Remote Sensing*, 24, 1601–1620, 2003.
4. Chuvieco, E., Cocero, D., Aguado, I., Palacios-Orueta A., & Prado, E., Improving burning efficiency estimates through satellite assessment of fuel moisture content. *Journal of Geophysical Research-Atmospheres*, 109, 1–8, 2004.

5. Carter, G. A., Responses of leaf spectral reflectance to plant stress. *American Journal of Botany*, 80, 239–243, 1993.

6. Peñuelas, J., Filella, I., Biel, C., Serrano, L., & Save, R., The reflectance at the 950–970 nm region as an indicator of plant water status. *International Journal of Remote Sensing*, 14, 1887–1905, 1993.

7. Peñuelas, J., Gamon, J. A., Fredeen, A. L., Merino, J., & Field, C. B., Reflectance indices associated with physiological changes in nitrogen and water limited sunflower leaves. *Remote Sensing of Environment*, 48, 135–146, 1994.

8. Stimson, H. C., Breshears, D. D., Ustin, S. L., & Kefauver, S. C., Spectral sensing of foliar water conditions in two co-occurring conifer species: *Pinus edulis* and *Juniperus monosperma*. *Remote Sensing of Environment*, 96, 108–118, 2005.

9. Begg, J. E., & Turner, N.C., Crop water deficits. *Advances in Agronomy*, 28, 161–217, 1976.

10. Strachan, I. B., Pattey, E., & Boisvert, J. B., Impact of nitrogen and environmental conditions on corn as detected by hyperspectral reflectance. *Remote Sensing of Environment*, 80, 213–224, 2002.

11. Dzikiti, S., Verreynne, J. S., Stuckens, J., Strever, A., Verstraeten, W. W., Swennen, R., & Coppin, P., Determining the water status of Satsuma mandarin trees [Citrus Unshiu Marcovitch] using spectral indices and by combining hyperspectral and physiological data. *Agricultural and Forest Meteorology*, 150(3), 369–379, 2010.

12. Hao, X., & Qu, J. J. Retrieval of real-time live fuel moisture content using MODIS measurements. *Remote Sensing of Environment*, 108(2), 130–137, 2007.

13. Chuvieco, E., Riaño, D., Aguado, I., & Cocero, D., Estimation of fuel moisture content from multitemporal analysis of Landsat Thematic Mapper reflectance data: Applications in fire danger assessment. *International Journal of Remote Sensing*, 23(11), 2145–2162, 2002.

14. Notarnicola, C., & Posa, F., Inferring vegetation water content from C and L band images. *IEEE Transactions on Geoscience and Remote Sensing*, 45(10), 3165–3171, 2007.

15. Danson, F. M., Steven, M. D., Malthus, T. J., & Clark, J. A., High-spectral resolution data for determining leaf water content. *International Journal of Remote Sensing*, 13(3), 461–470, 1992.

16. Rollin, E. M., & Milton, E. J., Processing of high spectral resolution reflectance data for the retrieval of canopy water content information. *Remote Sensing of Environment*, 65, 86–92, 1998.

17. Clevers J. G. P. W., Kooistra L., & Schaepman M. E., Using spectral information from the NIR water absorption features for the retrieval of canopy water content. *International Journal of Applied Earth Observation and Geoinformation*, 10(3), 388–397, 2008.

18. Anderson, M. C., Neale, C. M. U., Li, F., Norman, J. M., Kustas, W. P., Jayanthi, H. et al. Upscaling ground observations of vegetation water content, canopy height, and leaf area index during SMEX02 using aircraft and Landsat imagery. *Remote Sensing of Environment*, 92, 447–464, 2004.

19. Dawson, T. P., Curran, P. J., North, P. R. J., & Plummer, S. E., The propagation of foliar biochemical absorption features in forest canopy reflectance: A theoretical analysis. *Remote Sensing of Environment*, 67, 147–159, 1999.

20. Burgan, R. E., Use of remotely sensed data for fire danger estimation. EARSeL. *Advances in Remote Sensing*, 4(4), 1–8, 1996.

21. Ceccato, P., Flasse, S., Tarantola, S., Jacquemoud, S., & Gregoire, J. M., Detecting vegetation leaf water content using reflectance in the optical domain. *Remote Sensing of Environment*, 77, 22–33, 2001.

22. Kramer, P. J., *Plant and Soil Water Relationships: A Modern Synthesis*, McGraw-Hill, New York, 482 p, 1969.

23. Serrano, L., Ustin, S. L., Roberts, D. A., Gamon, J. A., & Peñuelas, J., Deriving water content of chaparral vegetation from AVIRIS data. *Remote Sensing of Environment*, 74, 570–581, 2000.

24. Feret, J. B., François, C., Asner, G.P., Gitelson, A. A., Martin, R. E., Bidel, L. P. R., Ustin, S. L., le Maire, G., & Jacquemoud, S., PROSPECT-4 and -5: Advances in the leaf optical properties model separating photosynthetic pigments. *Remote Sensing of Environment*, 112(6), 3030–3043, 2008.

25. Jacquemoud, S., & Baret, F., PROSPECT: A model of leaf optical properties spectra. *Remote Sensing of Environment*, 34, 75–91, 1990.

26. Knipling, E. B., Physical and physiological basis for the reflectance of visible and near-infrared radiation from vegetation. *Remote Sensing of Environment*, 1, 155–159, 1970.

27. Thomas, J. R., Namken, L. N., Oerther, G. F., & Brown, R. G., Estimating leaf water content by reflectance measurements. *Agronomy Journal*, 63, 845–847, 1971.

28. Ceccato, P., Gobron, N., Flasse, S., Pinty, B., & Tarantola, S., Designing a spectral index to estimate vegetation water content from remote sensing data: Part 1, Theoretical approach. *Remote Sensing of Environment*, 82, 188–197, 2002a.

29. Bowyer, P., & Danson, F. M., Sensitivity of remotely sensed spectral reflectance to variation in live fuel moisture content. *Remote Sensing of Environment*, 92, 297–308, 2004.

30. Seelig, H. D., Adams, W. W., Hoehn, A., Stodieck, L. S., Klaus, D. M., & Emery, W. J., Extraneous variables and their influence on reflectance-based measurements of leaf water content. *Irrigation Science*, 26(5), 407–414, 2008.

31. Aldakheel, Y. Y., & Danson, F. M., Spectral reflectance of dehydrating leaves: Measurements and modelling. *International Journal of Remote Sensing*, 18, 3683–3690, 1997.

32. Kokaly, R. F., & Clark, R. N., Spectroscopic determination of leaf biochemistry using band-depth analysis of absorption features and stepwise multiple linear regression. *Remote Sensing of Environment*, 67, 267–287, 1999.

33. Datt, B., Remote sensing of water content in eucalyptus leaves. *Australian Journal of Botany*, 47, 909–923, 1999.

34. Myneni, R.B., Ross, J., & Asrar, G., A review on the theory of photon transport in leaf canopies in slab geometry. *Agricultural and Forest Meteorology*, 45, 1–153, 1989.

35. Kokaly, R. F., Asner, G. P., Ollinger, S. V., Martin, M. E., & Wessman, C. A., Characterizing canopy biochemistry from imaging spectroscopy and its application to ecosystem studies. *Remote Sensing of Environment*, 113, 78–91, 2009.

36. Verhoef, W., Light scattering by leaf layers with application to canopy reflectance modeling: The SAIL model. *Remote Sensing of Environment*, 16, 125–141, 1984.

37. Kuusk, A., A fast, invertible canopy reflectance model. *Remote Sensing of Environment*, 51, 342–350, 1995.

38. Bull, C. R., Wavelength selection for near-infrared reflectance. *Journal of Agricultural Engineering Research*, 49, 113–125, 1991.

39. Lillesaeter, O. Spectral reflectance of partly transmitting leaves: laboratory measurements and mathematical modeling. *Remote Sensing of Environment*, 12, 247–254, 1982.

40. Goetz, A. F. H., Three decades of hyperspectral remote sensing of the Earth: A personal view. *Remote Sensing of Environment*, 113, 5–16, 2009.

41. Gao, B. C., NDWI, a normalized difference water index for remote sensing of vegetation liquid water from space. *Remote Sensing of Environment*, 58, 257–266, 1996.

42. Peñuelas, J., Piñol, J., Ogaya, R., & Filella, I., Estimation of plant water concentration by the reflectance water index (R900/R970). *International Journal of Remote Sensing*, 18, 2869–2875, 1997.

43. Zarco-Tejada, P. L. J., & Ustin, S. L., Modeling canopy water content for carbon estimates from MODIS data at land EOS validation sites. *International Geoscience and Remote Sensing Symposium*, 342–344, 2001.

44. Pu, R., Ge, S., Kelly, N. M., & Gong, P., Spectral absorption features as indicators of water status in coast live oak (*Quercus agrifolia*) leaves. *International Journal of Remote Sensing*, 24, 1799–1810, 2003.

45. Hardisky, M. A., Lemas, V., & Smart, R. M., The influence of soil salinity, growth form, and leaf moisture on the spectral reflectance of *Spartina alterniflora* canopies. *Photogrammetric Engineering and Remote Sensing*, 49, 77–83, 1983.

46. Elvidge, C. D., & Lyon, R. J. P., Estimation of the vegetation contribution to the 1.65/2.22 mm ratio in air-borne thematic-mapper imagery of the Virginia Range, Nevada. *International Journal of Remote Sensing*, 6, 75–88, 1985.

47. Hunt, E. R., Jr., Rock, B. N., & Nobel, P. S., Measurement of leaf relative water content by infrared reflectance. *Remote Sensing of Environment*, 22, 429–435, 1987.

48. Hunt, E. R., & Rock, B. N., Detection of changes in leaf water content using near and middle-infrared reflectances. *Remote Sensing of Environment*, 30, 43–54, 1989.

49. Zhao, C. J., Zhou, Q. F., Wang, J. H., & Huang, W. J., Band selection for relative depth indices (RDI) in analyzing wheat water status under field conditions. *International Journal of Remote Sensing*, 25, 2575–2584, 2004.

50. Chen, D., Huang, J. F., & Jackson, T.J., Vegetation water content estimation for corn and soybeans using spectral indices derived from MODIS near- and short-wave infrared bands. *Remote Sensing of Environment*, 98, 222–236, 2005.

51. Wang, L., & Qu, J. J., NMDI: A normalized multi-band drought index for monitoring soil and vegetation moisture with satellite remote sensing. *Geophysical Research Letters*, 34, L20405, 2007, doi:10.1029/2007GL031021

52. Colombo, R., Meroni, M., Marchesi, A., Busetto, L., Rossini, M., Giardino, C., & Panigada, C., Estimation of leaf and canopy water content in poplar plantations by means of hyperspectral indices and inverse modeling. *Remote Sensing of Environment*, 112(4), 1820–1834, 2008.

53. Kumar, L., High-spectral resolution data for determining leaf water content in Eucalyptus species: Leaf level experiments. *Geocarto International*, 22(1), 3–16, 2007.

54. Fava, F., Colombo, R., Bocchi, S., Meroni, M., Sitzia, M., Fois, N., & Zucca, C., Identification of hyperspectral vegetation indices for Mediterranean pasture characterization. *International Journal of Applied Earth Information and Geoinformation*, 11, 233–243, 2009.

55. Peñuelas, J., & Inoue, Y., Reflectance indices indicative of changes in water and pigment contents of peanut and wheat leaves. *Photosynthetica*, 36(3), 355–360, 1999.

56. Gamon, J. A., & Qiu, H.-L., Ecological applications of remote sensing at multiple scales. In F. I. Pugnaire & F. Valladares (Eds.), *Handbook of Functional Plant Ecology*, Marcel Dekker, New York, pp. 805–846, 1999.

57. Gamon, J. A., Qiu, H.-L., Roberts, D. A., Ustin, S. L., Fuentes, D. A., Rahman, A., Sims, D., & Stylinski, C., Water expressions from hyperspectral reflectance: Implications for ecosystem flux modeling. In R. O. Green (Ed.), *Summaries of the Eighth JPL Airborne Earth Science Workshop*, CD-ROM, JPL Publication, Pasadena, CA, 1999.

58. Claudio, H.C., Cheng, Y., Fuentes, D.A., Gamon, J.A., Luo, H., Oechel, W., Qiu, H.L., Rahman, A.F., & Sims, D.A., Monitoring drought effects on vegetation water content and fluxes in chaparral with the 970 nm water band index. *Remote Sensing of Environment*, 103(3), 304–311, 2006.

59. Eitel, J. U. H., Gessler, P. E., Smith, A. M. S., & Ronbrecht, R., Suitability of existing and novel spectral indices to remotely detect water stress in *Populus* spp. *Forest Ecology and Management*, 229, 170–182, 2006.

60. Piñol, J., Filella, I., Ogaya, R., & Peñuelas, J., Ground-based spectroradiometric estimation of life fine fuel moisture of Mediterranean plants. *Agricultural Forest Meteorology*, 90, 173–186, 1998.

61. Sims, D. A., & Gamon, J. A., Estimation of vegetation water content and photosynthetic tissue area from spectral reflectance: A comparison of indices based on liquid water and chlorophyll absorption features. *Remote Sensing of Environment*, 84, 526–537, 2003.

62. Zarco-Tejada, P. J., Rueda, C. A., & Ustin, S. L., Water content estimation in vegetation with MODIS reflectance data and model inversion methods. *Remote Sensing of Environment*, 85, 109–124, 2003.

63. De Santis, A., Vaughan, P., & Chuvieco, E., Foliage moisture content estimation from one-dimensional and two-dimensional spectroradiometry for fire danger assessment. *Journal of Geophysical Research*, 111, G04S03, 2006.

64. Wu, C., Niu, Z., Tang, Q., & Huang, W., Predicting vegetation water content in wheat using normalized difference water indices derived from ground measurements. *Journal of Plant Research*, 122(3), 317–326, 2009.

65. Zhang, J., Xu, Y., Yao, F., Wang, P., Guo, W., Li, L., & Yang, L., Advances in estimation methods of vegetation water content based on optical remote sensing techniques. *Science China Technological Sciences*, 53, 5, 2010.

66. Dennison, P. E., Roberts, D. A., Peterson, S. H., & Rechel, J., Use of normalized difference water index for monitoring live fuel moisture. *International Journal of Remote Sensing*, 26, 1035–1042, 2005.

67. Dasgupta, S., Qu, J. J., Hao, X., & Bhoi, S., Evaluating remotely sensed live fuel moisture estimations for fire behavior predictions in Georgia. *Remote Sensing of Environment*, 108, 138–150, 2007.

68. Tucker, C. J., Remote sensing of leaf water content in the near infrared. *Remote Sensing of Environment*, 10, 23–32, 1980.

69. Ripple, W. J., Spectral reflectance relationships to leaf water stress. *Photogrammetric Engineering and Remote Sensing*, 52(10), 1669–1675, 1986.

70. Cohen, W. B., Temporal versus spatial variation in leaf reflectance under changing water-stress conditions. *International Journal of Remote Sensing*, 12, 1865–1876, 1991.

71. Riggs, G. A., & Running, S. W., Detection of canopy water stress in conifers using the airborne imaging spectrometer. *Remote Sensing of Environment*, 35, 51–68, 1991.

72. Cibula, W. G., Zetka, E. F., & Rickman, D. L., Response of thematic mapper bands to plant water stress. *International Journal of Remote Sensing*, 13, 1869–1880, 1992.

73. Hunt, E. R., Jr., Airborne remote sensing of canopy water thickness scaled from leaf spectrometer data. *International Journal of Remote Sensing*, 12, 643–649, 1991.

74. Yilmaz, M. T., Hunt, E. R., Goins, L.D., Ustin, S. L., Vanderbilt, V. C., & Jackson, T. J., Vegetation water content during SMEX04 from ground data and Landsat 5 Thematic Mapper imagery. *Remote Sensing of Environment*, 112(2), 350–362, 2008a.

75. Yilmaz, M. T., Hunt, E. R., & Jackson T. J., Remote sensing of vegetation water content from equivalent water thickness using satellite imagery. *Remote Sensing of Environment*, 112, 2514–2522, 2008b.

76. Maki, M., Ishiahra, M., & Tamura, M., Estimation of leaf water status to monitor the risk of forest fires by using remotely sensed data. *Remote Sensing of Environment*, 90, 441–450, 2004.

77. Davidson, A., Wang, S., & Wilmshurst, J., Remote sensing of grassland—Shrubland vegetation water content in the shortwave domain. *International Journal of Applied Earth Observation and Geoinformation*, 8, 225–236, 2006.

78. Jackson, T. J., Chen, D., Cosh, M., Li, F., Anderson, M., Walthall, C. et al., Vegetation water content mapping using Landsat data derived normalized difference water index for corn and soybeans. *Remote Sensing of Environment*, 92, 475–482, 2004.

79. Filella I., & Peñuelas J., The red edge position and shape as indicators of plant chlorophyll content, biomass and hydric status. *International Journal of Remote Sensing*, 15, 1459–1470, 1994.

80. Liu, L., Wang, J., Huang, W., Zhao, C., Zhang, B., & Long, Q., Estimating winter wheat plant water content using red edge parameters. *International Journal of Remote Sensing*, 25, 3331–3342, 2004.

81. Roberts, D.A., Green, R.O., & Adams, J.B., Temporal and spatial patterns in vegetation and atmospheric properties from AVIRIS. *Remote Sensing of Environment*, 62, 223–240, 1997.

82. Hardy, C.C., & Burgan, R.E., Evaluation of NDVI for monitoring live moisture in three vegetation types of the western US. *Photogrammetric Engineering and Remote Sensing*, 65, 603–610, 1999.

83. Dilley, A. C., Millie, S., O'Brien, D. M., & Edwards, M., The relation between normalized difference vegetation index and vegetation moisture content at three grassland locations in Victoria, Australia. *International Journal of Remote Sensing*, 25, 3913–3928, 2004.

84. Verbesselt, J., Somers, B., Lhermitte, S., Jonckheere, I., Aardt, J. V., & Coppin, P., Monitoring herbaceous fuel moisture content with SPOT VEGETATION time-series for fire risk prediction in savanna ecosystems. *Remote Sensing of Environment*, 108, 357–368, 2007.

85. Cheng, Y. B., Wharton, S., Ustin, S. L., Zarco-Tejada, P. J., Falk, M., & Paw, U. K. T., Relationships between Moderate Resolution Imaging Spectroradiometer water indexes and tower flux data in an old growth conifer forest. *Journal of Applied Remote Sensing*, 1, 013513, 2007.

86. Curran, P. J., Dungan, J. L., & Peterson, D. L., Estimating the foliar biochemical concentration of leaves with reflectance spectrometry: Testing the Kokaly and Clark methodologies. *Remote Sensing of Environment*, 76, 349–359, 2001.

87. Tian, Q., Tong, Q., Pu, R., Guo, X., & Zhao, C., Spectroscopic determinations of wheat water status using 1650–1850 nm spectral absorption features. *International Journal of Remote Sensing*, 22, 2329–2338, 2001.

88. Butler, W. L., & Hopkins, W., Higher derivative analysis of complex absorption spectra. *Photochemistry and Photobiology*, 12, 439–450, 1970.

89. O'Haver, T. C., Derivative spectroscopy: Theoretical aspects. Analytical proceedings. *Proceedings of the Analytical Division of the Royal Society of Chemistry*, 19, 22–28, 1982.

90. Clevers, J. G. P. W., Kooistra, L., & Schaepman, M. E., Estimating canopy water content using hyperspectral remote sensing data. *International Journal of Applied Earth Observations and Geoinformation*, 12(2), 119–125, 2010.

91. Green, R. O., Conel, J. E., Margolis, J. S., Bruegge, C. J., & Hoover, G. L., An inversion algorithm for retrieval of atmospheric and leaf water absorption from AVIRIS radiance with compensation for atmospheric scattering. In R. O. Green (Ed.), *Proceedings of the 3rd Airborne Visible/Infrared Imaging Spectrometer Workshop*, 93–26, JPL Publication, Pasadena, CA, pp. 51–61, 1991.

92. Cheng, Y.B., Ustin, S.L., Riaño, D., Vanderbilt, V.C., Water content estimation from hyperspectral images and MODIS indexes in Southeastern Arizona. *Remote Sensing of Environment*, 112(2), 363–374, 2008.

93. Gao, B. C., & Goetz, A. F. H., Column atmospheric water vapor and vegetation liquid water retrievals from airborne imaging spectrometer data. *Journal of Geophysical Research*, 95, 3549–3564, 1990.

94. Gao, B. C., & Goetz, A. F. H., Retrieval of equivalent water thickness and information related to biochemical components of vegetation canopies from AVIRIS data. *Remote Sensing of Environment*, 52, 155–162, 1995.

95. Ustin, S. L., Roberts, D. A., Pinzón, J., Jacquemoud, S., Gardner, M., Scheer, G. C., Castaneda, M., & Palacios-Orueta, A., Estimating canopy water content of chaparral shrubs using optical methods. *Remote Sensing of Environment*, 65, 280–291, 1998.

96. Cheng, Y. B., Zarco-Tejada, P. J., Riaño, D., Carlos, A., & Ustin, S. L., Estimating vegetation water content with hyperspectral data for different canopy scenarios: Relationships between AVIRIS and MODIS indexes. *Remote Sensing of Environment*, 105(4), 354–366, 2006.

97. Roberts, D.A., Brown, K. J., Green, R., Ustin, S. L., & Hinckley, T., Investigating the relationship between liquid water and leaf area in clonal Populus. *Proc. 7th AVIRIS Earth Science Workshop*, JPL 97-21, Pasadena, CA, 91109, 10 p, 1998.

98. Roberts, D. A., Ustin, S. L., Ogunjemiyo, S., Greenberg, J., Dobrowski, S. Z., Chen, J. et al., Spectral and structural measures of northwest forest landscapes at leaf to landscape scales. *Ecosystems*, 7, 545–562, 2004.

99. Champagne, C. M., Staenz, K., Bannari, A., Mcnairn, H., & Deguise, J.C., Validation of a hyperspectral curve-fitting model for the estimation of plant water content of agricultural canopies. *Remote Sensing of Environment*, 87, 148–160, 2003.

100. Moreno, J. F., Baret, F., Leroy, M., Menenti, M., Rast, M., & Shaepman, M., Retrieval of vegetation properties from combined hyperspectral/multiangular optical measurements: results from the DAISEX campaigns. *Geoscience and Remote Sensing Symposium. Proceedings IGARSS '03 IEEE International*, IEEE, Toulouse, France, 2003.

101. Roberts, D. A., Dennison, P. E., Gardner, M. E., Hetzel, Y., Ustin, S. L., & Lee, C.T., Evaluation of the potential of Hyperion for fire danger assessment by comparison to the airborne visible/infrared imaging spectrometer. *IEEE Transactions on Geoscience and Remote Sensing*, 41, 1297–1310, 2003.

102. Dawson, T. P., Curran, P. J., & Plummer, S. E., LIBERTY—Modeling the effects of leaf biochemical concentration on reflectance spectra. *Remote Sensing of Environment*, 65, 50–60, 1998a.

103. Ganapol, B. D., Johnson, L. F., Hammer, P. D., Hlavka, C. A., & Peterson, D. L., LEAFMOD: A new within-leaf radiative transfer model. *Remote Sensing of Environment*, 63, 182–193, 1998.

104. Liang, S., *Quantitative Remote Sensing of Land Surfaces*, John Wiley and Sons, Hoboken, NJ, 534 p, 2004.

105. Asner, G. P., Biophysical and biochemical sources of variability in canopy reflectance. *Remote Sensing of Environment*, 64, 234–253, 1998.

106. Bacour, C., Jacquemoud, S., Tourbier, Y., Dechambre, M., & Frangi, J.-P., Design and analysis of numerical experiments to compare four canopy reflectance models. *Remote Sensing of Environment*, 79, 72–83, 2002.

107. Huemmrich, K. F., The GeoSAIL model: A simple addition to the SAIL model to describe discontinuous canopy reflectance. *Remote Sensing of Environment*, 75(3), 423–431, 2001.

108. Ceccato, P., Flasse, S., & Gregoire, J.M., Designing a spectral index to estimate vegetation water content from remote sensing data: Part 2. Validation and applications. *Remote Sensing of Environment*, 82, 198–207, 2002b.

109. North, P. R. J., Three-dimensional forest light interaction model using a Monte Carlo method. *IEEE Transactions on Geoscience and Remote Sensing*, 34, 946–956, 1996.

110. Shen, Y., Shi, R., Niu, Z., & Yan, C., Estimation models for vegetation water content at both leaf and canopy levels. *Geoscience and Remote Sensing Symposium, Proceedings. IEEE International*, IEEE, Seoul, South Korea, 2005.

111. Ghulam, A., Li, Z. L., Qin, Q. M., Yimit, H., & Wang, J., Estimating crop water stress with ETM+NIR and SWIR data. *Agr for Meteor*, 148, 1679–1695, 2008.

112. Kimes, D. S., Knyazikhin, Y., Privette, J. L., Abuelgasim, A. A., & Gao, F., Inversion of physically-based models. *Remote Sensing Reviews*, 18, 381–439, 2000.

113. Jacquemoud, S., & Baret, F. Inversion of the PROSPECT + SAILH canopy reflectance model from AVIRIS equivalent spectra: Theoretical study. *Remote Sensing of Environment*, 44, 281–292, 1993.

114. Jacquemoud, S., Baret, F., Andrieu, B., Danson F. M., & Jaggard, K., Extraction of vegetation biophysical parameters by inversion of the PROSPECT + SAIL models on sugar beet canopy reflectance data. Application to TM and AVIRIS sensors. *Remote Sensing of Environment*, 52, 163–172, 1995.

115. Jacquemoud, S., Ustin, S. L., Verdebout, J., Schmuck, G., Andreoli, G., & Hosgood, B., Estimating leaf biochemistry using the PROSPECT leaf optical properties model. *Remote Sensing of Environment*, 56, 194–202, 1996.

116. Fourty, T., & Baret, F., On spectral estimates of fresh leaf biochemistry. *International Journal of Remote Sensing*, 19, 1283–1297, 1997a.

117. Kötz, B., Schaepman, M., Morsdorf, F., Itten, K., & Allgöwer, B., Radiative transfer modeling within a heterogeneous canopy for estimation of forest fire fuel properties. *Remote Sensing of Environment*, 92, 332–344, 2004.

118. Asner, G. P., & Vitousek, P. M., Remote analysis of biological invasion and biogeochemical change. *Proceedings of the National Academy of Sciences (USA)*, 102, 4383–4386, 2005.

119. Li, J., & Goodenough, D. G., Mapping relative water content in Douglas-fir with AVIRIS and a canopy model. *Proc. IGARSS'05.*, IEEE, Seoul, Korea, pp. 3572–3574, 2005.

120. Toomey, M. P., & Vierling, L. A., Estimating equivalent water thickness in a conifer forest using Landsat TM and ASTER data: A comparison study. *Canadian Journal of Remote Sensing*, 32, 288–299, 2006.

121. Combal, B., Baret, F., Weiss, M., Trubuil, A., Macé, D., Pragnère, A., Myneni, R., Knyazikhin, Y., & Wang, L., Retrieval of canopy biophysical variables from bidirectional reflectance using prior information to solve the ill-posed inverse problem. *Remote Sensing of Environment*, 84, 1–15, 2002.

122. Riaño, D., Vaughan, P., Chuvieco, E., Zarco-Tejada, P. J., & Ustin, S. L., Estimation of fuel moisture content by inversion of radiative transfer models to simulate equivalent water thickness and dry matter content: Analysis at leaf and canopy level. *IEEE Transaction on Geoscience and Remote Sensing*, 43, 819–826, 2005a.

123. Kötz, B., Sun, G., Morsdorf, F., Ranson, K. J., Kneubühler, M., Itten, K., & Allgöwe, B., Fusion of imaging spectrometer and LIDAR data over combined radiative transfer models for forest canopy characterization. *Remote Sensing of Environment*, 106(4), 449–459, 2007.

124. Danson, F. M., & Bowyer, P., Estimating live fuel moisture content from remotely sensed reflectance. *Remote Sensing of Environment*, 92, 309–321, 2004.

125. Yebra, M., & Chuvieco, E., Linking ecological information and radiative transfer models to estimate fuel moisture content in the Mediterranean region of Spain: Solving the ill-posed inverse problem. *Remote Sensing of Environment*, 113, 2403–41, 2009.

126. Fourty, T., & Baret, F., Vegetation water and dry matter contents estimated from top-of-the-atmosphere reflectance data: A simulation. *Remote Sensing of Environment*, 61, 34–45, 1997b.

127. Riaño, D., Ustin, S., Usero, L., & Patricio, M. A., Estimation of fuel moisture content using neural networks. *Lecture Notes in Computer Science*, 3562, 489–498, 2005b.

128. Rubio, M. A., Riaño, D., Cheng, Y. B., & Ustin, S. L., Estimation of canopy water content from MODIS using artificial neural networks trained with radiative transfer models.ECAC Ljubljana, Slovenia, European Meteorological Society, 2006.

129. Trombetti, M., Riaño, D., Rubio, M. A., Cheng, Y. B., & Ustin, S. L., Multi-temporal vegetation canopy water content retrieval and interpretation using artificial neural networks for the continental USA. *Remote Sensing of Environment*, 112, 203–215, 2008.

130. Dawson, T. P., Curran, P. J., & Plummer, S. E., The biochemical decomposition of slash pine needles from reflectance spectra using neural networks. *International Journal of Remote Sensing*, 19, 1433–1438, 1998b.

131. Pinzon, J. E., Ustin, S. L., Canstenada, C. M., & Smith, M. O., Investigation of leaf biochemistry by hierarchical foreground/background analysis. *IEEE Transactions on Geoscience and Remote Sensing*, 36, 1–15, 1998.

132. Li, L., Cheng, Y. B., Ustin, S., Hu, X. T., & Riaño, D., Retrieval of vegetation equivalent water thickness from reflectance using genetic algorithm (GA)-partial least squares (PLS) regression. *Advances in Space Research*, 41(11), 1755–1763, 2008.

8 Estimation of Nitrogen Content in Herbaceous Plants Using Hyperspectral Vegetation Indices

D. Stroppiana, F. Fava, M. Boschetti, and P. A. Brivio

CONTENTS

8.1 INTRODUCTION

The nitrogen cycle is of particular interest to ecologists because nitrogen availability can affect the rate of key ecosystem processes, including primary production. However, the excessive use of artificial nitrogen fertilizers, together with fossil fuel combustion and the release of nitrogen in wastewater, has dramatically altered the global nitrogen cycle [1]. The extent and effects of the anthropogenically induced doubling of biologically available nitrogen in the soils, waters, and air of the Earth during the past century are still poorly understood. Nitrogen is an essential nutrient for crop and pasture growth. Crop N status is key information for crop management in precision agriculture, which aims to maximize productivity by limiting, at the same time, environmental impact of excessive fertilization. Since management and nutrient supply are intrinsically dependent on the growth stage of crops and nutrient availability in the soil, timely, spatially detailed, and accurate information on plant status is essential for sustainable agro-practices [2]. Variable rate (VR) technology is considered a promising approach to face issues related to N use efficiency, and it represents the basis for the implementation of rational top-dressing fertilization.

For pastures, nitrogen content is a key determinant of the nutritive value for livestock, being directly related to the crude protein concentration in herbage or forage [3].

Traditionally, N content has been estimated through soil testing and plant tissue analysis [4], which involves destructive field sampling and laboratory chemical analyses [5]. In addition to the effort required, destructive measurements prevent subsequent study of successive developmental stages. Moreover, they create a lag between sampling, laboratory analyses, and N treatments [6] or livestock grazing management [7].

Optical properties of leaves are driven by their biophysical and biochemical properties. Early works described the basic theory of the relationship between leaf optical and morphological properties (e.g., [8]) and investigated the effects on leaf spectra of environmental stresses (e.g., water stress, nutrient deficiency, and pests) [9,10]. Laboratory applications of reflectance spectroscopy clearly showed the potential of using optical properties for the assessment of N content [11]. However, since N is not an optically active parameter, measurements acquired with optical sensors rely on the relationship between chlorophyll and nitrogen. For example, clip sensors, such as the SPAD meter (Minolta Osaka Co., Ltd, Japan) and Dualex Scientific (Force-A, Orsay, France), were shown to provide accurate estimates of leaf chlorophyll content [12,13]. However, these sensors still require laborious field measurements on single leaves and might poorly represent high field variability unless carried out for an impracticable number of leaves and plants. Furthermore, measurements have been shown to be affected by growth stage, nutrient deficiency other than N, environmental conditions, and measurement position on leaves [7]. Ideally, a method is required that is accurate, not destructive, simple to use, and taken from the canopy as opposed to the leaf point of view [14].

Portable technology (smartphone, tablet) and flexible mounting of cameras and sensors make feasible a quick on-site assessment of crop biophysical parameters (e.g., LAI, N and P content) through smart applications [15–17]. These tools for N assessment are based on the capability of fast acquisition of color information from digital photography, thus representing an evolution of traditional approaches based on leaf color charts (LCCs) [18]. For example, the Pocket Nitrogen smart application [15] provides a proxy indicator of plant N concentration by processing digital photographs to derive the dark green color index (DGCI) [19]; the retrieval of N content depends on the availability of predefined crop-specific calibration functions.

Remote sensing techniques have significantly pushed forward the progress in nitrogen content assessment in agricultural crops [20] based on scaling up the relationship between optical and biophysical and biochemical properties from the leaf to canopy level [10]. Radiometric canopy data have, in fact, the potential of measuring reflected radiation from many plants within the field of view of the sensor [4], thus making N canopy assessment feasible, not destructive, and cost effective.

Spectral sensors can nowadays be mounted on tractors, unmanned aerial vehicles (UAVs), airplanes, and satellites, thus providing flexible solutions for multiscale and large-area assessment of surface conditions [10]. Although multispectral sensors have been used for a long time, the development and implementation of hyperspectral instruments has really boosted the use of remote sensing data for nitrogen assessment, importantly since reflectance features related to nitrogen content can only be measured by using contiguous and narrow bands [21,22]. The most widely used approach with hyperspectral data is based on empirical relationships between N content and reflectance in narrow bands and vegetation indices (VIs) [e.g., 23–28]. Empirical approaches do not attempt to directly relate changes in reflectance to N variability through the description of a causal relationship, but they measure the strength of the observed statistical correlation.

Here we review the use of hyperspectral vegetation indices (HVIs) for the estimation of N content in herbaceous plants (crops and pastures) by focusing on most common/recent advancements, describing and discussing their theoretical basis and methodological approaches.

8.2 STATE OF THE ART

Literature on crops (Table 8.1) has focused mainly on N status assessment of rice (*Oriza sativa*), maize (*Zea mays L. ssp*), and wheat (*Triticum ssp.*), as they are the most produced cereals worldwide [22,23,29–33], although other crops have also been studied [34–39]. Leaf-scale measurements have

TABLE 8.1

Summary of Recent Publications for Crops and Pastures

Herbaceous Vegetation	Sensor	Range [nm]	Scale		Parameter		Spectral Transform			R_λ	HVI						Reference
CROP			L	C	PP	N	CR	De	PC		SR	ND	Ch	S	RE	Ot	
Cotton	FS	350–1050								a	a						[34]
Rice	FS	447–1752								a	a	a					[23]
Rice	FS	300–1100								b		b		b			[29]
Rice	FS	350–2500								c,d							[7]
Rice	FS	447–1752								a	a	a					[31]
Rice	AS	400–1000								b							[44]
Rice	FS,AS	400–1100															[20]
Rice	FS,US	327–763											a		a		[51]
Sorghum	FS	350–2500								a	a						[105]
Sugarcane	FS	350–2500									a	a					[37]
Sunflowers	FS	390–1100									a	a					[35]
Wheat	FS	400–900										b					[22]
Wheat	FS	350–2500									a		a	a			[122]
Wheat	FS	350–2500									e	e				e	[124]
Wheat	FS	350–2500									a	a					[30]
Wheat	FS	350–2500									a	a				a	[32]
Wheat	FS	350–2500										b				b	[123]
Wheat	FS, AS	400–850									a	a	a	a	a		[48]
Wheat, corn	FS,AS	325–1075								a	a	a	a	a		a	[33]
Corn	AS	407–949								e							[46]
Corn	AS	400–885									a	a	a	a		a	[10]
Corn	AS	394–968											a	a		a	[45]
Corn	FS,AS	400–885										a	a	a			[57]
Spinach	FS	339–1094									b	b	b				[39]
Potato	FS, AS	401–982									b	b	b	b	b	b	[38]

(Continued)

TABLE 8.1 (*Continued*)

Summary of Recent Publications for Crops and Pastures

Herbaceous Vegetation	Sensor	Range [nm]	Scale		Parameter		Spectral Transform			R_λ	HVI						Reference
			L	C	PP	N	CR	De	PC		SR	ND	Ch	S	RE	Ot	
PASTURE																	
Bermudagrass (M)	FS	280–1100								a	a	a					[61]
Bermudagrass (M)	FS	368–1100									a					b	[6]
Bermudagrass (M)	FS	350–2500									a						[62]
Bermudagrass (M)	FS	350–2500								b,c	a						[26]
Bermudagrass (M)	FS	350–2500								a,c	a					a	[62]
Bermudagrass (M) and mixed grasslands (MG)	FS	350–2500								b,c	a						[63]
Ryegrass (M)	FS	400–900													c		[65]
Ryegrass (M)	FS	404–1650								b							[66]
Ryegrass (M)	FS	350–2500															[64]
Blue grama and Sandburg bluegrass (M)	FS	350–2500															[138]
Ryegrass (M), clover (M) and mixed grasslands (MG)	FS	350–2500								b, c	a						[67]
Sainfoin pastures (M)	FS	325–1150								c							[25]
Blue buffalo grass (G)	FS	350–2500								c	a	a					[69]
Blue buffalo grass (G)	AS	350–2500								c,a					a		[70]
Tropical grass (G)	FS	350–2500								c							[71]
Mixed grassland (MG)	AS	423–2507								c	a	a					[79]
Mixed grassland (MG)	FS	500–2400								b							[83]
Mixed grassland (MG)	AS	380–2500								b f							[82]
Mixed grassland (MG)	FS	350–2500								b							[84]
Mixed grassland (MG)	FS	350–2500								a	a	a					[81]
Mixed grassland (MG)	FS	305/1700								b	a	a					[87]
Mixed grassland (M)	FS/AS	350–2500								b							[86]
Natural prairie (MG)	FS,AS	350–2500															[80]
Alpine meadow (MG)	FS	325–1075								b	a	a					[24]
Mediterranean pastures (MG)	FS	325–1075									a						[28]

(Continued)

TABLE 8.1 (Continued)

Summary of Recent Publications for Crops and Pastures

Herbaceous Vegetation	Sensor	Range [nm]	Scale		Parameter		Spectral Transform			R_λ	HVI						Reference
			L	C	PP	N	CR	De	PC		SR	ND	Ch	S	RE	Ot	
PASTURE																	
Meadow (MG)	FS	360–1010								b,c						b	[138]
Temperate pastures (MG)	FS	350–2500								b						b,c	[85]
Temperate pastures (MG)	AS/SS	400–2500															[85]
Mixed grasslands (MG) and Ryegrass	FS	350–2500													a		[69]
Savanna (MG)	FS	350–2500								a,c							[72]
Savanna (MG)	AS	500–2450								c,d					b,c		[73]
Savanna (MG)	AS	500–2400								d							[74]
Savanna (MG)	FS	350–2500								b	a	a	a	a	a		[75]
Savanna (MG)	FS	350–2500								f							[77]
Savanna (MG)	FS	400–2500								b							[76]

Note: Gray shades represent the conditions of application.

Abbreviations: M, Monoculture; G, Greenhouse; MG, Mixed Grassland; FS, Field Sensor; US, UAV Sensor; AS, Airborne Sensor; SS, Satellite Sensor; L, Leaf spectra; C, Canopy spectra; PP, Photosynthetic Pigments; N, Nitrogen; CR, Continuum Removal; De, Derivative spectra; PC, Principal Components; R, Band Reflectance; SR, Simple Ratio; ND, Normalized Difference; Ch, Chlorophyll-sensitive indices; S, Soil indices; RE, Red edge; Ot, Other indices.

[a] Simple linear/nonlinear regression.
[b] Partial least-squares regression.
[c] Multilinear regression.
[d] Artificial neural networks.
[e] Discriminant analysis.
[f] Random forest/support vector regression.

been used to understand the physiological processes governing plant growth and to estimate leaf N content as an indicator of nutrition status [7,11,30,34,35,40], while canopy reflectance has been exploited for deriving crop N status for precision agriculture and yield estimation [23,31,32]. Some authors have more specifically addressed the issue of the estimation of photosynthetic pigment concentrations as indicators of vegetation physiological conditions due to their direct relationship with reflectance [41–43].

Although most studies rely on proximal sensing techniques (i.e., field spectrometers), the use of airborne hyperspectral sensors has largely increased in the last decade by exploiting, for example, Airborne Imaging Spectrometer for Applications (AISA) [44,45], Compact Airborne Spectrographic Imager (CASI) [46,47], and Pushbroom Hyperspectral Imager (PHI) [48] sensors. A few experiments have been carried out with satellite sensors such as the Hyperion Earth Observing 1 (EO-1), mainly by combining satellite images and proximal sensing hyperspectral data [49].

In the last few years, UAVs have been largely exploited as remote sensing platforms, mounting red-green-blue (RGB) and multispectral sensors to assess crop structural and biophysical parameters [e.g., 50–53]. UAVs mounting hyperspectral sensors are still underexploited due to the limited payload and the high cost of micro-hyperspectral sensors compatible with current payloads [51]; a few works have focused on nitrogen assessment [51,54–56]. It is noteworthy that micro-hyperspectral imaging sensors, which were developed for UAV applications, can be used as a cost-effective solution for airborne platforms; indeed, light sensors can be carried on different UAV airplanes, thus reducing the requirements related to those traditionally used for photogrammetric acquisition [57].

Active hyperspectral light detection and ranging (LiDAR) sensors have also been used for retrieving plant chlorophyll and nitrogen content, although the use of LiDAR intensity needs further exploration [58].

Principally, empirical statistical approaches, such as simple and multiple linear and nonlinear regression [7,27,31,34,45] and least-squares regression [29], have been used with HVIs. Image segmentation/classification as well as machine learning techniques have been applied to hyperspectral images, although they find little application in the specific topic of nitrogen content assessment [39,59,60].

Concerning pastures, in the last 20 years (Table 8.1), pioneer studies have mostly been performed in monoculture pastures, such as bermudagrass (*Cynodon Dactylon* L.) [6,26,61–63] and ryegrass (*Lolium* sp.) [64–67], or in greenhouse experiments at the leaf and canopy level [68–71]. In addition to their economic relevance, monoculture pastures are particularly suited for reflectance spectroscopy experiments since their canopy structure is vertically and horizontally homogeneous compared to mixed-species canopies. From the theoretical point of view of radiative transfer, these pastures represent a turbid medium similar to several crop canopies.

Estimating nitrogen content in mixed grassland systems and natural and seminatural pastures composed of mixed species and with a complex canopy structure is more challenging. Nevertheless, in the last decade, the progress in airborne and field hyperspectral spectroscopy has supported a steady increase of studies targeting these systems worldwide, such as African savannas [72–77], mountain and Mediterranean pastures [24,28,78], north American grasslands [79–81], and mixed grasslands in Australia and New Zealand [82–85].

Leaf-level experiments for nitrogen assessment in pastures are limited to a few example [65,69], as studies on leaves have been conducted mainly on crops. Most works have been performed at the canopy level, mainly by means of field spectrometric data (Table 8.1). However, for agricultural monitoring, the use of airborne sensors for nitrogen assessment of pastures has been increasing over the last few years in the context of precision grassland management applications. Mutanga and Skidmore [73] and Skidmore et al. [74] analyzed airborne HYMAP data (450–2500 nm) to map nitrogen concentration (nitrogen on a dry weight basis [mg N g^{-1} dry weight] and generally expressed as a percentage) in African savannas. Beeri et al. [80] used HYMAP to accurately map the carbon/nitrogen ratio in the American northwestern Great Plains, while Mirik et al. [79] acquired PROBE-1 (423–2507 nm) data from a helicopter to map forage quality in different grassland systems in

Yellowstone National Park. Most recently, pasture nutrient mapping applications have been proposed in New Zealand using the AISA-Fenix (380–2500 nm) [83], in New Hampshire (USA) using the ProSpecTIR VNIR/SWIR (VS4) imager [86], and in Australia using HYMAP [85]. The latter study also tested the potential of Hyperion (EO-1) satellite imagery for nitrogen retrieval, presenting one of the few space-borne applications in this context.

As for crops, most research has been based on empirical statistical approaches. Studies differ either for the type of spectral transformation applied (e.g., derivative, continuum removal and band depth, vegetation indices, etc.) and/or statistical techniques used (e.g., ordinary least-squares regression, multiple linear regression, partial least-squares [PLS] regression), and several included comparative analyses between retrieval methods [24–26,62,63,67–69,87]. The research works of Mutanga and co-authors [64,68,72,73], as an example, emphasized the advantages of using continuum removal [88] as a spectral transformation to enhance nitrogen absorption features, minimizing the spectral variability, which was independent of the nitrogen content. The use of vegetation indices and red-edge indices has been also widely explored compared to other methodologies.

In most recent literature, the increased use of hand-held and low-altitude platform (i.e., UAVs, airborne) hyperspectral sensors covering the whole VIS-NIR-SWIR domain has advanced the investigation of retrieval approaches exploring the whole spectral information content of hyperspectral data. In particular, the use of narrow bands in the SWIR domain for the assessment of nutrient content has been recently investigated with noteworthy results [89–91]. These findings suggest that this region of the electromagnetic spectrum could be used to develop new approaches/indices, as well as other key macro and micro nutrient indicators of herbaceous vegetation. It is, however, important to take into account the presence of water absorption features in the SWIR wavelength domain, which could lead to erroneous results, especially with high-altitude platforms (i.e., satellites) [89].

8.3   PHYSICAL AND PHYSIOLOGICAL BASIS

The domain of optical remote sensing use in vegetation monitoring (400–2500 nm) is usually divided into three spectral regions (VIS-NIR 400–700 nm, NIR 700–1100 nm, and SWIR 1100–2500 nm) characterized by specific light-vegetation interactions influenced by leaf biochemicals and canopy structural characteristics. The amount of radiation reflected from vegetation in the photosynthetically active radiation (PAR; 400–700 nm) region of the electromagnetic spectrum is regulated by pigment absorption within leaves. Chlorophyll a and chlorophyll b absorb the greatest proportion of radiation and provide energy for the reactions of photosynthesis, while carotenoids protect the reaction centers from excess light and help by intercepting PAR as auxiliary pigments of chlorophyll a [88]. Chlorophylls absorb radiation mainly in the blue (~450 nm) and red (~680 nm) wavelengths, whereas carotenoids have absorption features in the blue overlapping with chlorophyll. The red absorption peak is solely due to the presence of chlorophylls, but low chlorophyll content might saturate the 660–680 nm region, thus making it poorly sensitive to high chlorophyll contents [10,41]. Longer (~700 nm, red-edge) or shorter (~550 nm, green) wavelengths are therefore preferred because reflectance is more sensitive to moderate to high chlorophyll content [10].

The amount of absorbed PAR (APAR), which is not used by chlorophyll molecules for photosynthesis (photochemistry), is dissipated as heat and/or fluorescence. Remote sensing of plant characteristics and physiology is therefore the evaluation of the partitioning of excitation energy between photochemistry, fluorescence, and thermal dissipation [92]. The chlorophyll fluorescence spectra is the energy re-emitted in the red (650–780 nm, peak 685 nm) and far red wavelengths (>700 nm, peak 740 nm), and it represents 2%–5% of total APAR [93]. Measurement of this emitted radiation can be used to identify the dynamic status of vegetation and provide information complementary to spectral reflectance. The capacity to measure the amount of fluorescence emission can be an indirect indicator of the photosynthetic activity of a plant, and it has been experimentally demonstrated how hyperspectral measurements of chlorophyll fluorescence are a diagnostic assessment of occurring stress [94].

In the near-infrared (NIR) wavelengths (700–1100 nm), the high reflectance factor is caused by the internal leaf cellular structure. Between the red and NIR wavelengths, leaf reflectance is associated with the transition from chlorophyll absorption to leaf scattering; this position, referred to as red edge, is identified by the wavelength of the maximum slope of the reflectance spectrum between 650 and 800 nm [95]. An increase in the amount of chlorophyll in the canopy, either due to increases in the chlorophyll concentration or to LAI, results in a broadening of the red absorption feature and consequently in the shift of the red-edge position toward longer wavelengths [96].

The shortwave infrared region (SWIR; 1100–2500) is dominated by water absorptions and minor absorption features related to other foliar biochemicals, including nitrogen [11]. The potential of SWIR wavelengths for the assessment of nutrition content of herbaceous vegetation (crops and pastures) has been investigated, with promising results [89,91], although the influence of water absorption features remains an issue.

The relationship between N supply and chlorophyll formation has long been observed [97]. Since part of leaf N is contained in chlorophyll molecules, the amount of available N largely determines the amount of chlorophyll formed in plants, provided that other requirements for chlorophyll formation, such as light, iron supply, and magnesium, are present in sufficient quantities [98]. However, the N/chlorophyll relationship can be influenced by environmental conditions (nutrients and water stress), leaf position in the canopy, genotype, temperature, and leaf growth stage [99,100]. Since N stress induces a physiological change (pigment concentration), which in turn produces changes in leaf spectra, reflectance can be used to assess N status [34,101].

Also, the relation between leaf N and canopy spectra is indirectly due to the association with chlorophyll [102] since canopy spectra are determined by leaf optical properties besides the density and geometry of the canopy (LAI and leaf angle distribution [LDA]) and background reflectivity [103]. Canopy spectra can change dramatically during the season as a consequence of changes of the architecture and arrangement of plant components and changes in the proportion of soil and vegetation [9]. Reflectance from a canopy is considerably less than that from an individual leaf, although in the NIR wavelengths, attenuation is less pronounced. In fact, radiation is transmitted up to enhance the reflectivity of the upper leaves.

In conclusion, the use of leaf and canopy spectra for nitrogen assessment generally relies on the close relation between N and chlorophylls in the cell metabolism, although the experimental relationship established at the canopy scale remains purely empirical. Kokaly et al. [104], in fact, state that the two variables are only moderately correlated within and across ecosystems.

8.4 RETRIEVAL APPROACHES

The most widely used approaches to estimating N content are based on regressive models relating *in situ* measurements and vegetation indices. Simple linear and nonlinear regressions have been widely applied [27,31,34,105], although multivariate techniques, such as multiple linear regression (MLR), principal component regression (PCR), or partial least-squares regression, generally allowed slightly better nitrogen prediction [24,26,29,63]. PLS regression, with respect to MLR, was able to reduce the large number of measured collinear spectral variables to a few noncorrelated principal components (PCs) [22]. Stepwise linear regression has been exploited to identify significant wavebands for MLR where correlations between reflectance and chemical concentrations were high; these techniques generally provided good results for N estimation in crop and pastures [7,26,63], although the selected wavelengths were found to be only indirectly related to N content.

Some major drawbacks of multivariate techniques can, however, be pointed out: limited exportability of statistical models that use many predictors, overfitting of wavebands in the calibration equation, intercorrelation of leaf chemicals, and omission of expected wavelengths where absorption features are well known to be present [11].

Classification techniques have also been applied for N assessment, including artificial neural networks (ANNs) to derive biochemical characteristics of crop canopies [6,64,74]. These methods are widely used with hyperspectral data for their capability of handling large amounts of data and describing nonlinear relationships that may exist between observations [106]. However, training ANNs presents limitations similar to regression analysis due to the low generalization power. Goel et al. [60] compared performance of ANNs and decision trees (DTs), finding slightly better results for the former; however, they suggested that DTs could be preferred for their explicit formalization of the classification rules. Abdel-Rahman et al. [37] proposed a combined approach based on a random forest for feature selection and stepwise multiple linear (SML) regression for estimating sugarcane leaf N concentration from satellite Hyperion imagery.

The above-mentioned techniques have been applied directly to the acquired spectra and/or after transformation by means of derivatives, principal components, smoothing aimed at reducing noise, extracting nonredundant information, solving multicollinearity issues, and enhancing specific features of interest.

Physically based canopy reflectance models deserve to be mentioned. By means of radiative transfer equations, they describe the interactions between the incident radiation and the biophysical and biochemical vegetation parameters. They can be numerically inverted to retrieve canopy parameters from leaf/canopy radiometric measurements. However, models can be applied only to retrieve those parameters that are directly involved in the physical processes of radiative transfer, in this case photosynthetic pigments rather than N. Since models can be very complex, numerical inversion could be computationally intensive, especially when applied at the pixel scale. Moreover, since no universally applicable canopy reflectance model for all vegetation types has yet been defined, model selection is a compromise between complexity, invertibility, and computational efficiency [10,107]. Neither should we forget issues related to numerical inversion such as lack of convergence, sensitivity to parameter initialization, and difficulty of estimating all model parameters. However, radiative transfer models could help us to gain further insight into the spectral features directly influenced by nitrogen and describe the influence of structural canopy parameters on nitrogen estimates more effectively.

Radiative transfer models have been extensively used for the development of HVIs as tools for simulating spectral signals. This is a function of several input parameters, among which leaf chlorophyll content, covering different potential conditions of plant status that can occur in real-case scenarios, is important. Clevers and Kooistra [108] provide an example where both grassland and crops are addressed.

In general, statistical approaches are the simplest way to predict N content, but they provide relationships that are significantly space, time, and species dependent. The regression equations established by different studies do not extrapolate to other sites and times, as they depend on variable viewing and radiation geometries, canopy morphology, soil background, and the spectral characteristics of plant parts. The simplified approach of regressive techniques is suitable when the goal is prediction of *in-situ* quantities rather than understanding radiative transfer processes. In this framework, regressive models based on spectral vegetation indices are preferred to physically based models that are challenging as they are complex to be designed and parameterized [109].

8.5 REVIEW OF COMMON VEGETATION INDICES

VIs, in more or less complex and physically driven ways, enhance the signal from vegetation while minimizing the effect of factors such as solar irradiance, canopy architecture, and background [42]. VIs can be classified on the basis of the spectral attributes employed: reflectance in individual narrow bands, band ratio and (normalized) difference, derivative spectra, and chlorophyll-/soil-sensitive indices [110]. Some of these indices have been designed for broadband sensors, while others have specifically been developed for hyperspectral data such as the use of derivative analysis, which requires continuous spectra. This review can be further enhanced by reading Hatfield et al. [10], Pinter et al. [9], Zarco-Tejada et al. [111], and Tian et al. [112].

8.5.1 INDIVIDUAL/MULTIPLE WAVEBANDS REFLECTANCE

Nitrogen availability and growth stage significantly affect canopy reflectance (R_λ), although the correlation between R_λ and N content is a function of the wavelength (Figure 8.1). According to Stroppiana et al. [27], in the visible wavelengths, R_λ has a significant negative correlation to N concentration: an increase of LAI decreases N concentration and hence reflectance due to the greater proportion of absorbed solar radiation (Figure 8.1). The opposite can be observed for NIR wavelengths, while correlation drastically drops to zero in correspondence of the red edge; it is, in fact, the position of the red edge rather than the reflectance that has been found to be sensitive to N stress [32].

Visible wavelengths appear to be more suitable for estimating N concentration, although the high extinction coefficients of photosynthetic pigments result in very low reflectance, especially in correspondence of pigment absorption spectra (<5%), leading to a loss of detail. To enhance features of pigment absorption, Chapelle et al. [40] used the ratio analysis of reflectance spectra (RARS), and Mutanga et al. [68] successfully tested the continuum removal technique [88].

Other authors have investigated the correlation between R_λ and N content, and results are inconsistent. Some studies have found negative correlation in the visible and positive in the NIR wavelengths [7,24,30–32]. Fava et al. [28] found low positive correlation (<0.5) for visible and NIR wavelengths. Hansen and Schjoerring [22] found low (either positive and negative) correlation coefficients in the blue, red, and NIR regions for chlorophyll and N concentration and high correlation (either positive or negative in the NIR and visible regions, respectively) between reflectance and area-dependent variables such as LAI and chlorophyll/N density (nitrogen on an area basis [g N m^{-2} soil]). For corn leaves, Alchanatis et al. [113] found positive correlation in the visible and no correlation in the NIR range. This discrepancy may arise from the different variables used to quantify N status in plants: if N content is expressed on a unit area basis (N density) and therefore correlated to LAI, then the expected correlation is very similar to the R_λ-LAI correlation shown in Figure 8.1.

Clearly, common protocols for field measurements are necessary to make studies comparable. Although correlated to N content, the use of reflectance in individual bands has limitations due to

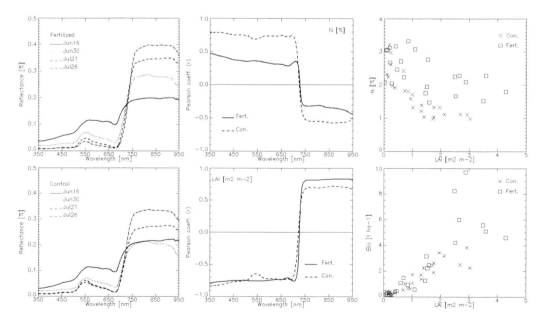

FIGURE 8.1 From the left: changes in canopy spectra with rice growth for fertilized (Fert.) and control conditions (Con.), Pearson correlation (r) of rice N concentration and LAI to canopy reflectance (R_λ) in the VIS-NIR, scatters of N-LAI and biomass-LAI measures (sample size = 48) [26].

the overlap of the effect of different nutrients [114] and the influence of leaf and canopy structural parameters; hence, the scientific focus has moved toward the use of vegetation indices [110] or multivariate approaches [11]. Table 8.2 summarizes the major indices proposed in the literature.

8.5.2 SIMPLE RATIO AND NORMALIZED DIFFERENCE INDICES

Simple ratio (SR; Equation 8.1) and normalized difference (ND; Equation 8.2) indices have been used for almost 45 years and are still widely used for many applications, mainly due to their simple formalization and utility:

$$SR = R_{\lambda 2} / R_{\lambda 1} \tag{8.1}$$

$$ND = (R_{\lambda 2} - R_{\lambda 1})/(R_{\lambda 2} + R_{\lambda 1}) \tag{8.2}$$

where $R_{\lambda 1}$, $R_{\lambda 2}$ is the spectral reflectance measured at wavelengths $\lambda 1$ and $\lambda 2$. Initially designed for multispectral instruments, with $\lambda 1 = red$ and $\lambda 2 = NIR$, they have been tested with almost all hyperspectral narrow bands [22,27–29,31,62]. The SR combines nutrient stress sensitive and insensitive bands to normalize for variations of exogenous factors such as irradiance, leaf orientation, irradiance angles, and shadowing. To estimate chlorophyll and N contents, Kim et al. [115] proposed $\lambda 1 = 550$ nm and $\lambda 2 = 700$ nm since the use of green wavelengths, which are more sensitive to N rates, instead of red could improve the precision and accuracy of the predictions [23,24,116]. For rice, Xue et al. [23] suggested $\lambda 1 = 560$ nm and $\lambda 2 = 810$ nm, while Zhu et al. [31] found $\lambda 1 = 660/680$ nm, $\lambda 2 = 950$ nm the best wavelengths for estimating leaf N accumulation (i.e., the product of leaf N concentration per unit dry weight and leaf dry weight per unit ground area).

Zhao et al. [36] proposed $\lambda 1 = 715/735$ nm, $\lambda 2 = 405/1075$ nm for a ratio index well correlated to sorghum leaf N concentration. Fava et al. [28] found that SR involving NIR bands (775–820 nm) and longer wavelengths of the red edge (740–770 nm) yielded the best correlation with N concentration in pasture (Figure 8.2). Abdel-Rahman et al. [37] used $R''_{\lambda 1} = 1316$ nm and $R''_{\lambda 2} = 743$ nm calculated from first-order derivative spectra to estimate sugarcane leaf N concentration. Since SR can be influenced by cloud and soil [117], the modified simple ratio (MSR) was introduced for boreal forests with $\lambda 2 = 800$ nm and $\lambda 1 = 670$ nm (Equation 8.3) and tested over other land covers [118,119].

$$MSR = (R_{\lambda 2} / R_{\lambda 1} - 1)/\sqrt{(R_{\lambda 2}/R_{\lambda 1}) + 1} \tag{8.3}$$

The normalized difference vegetation index (NDVI) is certainly a good indicator of vegetation greenness, but, in its initial form with $\lambda 1 = red$ ($\sim$670 nm) and $\lambda 2 = NIR$ ($\sim$800 nm) [120], it cannot follow physiological changes determined by stress conditions that produce changes at specific wavebands [34]. Changes in photosynthetic activity do not always determine changes in canopy structure [121], so indices sensitive to canopy characteristics may not be the best indicators of nutrient deficiency.

However, hyperspectral narrow bands sensitive to physiological changes can be used for defining an ND-type index: for example, the photochemical reflectance index (PRI) [122] and the normalized pigments chlorophyll ratio index (NPCI), which were found to be sensitive to physiological leaf status and N stress in sunflower leaves [34]. Wang et al. [123] indicated that a chlorophyll-based index, the plant pigment ratio (PPR), yielded the best predictions of leaf N concentration from canopy spectra of winter wheat. Hansen and Schjoerring [22] found the greatest correlation with $N_{density}$ when wavebands of the normalized index were selected in the NIR region ($\lambda 2$) and at the red shift ($\lambda 1$). When the purpose is to estimate N concentration (N_{conc}), the best wavebands were found to be in the visible domain and, in particular, in the blue (440–501 nm), characterized by a strong

TABLE 8.2

Summary of Spectral Indices Proposed for Chlorophyll/N Estimation

Vegetation Index	Name (Acronym)	Equation	Herbaceous Plant	Parameter	Reference
Simple Ratio (SR) and Modified Simple Ratio (MSR)	SR	R_{810}/R_{560}	Rice	Leaf N accumulation	[23]
	SR	R_{950}/R_{660} R_{950}/R_{680}	Rice	Leaf N accumulation	[31]
	SR	R_{405}/R_{715} R_{1075}/R_{735}	Sorghum	Leaf N concentration	[36]
	SR	$R_{775-820}/R_{740-770}$	Pasture	Canopy N concentration	[28]
	SR	R'_{743}/R'_{1316}	Sugarcane	Leaf N concentration	[37]
Normalized Difference Indices	Photochemical Reflectance Index (PRI)	$\dfrac{R_{570}-R_{531}}{R_{570}+R_{531}}$	Sunflower	Leaf status and N stress	[121]
	Normalized Pigments Chlorophyll Ratio Index (NPCI)	$\dfrac{R_{680}-R_{430}}{R_{680}+R_{430}}$	Sunflower	Leaf status and N stress	[34]
	Plant Pigment Ratio (PPR)	$\dfrac{R_{550}-R_{450}}{R_{550}+R_{450}}$	Winter wheat	Leaf N concentration	[122]
	Normalized Difference (Visible wavelengths)	$\dfrac{R_{503}-R_{483}}{R_{503}+R_{483}}$	Paddy rice	Canopy N concentration	[27]
	Normalized Difference	$\dfrac{R_{770}-R_{717}}{R_{770}+R_{717}}$ $\dfrac{R_{839}-R_{720}}{R_{839}+R_{720}}$	Wheat	Canopy N density	[22]
	Normalized Difference	$\dfrac{R_{573}-R_{440}}{R_{573}+R_{440}}$	Wheat	Canopy N concentration	[22]

(Continued)

TABLE 8.2 (*Continued*)
Summary of Spectral Indices Proposed for Chlorophyll/N Estimation

Vegetation Index	Name (Acronym)	Herbaceous Plant	Equation	Parameter	Reference
	Normalized Difference	Rice	$\dfrac{R_{1220} - R_{710}}{R_{1220} + R_{710}}$	Leaf N content	[31]
	Normalized Difference Spectral Index (NDSI)	Rice	$\dfrac{R_{825} - R_{735}}{R_{825} + R_{735}}$	Canopy N content	[20]
Chlorophyll and Soil-Sensitive Indices	Canopy Chlorophyll Content Index (CCCI)	Wheat	$\dfrac{NDRE - NDRE_{min}}{NDRE + NDRE_{max}}$	Plant N content	[123]
	Modified Chlorophyll Absorption in Reflectance Index (MCARI)	Corn	$[(R_{700} - R_{670}) - 0.2 \times (R_{700} - R_{550})] \times \left(\dfrac{R_{700}}{R_{670}}\right)$	N applied	[4]
	Transformed Chlorophyll Absorption Reflectance Index (TCARI)	Corn	$TCARI = 3*\left[\dfrac{(R_{700} - R_{670}) - 0.2*}{*(R_{700} - R_{550})}\left(R_{700}/R_{670}\right)\right]$	Canopy chlorophyll content	[42]
	Optimized Soil Adjusted Vegetation Index (OSAVI)	–	$OSAVI = (1+0.16)\dfrac{R_{800} - R_{670}}{R_{800} + R_{670} + 0.16}$	–	[126]
	TCARI/OSAVI	Corn	$TCARI/OSAVI$	N nutritional status	[12]
	Red Edge Position Linear	Crops and pastures	$REP = 700 + 40\left(\dfrac{R_{RE} - R_{700}}{R_{740} - R_{700}}\right)$ $R_{RE} = \left(R_{760} + R_{780}\right)/2$	Plant N content/ nutritional status	[130]

Abbreviations: R_λ, Reflectance at wavelength λ; R'_λ, First-order derivative reflectance at wavelength λ.

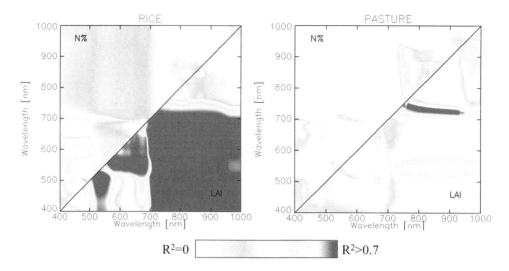

FIGURE 8.2 Linear correlation (R^2) between N concentration/LAI and ND for rice. (Left, from Starks, P.J. et al. *Grass and Forage Science*, 63, 168–178 [26]. and SR for pasture (right, From Stroppiana et al. *Field Crops Research*, 111, 119–129 [27].)

light absorption due to chlorophyll-a and chlorophyll-b, combined with a green (573–586 nm) or red (692 nm) band (Table 8.2). Indeed, Hansen and Schjoerring [22] found the red edge almost absent from band combinations in contrast to Gianelle and Guastella [24] and Fava et al. [28]. Note that due to a low signal-to-noise ratio at both the ends of the spectrum, Hansen and Schjoerring [22] restricted their analyses to the 438–883 nm range, discarding the longer NIR wavelengths. Zhu et al. [31] found that the best NDVI for rice leaf N content estimation involved near-infrared wavelengths (1220, 710). Stroppiana et al. [27] found that, for paddy rice, the combination of reflectance in two visible wavelengths ($\lambda 1 = 483$ nm, $\lambda 2 = 503$ nm) provided the highest correlation with field measurements of N concentration ($R^2 = 0.65$, ***$p < 0.001$), whereas visible/NIR combinations determined a high correlation with LAI rather than N concentration (Figure 8.2). With the same processing technique and hyperspectral imagery from ground-based and the CASI-3 sensors, Inoue et al. [20] showed that simple ratio and normalized difference indices containing a red-edge wavebands (R_{725}) and R_{835} were significantly correlated to canopy nitrogen content (CNC) of rice. The canopy chlorophyll content index (CCCI), which is derived from the minimum and maximum values of the normalized difference red edge [NDRE $= ((\rho_{790}-\rho_{720})/(\rho_{790}+\rho_{720}))$], as observed in the data, has proven to relate to plant N in wheat [124,125].

Other authors have attempted to include SWIR wavelengths in the formalization of ND-type indices; for example, Mahajan et al. [89] found significant correlation between rice N and P at 1460 nm, thus proposing new indices. Notice that most of these works rely on reflectance measured with hand-held spectroradiometers, and the robustness of their findings, when applied to remotely sensed data (UAV, airborne, or satellite), is still under discussion due to the likely influence of atmospheric disturbance and the presence of water absorption features in the SWIR domain.

8.5.3 CHLOROPHYLL- AND SOIL-SENSITIVE INDICES

Chlorophyll reflectance indices have been developed to enhance sensitivity to photosynthetic pigments and to reduce the influence of canopy architecture and the background [111]. The chlorophyll absorption in reflectance index (CARI) was proposed by Kim et al. [115] to reduce the influence of nonphotosynthetic parts of the plants and incorporates the green and red-edge regions of the spectrum.

The modified chlorophyll absorption in reflectance index (MCARI; Table 8.2) was introduced by Daughtry et al. [4] and measures the depth of the chlorophyll absorption at 670 nm relative to the reflectance at 550 nm and 700 nm. To reduce the combined effect of nonphotosynthetic materials and soil background, they proposed the use of the (R_{700}/R_{670}) ratio: the slope of the spectrum when the canopy contains no green biomass. However, MCARI is poorly predictive at low LAI due to the influence of the background [4]. Haboudane et al. [42] found it is still influenced by nonphotosynthetic elements at low chlorophyll concentrations and proposed the transformed chlorophyll absorption in reflectance index (TCARI; Table 8.2): in TCARI formulation, the ratio (R_{700}/R_{670}) is used to compensate for the influence of the background on the difference (R_{700}/R_{550}) since the change of background reflectance mainly influences the slope between 550 and 700 nm [115].

The family of soil adjusted vegetation indices (SAVIs; [126]) should further reduce the contribution of background reflectance and includes the optimized SAVI (OSAVI) [127]) and transformed SAVI (TSAVI) [128]) (Table 8.2). The major difference between these two indices is that OSAVI requires no information on optical soil properties and therefore has a more direct application, whereas TSAVI requires knowledge of the soil line. The soil line is the linear relationship between red and near-infrared reflectance of bare soil (i.e., constants a and b in TSAVI formulation; Table 8.2), which, in practice, can be derived from measuring an unsown area within the field.

Since neither sensitivity to chlorophyll nor insensitivity to the background seemed to be achievable with either of the above, Daughtry et al. [4] and Haboudane et al. [42] proposed MCARI/OSAVI and TCARI/OSAVI. Recently, Gabriel et al. [10] confirmed the good performance of combined indices (i.e., TCARI/OSAVI) for maize N nutritional status assessment. Using model simulations of the reflectance of open-canopy tree crops, Zarco-Tejada et al. [111] found, however, MCARI/OSAVI was less affected by soil background variations than TCARI/OSAVI for total chlorophyll content estimation; tree crops are a challenge for remote sensing due to the effect of background and shadowing of a complex canopy architecture.

Figure 8.3 shows scatter plots of the correlation between N concentration/biomass and example HVIs ($NDI_{483,503}$, $NDVI_{710,1220}$, $NDVI_{670,800}$, $SR_{670,800}$, $SR_{700,750}$, MCARI, TCARI, and TCARI/OSAVI) derived for rice canopy reflectance data acquired with a field spectrometer (FieldSpec FR PRO) over experimental plots [27]. The coefficients of determination (R^2) for both linear and exponential regressive models are given in the figure. Interestingly, $NDI_{483,503}$ shows a good correlation with N concentration and, at the same time, a poor correlation with aboveground biomass, while the other indices are correlated with biomass rather than N. On the other hand, $NDVI_{670,800}$ and $SR_{670,800}$ are not informative for N assessment since they are correlated to biomass. The same indices derived with different red/NIR wavelength combinations provided similar results when applied to the same dataset.

For these reasons, Stroppiana et al. [27] identified $NDI_{483,503}$ as the most suited index for defining a regressive model to estimate rice N concentration. The regressive model was applied to field spectro-radiometric measurements acquired in a different year for two rice varieties (Gladio–Indica type and Selenio–Japonica type) and under a wide range of nitrogen condition generated by a combination of different fertilization levels. N estimates were spatialized over experimental rice plots for three acquisition dates to obtain the nitrogen concentration maps shown in Figure 8.4 [27]. Validation of N estimates by comparison with ground measurements values showed a robust predictive value of the $NDI_{483,503}$ regressive model.

Soil-sensitive indices were insufficiently investigated for N assessment in pastures, probably because they are often characterized by complete canopy cover in late stages of growth, unless high grazing pressure occurred.

Very few experiments have been carried out with hyperspectral data acquired with UAVs. Uto et al. [51] investigated correlation between chlorophyll indices based on red-edge and NIR spectral ranges and SPAD readings (i.e., chlorophyll density); they found good correlation and promising results for future micro hyperspectral sensors spanning longer wavelengths.

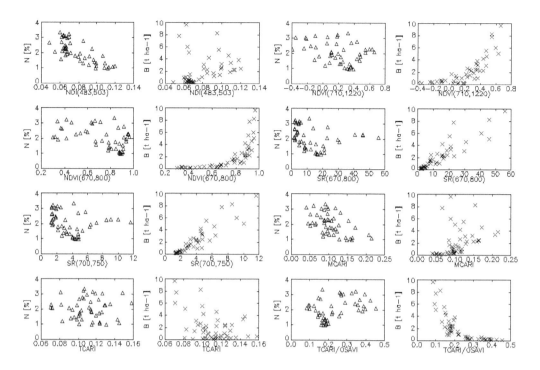

FIGURE 8.3   Scatter plots of VI-N [%] and VI-Biomass [tha-1] for VIs described above. Field data were collected for experimental paddy rice fields in Italy (sample size = 48) [26].

Recent works have explored the use of chlorophyll fluorescence and fluorescence-sensitive reflectance indices derived from airborne and UAV imagery [57,129].

8.5.4   THE RED EDGE

The red edge identifies the steep transition between the reflectance absorption feature in red wavelengths (~680 nm) and the high NIR reflectance (~740 nm) [130], and the point of the

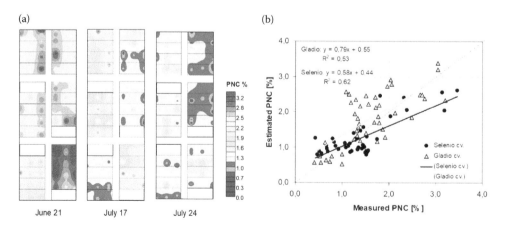

FIGURE 8.4   (a) Maps of nitrogen concentration over experimental rice plots for three dates of the 2006 growing season. Maps have been obtained by spatialization of estimates of N concentration derived with a regression (NDI$_{483,503}$ − N content); (b) validation of N estimates by comparison with field N measurements for two rice varieties. Grey blocks identify the plots where rice was not sown.

maximum slope (or inflection point) is defined as the red-edge position (REP) or red-edge index. The red edge has been extensively used to estimate plant N content and nutritional status in crops and pastures [32,65,70], and several techniques have been proposed to locate the REP. The linear interpolation technique assumes that the red-edge curve can be modeled as a straight line, and the REP is estimated by fitting a linear equation using the slope of the line.

The algorithm proposed by Guyot and Baret [131] uses four wavebands (670, 700, 740, and 780 nm) and calculates the REP as shown in Table 8.2.

The inverted Gaussian (IG) technique [132,133] consists of fitting a Gaussian model to the reflectance red edge by iterative fitting optimization procedure (Equation 8.4). The REP is calculated as the midpoint of the ascending edge of the Gaussian curve (REP = $\lambda_0+\sigma$).

$$R(\lambda) = R_s - \left(R_s - R_0\right)\exp\left(-\frac{(\lambda_0 - \lambda)^2}{2\sigma^2}\right)$$ (8.4)

where R_s is the maximum red-edge reflectance (NIR shoulder), R_0 and λ_0 are the minimum spectral reflectance and corresponding wavelength, and σ is the Gaussian function variance.

A third group of methods uses fitting of continuous high-order curves, such as cubic-spline or high-order polynomial, to the first derivative of the reflectance spectrum; REP is calculated as the maximum value [130,134,135].

The main drawback of the techniques based on the first derivative spectrum is that they do not account for the possibility of more than one maximum in the derivative of the red edge [136]. In contrast, several studies demonstrated a bimodal distribution of the derivative reflectance curve in the red edge, characterized by two maxima around 700 and 725 nm [65,69,130]. As a consequence, the red-edge position could represent whichever maximum was dominant, and therefore a red-edge shift could involve a jump between the two maxima, creating a discontinuity in the red-edge position [136].

To mitigate the destabilizing effect of the double-peak feature, and to predict leaf nitrogen content with high accuracy in herbaceous vegetation, Cho and Skidmore [69] proposed an alternative methodology called the linear extrapolation technique. This technique is based on linear extrapolation of two straight lines through two points on the far-red (680–700 nm) and two points on the NIR (725–760 nm) flanks of the first derivative reflectance spectrum of the red edge. The REP is identified by the wavelength value at the intersection of the straight lines.

Overall, comparative studies showed that the accuracy of different techniques to estimate the red edge is dependent on several factors related to the biological characteristics of the plant material [69], and no single extraction method could be considered superior in general [137]. Despite the good performances of red-edge indices [32,69,70], the effect of varying LAI on the red edge is still an open issue. Lamb et al. [65] demonstrated by radiative transfer model simulations that only high LAI values determine low sensitivity of the red edge to LAI, thus allowing an accurate chlorophyll/nitrogen estimation.

8.6 CONCLUSIONS

With the advent of hyperspectral sensors, remote sensing techniques have become particularly attractive for assessing crop and pasture N status. Compared to multispectral sensors, the availability of narrow and contiguous bands is fundamental for retrieving nitrogen content [21,24,28,33]. Indeed, spectral resolution becomes even more important than spatial resolution [138]. A vast literature is available, with increasing applications exploiting airborne sensors, whereas space-borne instruments still suffer from numerous issues that limit their use for N retrieval. Atmospheric correction, spectral and spatial resolution, and signal-to-noise ratio (SNR) are not always adequate for identifying the absorption features of the chemical compounds. Future space missions, such as Germany's

Environmental Mapping and Analysis Programme (EnMAP) and Italy's PRecursore IperSpettrale della Missione Applicativa (PRISMA), both scheduled to be launched in 2018, will offer new opportunities for the scientific community.

It is important to underline that the recent and rapid development of microtechnology for UAVs opened new opportunities and flexible solutions for vegetation monitoring; although, at present, the use of hyperspectral data is not widespread due to size/payload and economic limitations, UAV platforms will soon be operative platforms also carrying hyperspectral sensors.

The high empirical correlation between chlorophylls and nitrogen is the basis for using spectra for N assessment, although the relation between spectra changes and physiological processes has to be further investigated. Statistical regressive methods, based on the use of vegetation indices, have been useful approaches for the retrieval of N content from leaf and canopy spectra. However, there is still little agreement over which vegetation index has the strongest and more stable relationship with N.

Ratio and normalized difference indices have been shown to reduce the influence of exogenous factors (e.g., canopy, background signal) but little consensus has been reached on the narrow wavebands that should be used in their formulation. The visible and red-edge wavelengths are often reported as important spectral regions for nitrogen assessment and could be better suited than NIR wavebands, which are strongly influenced by structural canopy parameters. The decorrelation of the optimal HVI from factors other than N concentration has proven to be rather difficult to achieve. Canopy architecture can significantly influence changes in canopy spectra, and some indices have shown a correlation with N concentration only as a consequence of the correlation N-LAI. Since near-infrared wavelengths are mainly influenced by canopy structure, HVIs that use visible and red-edge wavebands should be preferred [27,28]. The role of vegetation phenology and growth stage in the retrieval of nitrogen content has also been insufficiently investigated [21,65], not to forget that other nutrient deficiencies might produce spectral changes that overlap with N stress features. Hence, the ability of predicting N status may depend on their availability, and a greater predictive capability can be achieved in conditions where other nutrients are not limiting or well known.

Increasing attention has recently been paid to SWIR wavelengths for formulating new indices, yet the feasibility of upscaling these findings to remotely sensed data and, in particular, satellite data is still under discussion.

Chlorophyll fluorescence, which is recognized as an indicator for early detection of vegetation stress conditions, also deserves further attention in relation to the assessment of nitrogen content.

Several studies have, however, shown that the performance of VIs is a function of site and vegetation characteristics, and none of the indices appear to be robust enough across sites. Empirical models therefore remain applicable mainly at local/regional scales. Therefore, there is a need not only for comparative analysis of different vegetation indices, which have been extensively pursued in the literature, but also for studies comparing methodological approaches in terms of their consistency and robustness in space and time. Yet comparisons are often hampered by the different field conditions/parameters used in the experiments (e.g., N concentration/density/accumulation).

In summary, there are some priorities to be addressed by future research:

1. Further understanding of the physiological processes involved in the relation between plant N content and leaf/canopy spectra, also by exploiting simulations from radiative transfer models

2. Comparison of methodologies on datasets acquired with common protocols of measurements to evaluate consistency and robustness

3. Setup of field experiments designed to evaluate the combined influence on vegetation indices of canopy architecture, other nutrients stress, and vegetation growth stage

4. Development of procedures for upscaling HVIs, including SWIR wavelengths to high-altitude platforms (i.e., satellite) to reduce the influence of water absorption features

REFERENCES

1. Holland, E.A., Dentener, F.J., Braswell, B.H., Sulzman, J.M. 1999. Contemporary and pre-industrial global reactive nitrogen budgets, *Biogeochemistry*, 46, 7–43.
2. Dobermann, A., Witt, C., Dawe, D., Abdulrachman, S., Gines, H.C., Nagarajan, R., Satawathananont, S. et al. 2002. Site-specific nutrient management for intensive rice cropping systems in Asia, *Field Crops Research*, 74, 37–66, 10.1016/S0378-4290(01)00197-6.
3. Pearson, C.J., Ison, R.L. 1997. *Agronomy of Grassland Systems*, Cambridge University Press, Cambridge, UK.
4. Daughtry, C.S.T., Waithall, C.L., Kim, M.S., de Colstoun, E.B., McMurtrey III, J.E. 2000. Estimating corn leaf chlorophyll concentration from leaf and canopy reflectance, *Remote Sensing of Environment*, 74, 229–239.
5. Wu, J., Wang, D., Rosen, C.J., Bauer, M.E. 2007. Comparison of petiole nitrate concentrations, SPAD chlorophyll readings, and QuickBird satellite imagery in detecting nitrogen status of potato canopies, *Field Crops Research*, 101, 96–103.
6. Starks, P.J., Coleman, S.W., Phillips, W.A. 2004. Determination of forage chemical composition using remote sensing, *Journal of Range Management*, 57, 635–640.
7. Yi, Q-X., Huang, J-F., Wang, F-M., Wang, X-Z., Liu, Z-Y. 2007. Monitoring rice nitrogen status using hyperspectral reflectance and artificial neural networks, *Environmental Science & Technology*, 41, 6770–6775.
8. Knipling, E.B. 1970. Physical and physiological basis for the reflectance of visible and near-infrared radiation from vegetation, *Remote Sensing of Environment*, 1, 155–159.
9. Pinter, P.J. Jr., Hatfield, L., Schepers, J.S., Barnes, E.M., Moran, M.S., Daughtry, C.S.T., Upchurch, D.R. 2003. Remote sensing for crop management, *Photogrammetric Engineering & Remote Sensing*, 69, (6), 647–664.
10. Hatfield, J.L., Gitelson, A.A., Schepers, J.S. and Walthall, C.L. 2008. Application of spectral remote sensing for agronomic decisions, *Agronomy Journal*, 100, 117–131.
11. Curran, P.J. 1989. Remote sensing of foliar chemistry, *Remote Sensing of Environment*, 30, 271–278.
12. Gabriel, J.L., Zarco-Tejada, P.J., López-Herrera, P.J., Pérez-Martín, E., Alonso-Ayuso, M., Quemada, M. 2017. Airborne and ground level sensors for monitoring nitrogen status in a maize crop, *Biosystems Engineeering*, 160, 124–133.
13. Reyes, J.F., Correa, C., Zúñiga, J. 2017. Reliability of different color spaces to estimate nitrogen SPAD values in maize, *Computers and Electronics in Agriculture*, 143, 14–22. ISSN 0168-1699, https://doi.org/10.1016/j.compag.2017.09.032.
14. Curran, P.J., Dungan, J.L., and Gholz, H.L. 1990. Exploring the relationship between reflectance red edge and chlorophyll content in slash pine, *Tree Physiology*, 7, 33–48.
15. Confalonieri, R., Foi, M., Casa, R., Aquaro, S., Tona, E., Peterle, M., Boldini, A. et al. 2013. Development of an app for estimating leaf area index using a smartphone. Trueness and precision determination and comparison with other indirect methods, *Computers and Electronics in Agriculture*, 96, 67–74.
16. Moonrungsee, N., Pencharee, S., Jakmunee, J. 2015. Colorimetric analyzer based on mobile phone camera for determination of available phosphorus in soil, *Talanta*, 136, 204–209.
17. Rigon, J.P.G., Capuani, S., Fernandes, D.M., Guimarães, T.M. 2016. A novel method for the estimation of soybean chlorophyll content using a smartphone and image analysis, *Photosynthetica*, 54, (4), 559–566.
18. Alam, M.M., Ladha, J.K., Khan, S.R., Foyjunessa, Harun-ur-rarun-ur-rashid Khan, A.H., Buresh, R.J. 2005. Leaf color chart for managing nitrogen fertilizer in lowland rice in Bangladesh, *Agronomy Journal*, 97, 949–959.
19. Karcher, D.E., Richardson, M.D. 2003. Quantifying turfgrass color using digital image analysis, *Crop Science*, 43, 943–951.
20. Inoue, Y., Sakaiya, E., Zhu, Y., Takahashi, W. 2012. Diagnostic mapping of canopy nitrogen content in rice based on hyperspectral measurements, *Remote Sensing of Environment*, 126, 210–221, doi.org/10.1016/j.rse.2012.08.026.
21. Thenkabail, P.S., Smith, R.B., Pauw, E.D. 2000. Hyperspectral vegetation indices and their relationships with agricultural crop characteristics, *Remote Sensing of Environment*, 71, 158–182.
22. Hansen, P.M., Schjoerring, J.K. 2003. Reflectance measurement of canopy biomass and nitrogen status in wheat crops using normalized difference vegetation indices and partial least squares regression, *Remote Sensing of Environment*, 86, 542–553.
23. Xue, L., Cao, W., Luo, W., Dai, T., Zhu, Y. 2004. Monitoring leaf nitrogen status in rice with canopy spectral reflectance, *Agronomy Journal*, 96, 135–142.

24. Gianelle, D., Guastella, F. 2007. Nadir and off-nadir hyperspectral field data: Strengths and limitations in estimating grassland biophysical characteristics, *International Journal of Remote Sensing*, 28, 1547–1560.

25. Albayrak, S. 2008. Use of reflectance measurements for the detection of N, P, K, ADF and NDF contents in Sainfoin Pasture, *Sensors*, 8, 7275–7286.

26. Starks, P.J., Zhao, D., Brown, M.A. 2008. Estimation of nitrogen concentration and *in vitro* dry matter digestibility of herbage of warm-season grass pastures from canopy hyperspectral reflectance measurements, *Grass and Forage Science*, 63, 168–178.

27. Stroppiana, D., Boschetti, M., Brivio, P.A., Bocchi, S. 2009. Plant nitrogen concentration in paddy rice from field canopy hyperspectral radiometry, *Field Crops Research*, 111, 119–129.

28. Fava, F., Colombo, R., Bocchi, S., Meroni, M., Sitzia, M., Fois, N., Zucca, C. 2009. Identification of hyperspectral vegetation indices for Mediterranean pasture characterization, *International Journal of Applied Earth Observation and Geoinformation*, 11, 233–243.

29. Nguyen, H.T., Lee, B-W. 2006. Assessment of rice leaf growth and nitrogen status by hyperspectral canopy reflectance and partial least square regression, *European Journal of Agronomy*, 24, 349–356.

30. Yao, X., Feng, W., Zhu, Y., Tian, Y.C., Cao, W.X. 2007. A non-destructive and real-time method of monitoring leaf nitrogen status in wheat, *New Zeland Journal of Agricultural Research*, 50, 935–942.

31. Zhu, Y., Zhou, D., Yao, X., Tian, Y., Cao, W. 2007. Quantitative relationships of leaf nitrogen status to canopy spectral reflectance in rice, *Australian Journal of Agricultural Research*, 58, 1077–1085.

32. Feng, W., Yao, X., Zhu, Y., Tian, Y.C., Cao, W.X. 2008. Monitoring leaf nitrogen status with hyperspectral reflectance in wheat, *European Journal of Agronomy*, 28, 394–404.

33. Chen, P., Haboudane, D., Trembaly, N., Wang, J., Vigneault, P., Li, B. 2010. New spectral indicator assessing the efficiency of crop nitrogen treatment in corn and wheat, *Remote Sensing of Environment*, 114, 1987–1997.

34. Peñuelas, J., Gamon, J.A., Fredeen, A.L., Merino, J., and Field, C.B. 1994. Reflectance indices associated with physiological changes in nitrogen- and water-limited sunflower leaves, *Remote Sensing of Environment*, 48, 135–146.

35. Tarpley, L., Reddy, K.R., Sassenrath-Cole, G.F. 2000. Reflectance indices with precision and accuracy in predicting cotton leaf nitrogen concentration, *Crop Science*, 40, 1814–1819.

36. Zhao, D., Reddy, K.R., Kakani, V.G., Reddy, V.R. 2005. Nitrogen deficiency effects on plant growth, leaf photosynthesis, and hyperspectral reflectance properties of sorghum, *European. J. Agronomy*, 22, 391–403.

37. Abdel-Rahman, E.M., Ahmed, F.B., van der Berg, M. 2010. Estimation of surgarcane leaf nitrogen concentration using *in situ* spectroscopy, *International Journal of Applied Earth Observation and Geoinformation*, 12S, S52–S57.

38. Nigon, T.J., Mulla, D.J., Rosen, C.J., Cohen, Y., Alchanatis, V., Knight, J., Rud, R. 2015. Hyperspectral aerial imagery for detecting nitrogen stress in two potato cultivars, *Computers and Electronics in Agriculture*, 112, 36–46.

39. Corti, M., Gallina, P.M., Cavalli, D., Cabassi, G. 2017. Hyperspectral imaging of spinach canopy under combined water and nitrogen stress to estimate biomass, water, and nitrogen content, *Biosystems Engineering*, 158, 38–50. ISSN 1537-5110, https://doi.org/10.1016/j.biosystemseng.2017.03.006.

40. Chapelle, E.W., Kim, M.S., McMurtrey III, J.E. 1992. Ratio analysis of reflectance spectra (RARS): An algorithm for the remote estimation of the concentration of chlorophyll a, chlorophyll b, and carotenoids in soybean leaves, *Remote Sensing of Environment*, 39, 239–247.

41. Sims, D., Gamon, J.A. 2002. Relationship between leaf pigment content and spectral reflectance across a wide range of species, leaf structure and development stages, *Remote Sensing of Environment*, 81, 337–354.

42. Haboudane, D., Miller, J.R., Tremblay, N., Zarco-Tejada, P.J., Dextraze, L. 2002. Integrated narrow-band vegetation indices for prediction of crop chlorophyll content for application to precision agriculture, *Remote Sensing of Environment*, 81, 416–426.

43. Gitelson, A.A., Keydan, G.P., Merzlyak, M.N. 2006. Three-band model for non invasive estimation of chlorophyll, carotenoids, and anthocyanin contents in higher plant leaves, *Geophysical Research Letters*, 33, (L11140), doi:10.1029/2006Gl026457.

44. Ryu, C., Suguri, M., Umeda, M. 2009. Model for predicting the nitrogen content of rice at panicle initiation stage using data from airborne hyperspectral remote sensing, *Biosystems Engineering*, 104, 465–475.

45. Cilia, C., Panigada, C., Rossini, M., Meroni, M., Busetto, L., Amaducci, S., Boschetti, M., Picchi, V., Colombo, R. 2014. Nitrogen status assessment for variable rate fertilization in maize through hyperspectral imagery, *Remote Sensing*, 6(7), 6549–6565, doi:10.3390/rs6076549.

46. Goel, P.K., Prasher, S.O., Landry, J.A., Patel, R.M., Bonnell, R.B., Viau, A.A., Miller, J.R. 2003. Potential of airborne hyperspectral remote sensing to detect nitrogen deficiency and weed infestation in corn, *Computers and Electronics in Agriculture*, 38, 99–124.

47. Chen, P., Haboudane, D., Tremblay, N., Wang, J., Vigneault, P., Li, B. 2010. New spectral indicator assessing the efficiency of crop nitrogen treatment in corn and wheat, *Remote Sensing of Environment*, 114, 1987–1997, doi:10.1016/j.rse.2010.04.006.

48. Zhou, X., Huang, W., Kong, W., Ye, H., Luo, J., Chen, P. 2016. Remote estimation of canopy nitrogen content in winter wheat using airborne hyperspectral reflectance measurements, *Advances in Space Research*, 58, 1627–1637.

49. Moharana, S., Dutta, S. 2016. Spatial variability of chlorophyll and nitrogen content of rice from hyperspectral imagery, *ISPRS Journal of Photogrammetry and Remote Sensing*, 122, 17–29. ISSN 0924-2716, https://doi.org/10.1016/j.isprsjprs.2016.09.002.

50. Zhang, C., Kovacs, J.M. 2012. The application of small unmanned aerial systems for precision agriculture: A review, *Precision Agriculture*, 13, 693–712.

51. Uto, K., Seki, H., Saito, G., Kosugi, Y. 2013. Characterization of rice paddies by a UAV-mounted miniature hyperspectral sensor system, *IEEE Journal of Selected Topics in Applied Earth Observations and Remote Sensing*, 6(2), 851–860.

52. Verger, A., Vigneau, N., Chéron, C., Gilliot, J.M., Baret, F. 2014. Green area index from unmanned aerial system over wheat and rapeseed crops, *Remote Sensing of Environment*, 152, 654–664.

53. Bendig, J., Yu, K., Aasen, H., Bolten, A., Bennertz, S., Broscheit, J., Gnyp, M.L., Bareth, G. 2015. Combining UAV-based plant height from crop surface models, visible, and near infrared vegetation indices for biomass monitoring in barley. *International Journal of Applied Earth Observation and Geoinformation*, 39, 79–87.

54. Zarco-Tejada, P.J., Guillén-Climent, M.L., Hernández-Clemente, R., Catalina, A., González, M.R., Martín, P. 2013. Estimating leaf carotenoid content in vineyards using high resolution hyperspectral imagery acquired from an unmanned aerial vehicle (UAV), *Agricultural and Forest Meteorology*, 281–294. ISSN 0168-1923, https://doi.org/10.1016/j.agrformet.2012.12.013m.

55. Caturegli, L., Corniglia, M., Gaetani, M., Grossi, N., Magni, S., Migliazzi, M., Angelini, L. et al. 2016. Unmanned aerial vehicle to estimate nitrogen status of turfgrasses, *PLOS ONE*, 11, (6), e0158268. https://doi.org/10.1371/journal.pone.0158268.

56. Pölönen, I., Saari, H., Kaivosoja, J., Eija, H., Liisa, P. 2013. Hyperspectral imaging based biomass and nitrogen content estimations from light-weight UAV. *Proceedings Volume 8887, Remote Sensing for Agriculture, Ecosystems, and Hydrology XV*; 88870J (2013); doi:10.1117/12.2028624.

57. Quemada, M., Gabriel, J.L., Zarco-Tejada, P. 2014. Airborne hyperspectral images and ground-level optical sensors as assessment tools for maize nitrogen fertilization, *Remote Sensing*, 6, (4), 2940–2962, doi:10.3390/rs6042940.

58. Sun, J., Shi, S., Gong, W., Yang, J., Du, L., Song, S., Chen, B., Zhang, Z. 2017. Evaluation of hyperspectral LiDAR for monitoring rice leaf nitrogen by comparison with multispectral LiDAR and passive spectrometer, *Scientific Reports*, 7, doi:10.1038/srep40362.

59. Plaza, A., Benediktsson, J.A., Boardman, J.W., Brazile, J., Bruzzone, L., Camps-Valls, G., Chanussot, J. et al. 2009. Recent advances in techniques for hyperspectral image processing, *Remote Sensing of Environment*, 113, (1), S110–S122. ISSN 0034-4257, https://doi.org/10.1016/j.rse.2007.07.028.

60. Goel, P.K., Prasher, S.O., Patel, R.M., Landry, J.A., Bonnell, R.B., Viau, A.A. 2003. Classification of hyperspectral data by decision trees and artificial neural networks to identify weed stress and nitrogen status of corn, *Computers and Electronics in Agriculture*, 39, (2), 67–93. ISSN 0168-1699, https://doi.org/10.1016/S0168-1699(03)00020-6.

61. Sembiring, H., Raun, W.R., Johnson, G.V., Stone, M.L., Solie, J.B., Phillips, S.B. 1998. Detection of nitrogen and phosphorus nutrient status in bermudagrass using spectral radiance, *Journal of Plant Nutrition*, 21, 1189–1206.

62. Starks, P.J., Zhao, D.L., Phillips, W.A., Coleman, S.W. 2006. Development of canopy reflectance algorithms for real-time prediction of bermudagrass pasture biomass and nutritive values, *Crop Science*, 46, 927–934.

63. Zhao, D., Starks, P.J., Brown, M.A., Phillips, W.A., Coleman, S.W. 2007. Assessment of forage biomass and quality parameters of bermudagrass using proximal sensing of pasture canopy reflectance, *Grassland Science*, 53, 39–49.

64. Mutanga, O., Skidmore, A.K., van Wieren, S. 2003. Discriminating tropical grass (*Cenchrus ciliaris*) canopies grown under different nitrogen treatments using spectroradiometry, *ISPRS Journal of Photogrammetry and Remote Sensing*, 57, 263–272.

65. Lamb, D.W., Steyn-Ross, M., Schaare, P., Hanna, M.M., Silvester, W., & Steyn-Ross, A. 2002. Estimating leaf nitrogen concentration in ryegrass (*Lolium* spp.) pasture using the chlorophyll red-edge: Theoretical modelling and experimental observations, *International Journal of Remote Sensing*, 23, 3619–3648.

66. Schut, A.G.T., Lokhorst, C., Hendriks, M., Kornet, J.G., Kasper, G. 2005. Potential of imaging spectroscopy as tool for pasture management, *Grass and Forage Science*, 60, 34–45.

67. Biewer, S., Fricke, T., Wachendorf, M. 2009. Development of canopy reflectance models to predict forage quality of legume-grass mixtures, *Crop Science*, 49, 1917–1926.

68. Mutanga, O., Skidmore, A.K., Kumar, L., Ferwerda, J. 2005. Estimating tropical pasture quality at canopy level using band depth analysis with continuum removal in the visible domain, *International Journal of Remote Sensing*, 26, 1093–1108.

69. Cho, M.A. Skidmore, A.K. 2006. A new technique for extracting the red edge position from hyperspectral data: The linear extrapolation method, *Remote Sensing of Environment*, 101, 181–193.

70. Mutanga, O., Skidmore, A.K. 2007. Red edge shift and biochemical content in grass canopies, *ISPRS Journal of Photogrammetry and Remote Sensing*, 62, 34–42.

71. Knox, N.M., Skidmore, A.K., Schlerf, M., de Boer, W.F., van Wieren, S.E., van der Waal, C., Prins, H.H.T., Slotow, R. 2010. Nitrogen prediction in grasses: Effect of bandwidth and plant material state on absorption feature selection, *International Journal of Remote Sensing*, 31, 691–704.

72. Mutanga, O., Skidmore, A.K., Prins, H.H.T. 2004. Predicting *in situ* pasture quality in the Kruger National Park, South Africa, using continuum-removed absorption features, *Remote Sensing of Environment*, 89, 393–408.

73. Mutanga, O., Skidmore, A.K. 2004. Integrating imaging spectroscopy and neural networks to map grass quality in the Kruger National Park, *South Africa, Remote Sensing of Environment*, 90, 104–115.

74. Skidmore, A.K., Ferwerda, J.G., Mutanga, O., Van Wieren, S.E., Peel, M., Grant, R.C., Prins, H.H.T., Balcik, F.B., Venus, V. 2010. Forage quality of savannas—Simultaneously mapping foliar protein and polyphenols for trees and grass using hyperspectral imagery, *Remote Sensing of Environment*, 114, 64–72.

75. Ramoelo, A., Skidmore, A.K., Cho, M.A., Mathieu, R., Heitkönig, I.M.A., Dudeni-Tlhone, N., Schlerf, M., Prins, H.H.T. 2013. Non-linear partial least square regression increases the estimation accuracy of grass nitrogen and phosphorus using *in situ* hyperspectral and environmental data, *ISPRS journal of Photogrammetry and Remote Sensing*, 82, 27–40.

76. Adjorlolo, C., Mutanga, O., Cho, M.A. 2015. Predicting C3 and C4 grass nutrient variability using *in situ* canopy reflectance and partial least squares regression, *International Journal of Remote Sensing*, 36, (6), 1743–1761.

77. Singh, L., Mutanga, O., Mafongoya, P. and Peerbhay, K. 2017. Remote sensing of key grassland nutrients using hyperspectral techniques in KwaZulu-Natal, *South Africa, I*, 36005.

78. Boschetti, M., Bocchi, S. and Brivio, P.A. 2007. Assessment of pasture production in the Italian Alps using spectrometric and remote sensing information, *Agriculture Ecosystems & Environment*, 118, 267–272.

79. Mirik, M., Norland, J.E., Crabtree, R.L., Biondini, M.E. 2005. Hyperspectral one-meter-resolution remote sensing in Yellowstone National Park, Wyoming: I. Forage nutritional values, *Rangeland Ecology & Management*, 58, 452–458.

80. Beeri, O., Phillips, R., Hendrickson, J., Frank, A.B., Kronberg, S. 2007. Estimating forage quantity and quality using aerial hyperspectral imagery for northern mixed-grass prairie, *Remote Sensing of Environment*, 110, 216–225.

81. Ling, B., Goodin, D.G., Mohler, R.L., Laws, A.N. and Joern, A. 2014. Estimating canopy nitrogen content in a heterogeneous grassland with varying fire and grazing treatments: Konza Prairie, Kansas, USA, *Remote Sensing*, 6, (5), 4430–4453.

82. Pullanagari, R.R., Yule, I.J., Tuohy, M.P., Hedley, M.J., Dynes, R.A., King, W.M. 2012. In-field hyperspectral proximal sensing for estimating quality parameters of mixed pasture, *Precision Agriculture*, 13, (3), 351–369.

83. Pullanagari, R.R., Kereszturi, G., Yule, I.J. 2016. Mapping of macro and micro nutrients of mixed pastures using airborne AisaFENIX hyperspectral imagery, *ISPRS Journal of Photogrammetry and Remote Sensing*, 117, 1–10.

84. Sanches, I.D., Tuohy, M.P., Hedley, M.J., Mackay, A.D. 2013. Seasonal prediction of *in situ* pasture macronutrients in New Zealand pastoral systems using hyperspectral data, *International Journal of Remote Sensing*, 34, (1), 276–302.

85. Thulin, S., Hill, M.J., Held, A., Jones, S., Woodgate, P. 2014. Predicting levels of crude protein, digestibility, lignin and cellulose in temperate pastures using hyperspectral image data, *American Journal of Plant Sciences*, 5, (7), 997.

86. Pellissier, P.A., Ollinger, S.V., Lepine, L.C., Palace, M.W., McDowell, W.H. 2015. Remote sensing of foliar nitrogen in cultivated grasslands of human dominated landscapes, *Remote Sensing of Environment*, 167, 88–97.

87. Safari, H., Fricke, T., Wachendorf, M. 2016. Determination of fibre and protein content in heterogeneous pastures using field spectroscopy and ultrasonic sward height measurements, *Computers and Electronics in Agriculture*, 123, 256–263.

88. Kokaly, R.F., Clark, R.N. 1999. Spectroscopic determination of leaf biochemistry using band-depth analysis of absorption features and stepwise multiple linear regression, *Remote Sensing of Environment*, 67, 267–287.

89. Mahajan, G.R., Pandey, R.N., Sahoo, R.N., Gupta, V.K., Datta Dinesh Kumar, S.C. 2017. Monitoring nitrogen, phosphorus and sulphur in hybrid rice (*Oryza sativa* L.) using hyperspectral remote sensing, *Precision Agriculture*, 18(5), 736–761.

90. Pellissier, P.A., Ollinger, S.V., Lepine, L.C., Palace, M.W., McDowell, W.H. 2015. Remote sensing of foliar nitrogen in cultivated grasslands of human dominated landscapes. *Remote Sensing of Environment*, 167, 88–97.

91. Pimstein, A., Karnieli, A., Bansal, S.K., Bonfil, D. J. 2011. Exploring remotely sensed technologies for monitoring wheat potassium and phosphorus using field spectroscopy, *Field Crop Research*, 121, 125–135.

92. Tremblay, N., Wang, Z., Cerovic, Z.G. 2012. Sensing crop nitrogen status with fluorescence indicators. A review, *Agronomy for Sustainable Development*, 32, 451–464, doi:10.1007/s13593-011-0041-1.

93. ESA. 2015. Report for Mission Selection: FLEX. ESA SP-13330/2 (2 volume series), European Space Agency, Noordwijk, The Netherlands. http://esamultimedia.esa.int/docs/EarthObservation/SP1330-2_FLEX.pdf.

94. Rossini, M., Nedbal, L., Guanter, L., Ač, A., Alonso, L., Burkart, A., Cogliati, S. et al. 2015. Red and far red sun-induced chlorophyll fluorescence as a measure of plant photosynthesis, *Geophysical Research Letters*, 42, 1632–1639, doi:10.1002/2014GL062943.

95. Curran, P.J., Dungan, J.L., Macler, B.A., Plummer, S.E. 1991. The effect of a red leaf pigment on the relationship between red edge and chlorophyll concentration, *Remote Sensing of Environment*, 35, 69–76.

96. Dawson, T.P. Curran, P.J. 1998. A new technique for interpolating red edge position, *International Journal of Remote Sensing*, 19, (11), 2133–2139.

97. Schertz, F.M. 1921. A chemical and physiological study of mottling of leaves, *Botanical Gazette*, 71, 81–130.

98. Tam, R.K., Magistad, O.C. 1935. Relationship between nitrogen fertilization and chlorophyll content in pineapple plants, *Plant Physiology*, 10, 159–168.

99. Houlès, V., Guérif, M., Mary, B. 2007. Elaboration of a nitrogen nutrition indicator for winter wheat based on leaf area index and chlorophyll content for making nitrogen recommendations, *European Journal of Agronomy*, 27, 1–11.

100. Filella, I., Serrano, I., Serra, J., Penuelas, J. 1995. Evaluating wheat nitrogen status with canopy reflectance indices and discriminant analysis, *Crop Science*, 35, 1400–1405.

101. Thomas, J.R., Oerther, G.F. 1972. Estimating nitrogen content of sweet pepper leaves by reflectance measurements, *Agronomy Journal*, 64, 11–13.

102. Yoder, B.J., Pettigrew-Crosby, R.E. 1995. Predicting nitrogen and chlorophyll content and concentrations from reflectance spectra (400–2500 nm) at leaf and canopy scales, *Remote Sensing of Environment*, 53, 199–211.

103. Gausman, H.W., Gerbermann, A.H., Wiegand, C.L., Leamer, R.W., Rodriguez, R.R., Noriega, J.R. 1975. Reflectance differences between crop residues and bare soils, *Soil Science Society of America Journal.*, 39, 752–755.

104. Kokaly, R.F., Asner, G.P., Ollinger, S.V., Martin, M.E., Wessman, C.A. 2009. Characterizing canopy biochemistry from imaging spectroscopy and its application to ecosystem studies, *Remote Sensing of Environment*, 113, 578–591.

105. Zhao, D.H., Li, J.L., Qi, J.G. 2005. Identification of red and NIR spectral regions and vegetative indices for discrimination of cotton nitrogen stress and growth stage, *Computers and Electronics in Agriculture*, 48, 155–169.

106. Plaza, A., Benediktsson, J.A., Boardman, J.W., Brazile, J., Bruzzone, L., Camps-Valls, G., Chanussot, J. et al. 2009. Recent advances in techniques for hyperspectral image processing. *Remote Sensing of Environment*, 113(1), S110–S122.

107. Jacquemoud, S., Baret, F., Andrieu, B., Danson, F.M., Jaggard, K. 1995. Extraction of vegetation biophysical parameters by inversion of the PROSPECT+SAIL models on sugar beet canopy reflectance data. Application to TM and AVIRIS sensors, *Remote Sensing of Environment*, 52, 163–172.

108. Clevers, J.G.P.W., Kooistra, L. 2012. Using hyperspectral remote sensing data for retrieving canopy chlorophyll and nitrogen content, *IEEE Journal of Selected Topics in Applied Earth Observations and Remote Sensing*, 5, (2), 574–583, doi: 10.1109/JSTARS.2011.2176468.

109. Fernandes, R., Leblanc, S.G. 2005. Parametric (modified least square) and non-parametric (Theil-Sen) linear regression for predicting biophysical parameters in the presence of measurements errors, *Remote Sensing of Environment*, 95, 301–316.

110. Blackburn, G.A. 1999. Relationship between spectral reflectance and pigment concentrations in stacks of deciduous broadleaves, *Remote Sensing of Environment*, 70, 224–237.

111. Zarco-Tejada, P.J., Miller, J.R., Morales, A., Berjón, A., Agüera, J. 2004. Hyperspectral indices and model simulation for chlorophyll estimation in open-canopy tree crops, *Remote Sensing of Environment*, 90, 463–476.

112. Tian, Y.C., Yao, X., Yang, J., Cao, W.X., Hannaway, D.B., Zhu, Y. 2001. Assessing newly developed and published vegetation indices for estimating rice leaf nitrogen concentration with ground- and space-based hyperspectral reflectance, *Field Crops Research*, 120, (2), 299–310. ISSN 0378-4290, https://doi.org/10.1016/j.fcr.2010.11.002.

113. Alchanatis, V., Schmilvitch, Z. 2005. In-field assessment of single leaf nitrogen status by spectral response measurements, *Precision Agriculture*, 6, 25–39.

114. Masoni, A., Ercoli, L., Mariotti, M. 1996. Spectral properties of leaves deficient in iron, sulfur, magnesium, and manganese, *Agronomy Journal*, 88, 937–943.

115. Kim, M.S., Daughtry, C.S.T., Chapelle, E.W., McMurtrey, J.E. 1994. The use of high spectral resolution bands for estimating absorbed photosynthetically active radiation (APAR). *Proceedings of the 6th International Symposium on Physical Measurements and Signatures in Remote Sensing*, France: Val D'Isere, 229–306.

116. Hinzman, L.D., Bauer, M.E., Daughtry, C.S.T. 1986. Effects of N fertilization on growth and reflectance characteristics of winter wheat, *Remote Sensing of Environment*, 19, 47–61.

117. Slater, P.N., Jackson, R.D. 1982. Atmospheric effects on radiation reflected from soil and vegetation as measured by orbit sensors using various scanning directions, *Applied Optics*, 21, 3923–4.

118. Chen, J. 1996. Evaluation of vegetation indices and modified simple ratio for boreal applications, *Canadian Journal of Remote Sensing*, 22, 229–242.

119. Cundill, S.L., van der Werff, H.M.A., van der Meijde, M. 2015. Adjusting spectral indices for spectral response function differences of very high spatial resolution sensors simulated from field spectra, *Sensors*, 15, (3), 6221–6240, doi:10.3390/s150306221.

120. Rouse, J.W., Haas, R.H. Jr., Schell, J.A., Deering, D.W. 1974. Monitoring vegetation systems in the Great Plains with ERTS, NASA SP-351, *3rd ERTS-1 Symp*. Washington, DC, pp. 309–317.

121. Running, S.R., Nemani, R.R. 1998. Relating seasonal patterns of the AVHRR vegetation index to simulated photosynthesis and transpiration of forests in different climates, *Remote Sensing of Environment*, 24, 347–367.

122. Gamon, J.A., Peñuelas, J., Field, C.B. 1992. A narrow-waveband spectral index that tracks diurnal changes in photosynthetic efficiency, *Remote Sensing of Environment*, 41, 35–44.

123. Wang, Z.J., Wang, J.H., Liu, L.Y., Huang, W.J., Zhao, C.J., Wang, C.Z. 2004. Prediction of grain protein content in winter wheat (*Triticum aestivum* L.) using plant pigment ratio (PPR), *Field Crops Research*, 90, 311–321.

124. Fitzgerald, G.J., Rodriguez, D., Christensen, L.K., Belford, R., Sadras, V.O., Clarke, T.R. 2007. Spectral and thermal sensing for nitrogen and water status in rainfed and irrigated wheat environments, *Precision Agriculture*, 7, 233–248.

125. Tilling, A.K., O'leary, G.J., Ferwerda, J.G., Jones, S.D., Fitzgerald, G.J., Rodriguez, D., Belford, R. 2007. Remote sensing of nitrogen and water stress in wheat, *Field Crops Research*, 104, 77–85.

126. Huete, A.R., Jackson, R.D., Post, D.F. 1985. Spectral response of a plant canopy with different soil backgrounds, *Remote Sensing of Environment*, 17, 37–53.

127. Rondeaux, G., Steven, M., Baret, F. 1996. Optimization of soil-adjusted vegetation indices, *Remote Sensing of Environment*, 55, 95–107.

128. Baret, F., Guyot, G., Major, D.J. 1989. TSAVI: A vegetation index which minimizes soil brightness effects on LAI and APAR estimations, in *Quantitative Remote Sensing for the Nineties, Proc. IGARSS '89*, Vancouver, Canada, vol. 3, 1355–1358.

129. Zarco-Tejada, P.J., Morales, A., Testi, L., Villalobos, F.J. 2013. Spatio-temporal patterns of chlorophyll fluorescence and physiological and structural indices acquired from hyperspectral imagery as compared with carbon fluxes measured with eddy covariance, *Remote Sensing of Environment*, 133, 102–115.

130. Horler, D.N.H., Dockray, M., Barber, J. 1983. The red edge of plant leaf reflectance, *International Journal of Remote Sensing*, 4, 273–288.

131. Guyot, G., Baret, F. 1988. Utilisation de la haute résolution spectrale pour suivre l'état des couverts végétaux. *Proceedings of the 4th International Colloquium on Spectral Signatures of Objects in Remote Sensing. ESA SP-287*, Assois, France, 279–286.

132. Bonham-Carter, G.F. 1988. Numerical procedures and computer program for fitting an inverted Gaussian model to vegetation reflectance data, *Computers and Geosciences*, 14, (3), 339–356.

133. Miller, J.R., Hare, E.W., Wu, J. 1990. Quantitative characterization of the red edge reflectance. An inverted-Gaussian reflectance model, *International Journal of Remote Sensing, 11*, 10, 1755–1773.

134. Savitzky, A., Golay, M.J.E. 1964. Smoothing and differentiation of data by simplified least-squares procedures, *Analytical Chemistry, 36*, 8, 1627–1639.

135. Demetriades-Shah, T.H., Steven, M.D., Clark, J.A. 1990. High resolution derivative spectra in remote sensing, *Remote Sensing of Environment*, 33, 55–64.

136. Clevers, J.G.P.W., De Jong, S.M., Epema, G.F., Van der Meer, F., Bakker, W.H., Skidmore, A.K., Sholte, K.H. 2002. Derivation of the red edge index using MERIS standard band setting, *International Journal of Remote Sensing*, 23, (16), 3169–3184.

137. Baranoski, G.V.G., Rokne, J.G. 2005. A practical approach for estimating the red edge position of plant leaf reflectance, *International Journal of Remote Sensing*, 26, (3), 503–521.

138. Thenkabail, P.S., Enclona, E.A., Ashton, M.S., Van der Meer, B. 2004. Accuracy assessment of hyperspectral wavebands performance for vegetation analysis applications, *Remote Sensing of Environment*, 91, 354–376.

9 Hyperspectral Remote Sensing of Leaf Chlorophyll Content

From Leaf, Canopy, to Landscape Scales

Yongqin Zhang

CONTENTS

9.1   INTRODUCTION

A plant's physiological state is governed by its biochemical constituents, including photosynthetic and other enzyme systems, structural and nonstructural carbohydrates, chlorophyll and associated light harvesting complexes, and photoprotective and ancillary pigments. Many biochemical processes, such as photosynthesis, net primary production, and decomposition, are related to the content of biochemicals in leaves [1–3]. Of these biochemicals, leaf chlorophyll content stands out as being both sensitive to environmental conditions and having a very strong influence on leaf optical properties and canopy albedo [4–6]. All green leaves have major absorption features in the 400–700 nm range caused by electron transitions in chlorophyll and carotenoid pigments [7]. Most green vegetation shows absorption peaks near 420, 490, and 670 nm due to the strong absorption peaks of chlorophyll *a* and *b*. The absorption of other pigments in the leaf, such as carotenes and xanthophylls, is usually obscured by the chlorophyll *a* and *b* absorption. Differences in leaf and canopy reflectance between healthy and stressed vegetation due to changes in chlorophyll levels have been detected in the green peaks and along the red edge (690–750 nm) [8–12]. Changes of red-edge position and slope are associated with vegetation stress [9,13–15]. Canopy reflectance in the green and far-red regions is also sensitive to variations in chlorophyll concentration and can act as an indicator of vegetation stress [10]. Leaf chlorophyll content is of great use for forest health status evaluation and sustainable forest management [16,17]. Leaf chlorophyll content also serves as an input to photosynthesis and carbon cycle models. Changes in leaf optical properties and chlorophyll content, including responses to rising atmospheric CO_2 and other global change variables, may have important implications for climate forcing as well [18].

Quantitative estimates of leaf chlorophyll content may thus provide a useful indicator of important physiological processes in vegetation that can be readily assessed via hyperspectral remote sensing data. Cutting-edge hyperspectral technology provides simultaneous acquisition of information in narrow but contiguous spectral bands. Hyperspectral data are almost spectrally "continuous," which is sufficient to detect subtle absorption features in foliar spectra and to study the correlations of vegetation's minor absorption features to biochemical parameters. They are particularly useful for estimating vegetation structure and biochemistry, which are important for studying nutrient cycling, productivity, and vegetation stress and for ecosystem modeling [19–22].

Here, methods for estimating leaf and forest canopy chlorophyll content using hyperspectral remote sensing data are summarized, and the suitability and limitations of each method are analyzed. In particular, recent advances [23–25] of quantitative methods using physically based algorithms are presented in more detail. These descriptions follow a sequence of estimating broadleaf and needleleaf chlorophyll content from leaf-level hyperspectral measurements, retrieving leaf chlorophyll content of forest canopies from hyperspectral remote sensing imagery, and scaling of leaf-level chlorophyll estimation to forest canopy levels. The applications of estimating forest chlorophyll content using hyperspectral data are discussed.

9.2 ESTIMATING LEAF CHLOROPHYLL CONTENT FROM LEAF-LEVEL HYPERSPECTRAL MEASUREMENTS

Leaf optical properties (and thus leaf chemistry) and canopy structure determine the remote sensing signals originating from vegetated surfaces. Deriving leaf optical and canopy structural properties are the two main domains of remote sensing of terrestrial ecosystems.

Leaf optical properties depend on leaf structure, leaf biochemical composition, distribution of leaf biochemical constituents, and the complex refraction index of these constituents [26–31]. Radiation scattering by a leaf is caused by optical inhomogeneities in the leaf surface, leaf thickness, cell size, and the internal cell structure. Variability in leaf optical properties is wavelength dependent. Spectrally continuous leaf-level hyperspectral measurements are especially useful to detect subtle features and changes in the leaf optical spectra that correlate with leaf chlorophyll content. Accurate estimation of leaf chlorophyll content from leaf optical spectra is an essential step to derive canopy chlorophyll content from remotely sensed imagery.

There are two types of approach to estimat leaf chlorophyll content from remote sensing products: the empirical method and physically based method. The empirical approach uses statistical models derived from spectral indices [32] or through partial least-squares regression [33], wavelet analysis [34], and machine learning algorithms such as Bayesian optimization [35]. Empirical spectral indices are perhaps the most popular and straightforward means of retrieving chlorophyll content from reflectance factors. The physically based approach accounts for the underlying physics and the complexity of leaf biochemical properties. Various physically based models have been developed to simulate leaf spectral characteristics based on the interactions of incident radiation with leaves. Details of empirical and physically based approaches will be presented in Sections 9.3.1 and 9.3.2. Here, a comparison of the strength and limitations is listed in Table 9.1.

9.2.1 EMPIRICAL METHOD FOR LEAF CHLOROPHYLL CONTENT ESTIMATION

At the leaf level, empirical relationships between spectral indices and chlorophyll content measurements have been widely exploited for estimating leaf chlorophyll content. They have been

TABLE 9.1

Comparison of the Two Approaches in Estimating Leaf Chlorophyll Content

	Empirical Approach	Physically Based Approach
Strengths	• Straightforward. • Non-destructive. • Efficient.	• The approach is robust and produces accurate prediction of leaf pigment content, as physics laws are applied to consider interactions of light with leaf and canopy elements. • The approach is applicable for different species and can produce accurate prediction of leaf pigment content.
Limitations	• Species- and site-specific, statistical relationships often need to be recalibrated. • Most are applicable for broadleaves, not for needleleaves. • Lack of generality and transferability across different temporal and spatial scales [36].	• Complicated. • Calibration of model parameters presents difficulty.

proven effective for nondestructive estimations of leaf chlorophyll content from field and laboratory measurements of leaf spectral reflectance [37–42].

Many spectral indices have been developed through identifying relationships between the leaf reflectance and chlorophyll content. le Maire et al. [32] summarized the chlorophyll spectral indices published until 2002. In the red and blue regions of the electromagnetic spectrum, chlorophylls have strong absorbance peaks. Spectral indices employ ratios of narrow bands within spectral ranges that are sensitive to chlorophylls to those not sensitive and/or related to some other control on reflectance. To avoid the saturation of indices under low chlorophyll content, reflectances near, instead of exactly at, the maximum absorption wavelengths are generally selected to develop spectral indices. This method solves the problem of overlapping absorption spectra of different pigments, the effects of leaf surface interactions, and leaf structure. The majority of chlorophyll indices are based on ratios of narrow bands in the visible and near-infrared region [43–45]. Some indices use ratios of narrow bands only in the visible [46], red-edge [9], or red-edge and near-infrared shoulder regions [39]. Some indices use three bands [32,47–49] and can differentiate chlorophyll *a* and *b* [50]. Generally, indices using three bands are applicable for leaf-level chlorophyll estimation. Multiple narrow bands, or whole spectrum through principal component transformation [51], factor analysis [52], artificial neural networks [53], and stepwise multiple regression [54,55], have attempted to quantify leaf and canopy chlorophyll concentrations.

le Maire et al. [32] concluded that at leaf level, simple spectral indices give better estimations than indices related to the red-edge inflection point, derivative-based indices, or indices based on neural network analysis of empirical hyperspectral data. Zhang et al. [23] measured reflectance and transmittance spectral of 255 sugar maples (*Acer saccharum*) from 350 to 2500 nm at 1-nm intervals using a portable field spectroradiometer FieldSpec Pro FR (Analytical Spectral Devices, Inc., Boulder, CO, USA) attached via fiber optic to the Li-Cor 1800 integrating sphere (Li-Cor 1800–12S, Li-COR, Inc., Lincoln, NE, USA). Chlorophyll *a* and *b* content of the 255 leaves were measured subsequently to the leaf spectral measurements. Using the leaf spectral and chlorophyll content measurements, Zhang et al. [23] analyzed a number of simple chlorophyll indices for estimating broadleaf chlorophyll content. These indices have previously been shown to produce low deviation from measurements of chlorophyll content, including the modified simple ratio index mSR and the modified normalized difference index mND of Sims and Gamon [48], the double difference index DD, the first derivative-based index BmSR, the first derivative-based index BmND of le Maire et al. [32], and the red-edge normalized differences index NDI of Gitelson and Merzlyak

TABLE 9.2

Estimates of Leaf Chlorophyll Content Using Empirical Indices in Comparisons to Laboratory Chlorophyll Measurements

Spectral Index	Formula	RMSE (μg/cm^2)	R^2
mSR	$(R_{728} - R_{434})/(R_{720} - R_{434})$	3.94	0.875
mND	$(R_{728} - R_{720})/(R_{728} + R_{720} - 2*R_{434})$	3.97	0.872
BmSR	$(\delta R_{722} - \delta R_{502})/(\delta R_{701} - \delta R_{502})$	3.98	0.872
NDI	$(R_{750} - R_{705})/(R_{750} + R_{705})$	4.11	0.864
DD	$(R_{749} - R_{720}) - (R_{701} - R_{672})$	4.15	0.861
BmND	$(\delta R_{722} - \delta R_{699})/(\delta R_{722} + \delta R_{699} - 2*\delta R_{502})$	4.27	0.853

[39]. The performances of these indices for chlorophyll content estimations are listed in Table 9.2. Compared to the laboratory measurements of leaf chlorophyll content, these spectral indices gave good estimates of broadleaf chlorophyll content. Among the indices, the modified simple ratio mSR produced the best estimation with a root mean square error (RMSE) of 3.94 μg/cm^2 and an R^2 of 0.875, and the modified normalized difference index mND performed nearly as well.

Spectral indices provide nondestructive, efficient, and sensitive measurements of leaf chlorophyll content from leaf spectral reflectance. They can serve as indicators of vegetation stress, senescence, and disease. However, they are generally suitable for broadleaved species. Spectral indices may perform poorly for needleleaf plants, as needleleaves are smaller and narrower compared to broadleaves. Also, spectral indices were also developed for some specific plant species. As the size, shape, surface, and internal structure of leaves may vary from species to species, the application of optical indices to other vegetation types or vegetation communities needs to be reinvestigated. Efforts have been made to improve the robustness and generality of chlorophyll indices by testing them over a range of species and physiological conditions. Nevertheless, when applied to a specific species, spectral indices need calibration.

9.2.2 Physically Based Model Inversion Method to Estimate Leaf Chlorophyll

Considerable progress has been made in physically based models that simulate leaf spectral characteristics based on their interactions with incident radiation with foliar elements. These radiometric models account for the underlying physics and complexity of leaf biochemical properties and therefore are robust and applicable for different species. Four categories of models now exist: (i) plate models assume a leaf as one or several absorbing plates with rough surfaces giving rise to scattering of light [56], typical of which is PROSPECT [57]; (ii) N-flux models consider a leaf as a horizontally homogeneous parallel medium with downwelling and upwelling irradiance and isotropic scattering [58,59]; (iii) ray-tracing models describe the complexity of leaf internal structure with a detailed description of individual cells and their unique arrangement inside tissues [60–63]; and (iv) stochastic models such as the LEAFMOD simulate leaf optical properties by a Markov chain or basic radiative transfer equations [64–67].

Numerical inversion of leaf-level radiative transfer (RT) models, such as PROSPECT and LEAFMOD, has demonstrated success for predicting leaf chlorophyll content [58,67–70]. Numerical inversion techniques offer the potential of a generically superior approach to estimating leaf chlorophyll content from hyperspectral data than spectral indices and other approaches that are based on empirical calibrations. Of all the leaf optical models, PROSPECT is a simple but effective model for leaf biochemical content retrieval. The PROSPECT model has been widely validated with various broadleaf species.

9.2.2.1   Modeling Method for Broadleaf Chlorophyll Content Estimation

The PROSPECT model assumes a leaf is a stack of L identical elementary layers separated by $L - 1$ air spaces. The number of layers mimics the scattering within the leaf. Scattering is described by the refractive index (n) of leaf materials and by a parameter characterizing the leaf mesophyll structure (L). Layers are defined by their refractive index and absorption coefficient K_i. Absorption is the linear summation of the contents of the constituent chemicals and the corresponding specific absorption coefficients.

$$K(\lambda) = Ke(\lambda) + \sum K_i(\lambda)C_i/L \qquad (9.1)$$

where λ is the wavelength; $K_e(\lambda)$ is the absorption coefficient of elementary albino and dry layer; C_i is the content of layer constituent i (chlorophyll$_{a+b}$, water and dry matter) per unit area; $K_i(\lambda)$ is the corresponding specific absorption coefficients of the constituent i; and L is the leaf structure parameter, which is the number of compact layers specifying the average number of air/cell walls interfaces within the mesophyll.

Through numerical iteration, the chemical contents can be derived from the leaf spectra. First, an initial guess of the structure parameter L and the concentration of three constituents are input in the forward model to calculate the absorption coefficient $K(\lambda)$ and the hemispherical reflectance and transmittance. The estimated hemispherical reflectance and transmittance are then compared with the measured leaf reflectance and transmittance. Using an optimization algorithm, the i constituents can be numerically iterated by minimizing the merit function [71]:

$$\Delta = \sum_{\lambda}\{[R_{mes}(\lambda) - R_{mod}(\lambda)]^2 + [T_{mes}(\lambda) - T_{mod}(\lambda)]^2\} \qquad (9.2)$$

where R_{mes} and T_{mes} are the measured reflectance and transmittance, and R_{mod} and T_{mod} are the estimated reflectance and transmittance from the model.

Zhang et al. [23] found that the original PROSPECT model performed well for the overstory leaf samples collected in summer (July and August) that have high chlorophyll content (Figure. 9.1a). However, the variations of leaf chlorophyll content across the season and canopy height were not well captured. Specifically, understory leaf samples and samples collected in the early (May and early June) and late (middle to late September) growing season have low leaf chlorophyll content. The model

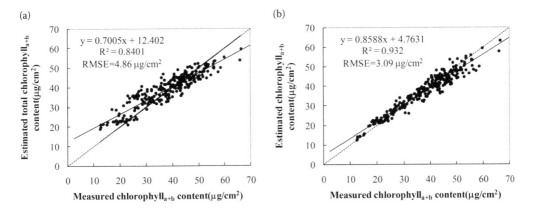

FIGURE 9.1   Comparison of leaf chlorophyll$_{a+b}$ content from measurements and from estimates from (a) leaf chlorophyll content from measurements and from the original PROSPECT model estimates (b) leaf chlorophyll content from the measurements and the PROSPECT model after incorporating leaf thickness factor.

TABLE 9.3

Leaf Thickness Factor Used in the Model for Overstory and Understory Leaves through the Growing Season

	May 27	Jun 10	Jul 1	Jul 27	Aug 16	Aug 30	Sept 11	Sept 30
Overstory leaves	1.45	1.07	1.02	1.00	1.00	1.00	1.05	1.30
Understory leaves	1.55	1.25	1.20	1.18	1.05	1.05	1.12	1.20

predictions tended to overestimate leaf chlorophyll content. The specific absorption coefficients of the biochemical constituents in the PROSPECT were calibrated using the data collected in summer [21,72]. Leaf structure, chlorophyll content, and optical properties vary both seasonally and with respect to canopy position. Assuming that leaf structural variables were across-season averages, the specific absorption coefficient of chlorophyll in summer or at upper canopies would tend to be low. Using the specific absorption coefficient calibrated in summer (overstory) would thus result in overestimation of leaf chlorophyll content for leaves in other seasons (understory).

To simulate the seasonal and canopy height variations of leaf chlorophyll content, Zhang et al. [23] defined and incorporated a leaf thickness factor in the PROSPECT model to consider the influence of seasonal and canopy-height variability in leaf structure on light absorption. The specific absorption coefficients of all constituents were adjusted using the same leaf thickness factor.

$$K(\lambda) = \text{Ke}(\lambda)/T + (\sum K_i(\lambda)C_i)/L * T \qquad (9.3)$$

where T is the thickness factor, which is the ratio of the leaf thickness in summer to that in other growing seasons, or the thickness ratio of overstory leaves to understory leaves. Based on the leaf thickness measurements for sugar maple leaves, the thickness factor was calculated for overstory and understory leaves through the whole season (Table 9.3).

Leaf chlorophyll content was estimated with leaf reflectance and transmittance measurements from 255 sugar maple leaves as input into the PROSPECT with consideration of leaf thickness. After incorporating the leaf thickness factor in the model for these samples, the estimation was improved compared with the estimation from the original PROSPECT, from RMSE = 4.86 μg/cm^2 and R^2 = 0.84 to RMSE = 3.09 μg/cm^2 and R^2 = 0.93 (Figure 9.1b). With the additional input of leaf thickness as a surrogate to capture the seasonal and location variations in leaf structure and nonchlorophyll light absorption, the model performed better in estimating the low chlorophyll content from understory and overstory leaves in the early and late growing season, though there remains some bias with low values of leaf chlorophyll content being slightly, but systematically, overestimated by the model inversion.

Féret et al. [73] recalibrated the physical and optical constants in the PROSPECT model using comprehensive datasets encompassing hundreds of leaves from various species. The chlorophyll content estimates were improved for leaves with low chlorophyll content.

9.2.2.2   Modeling Method for Needleleaf Chlorophyll Content Estimation

The relative small size, narrow width, and irregular shape of needleleaves impose challenges for both the measurement of leaf optical properties and estimation of chlorophyll content. Leaf-level models mentioned in 9.2.2.1 are intended for broadleaf species. An RT model named LIBERTY was developed for estimating leaf optical spectra of jack pine needles [74]. However, comparative studies show that LIBERTY produces poor estimates of needleleaf chlorophyll content [75].

Needleleaves are usually narrow and thick. Needle structures with adaxial and abaxial surfaces are neither parallel nor necessarily flat planes. The validity of the model for needleleaves needs investigation since the assumption of infinite plane layers as for broadleaves is violated. Figure 9.2 schematically shows the structural effects of a broadleaf and needleleaf on light transfer. By assumption, a broadleaf

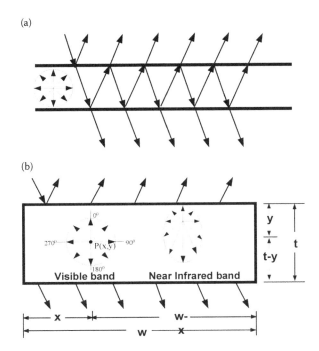

FIGURE 9.2 Schematic diagram showing the light transport in a monocotyledon broad leaf (a) based on the assumption of the PROSPECT model, and (b) in needle. For a broad leaf, the light transport in the leaf interior is assumed to be isotropic. For a needle, a phase function is introduced to characterize the directional scattering in the visible and near-infrared regions.

is composed of horizontally infinite layers. As the thickness of a broadleaf is much smaller than the leaf width, the incident light leaves the outer layers as reflected or transmitted light after multiple interactions with the internal layers. Light transmitted through leaf edges is negligible compared to the amount that exits the first (uppermost) and last (lowermost) layer. For thick and narrow black spruce needles (Figure 9.2b), needleleaf length can be assumed to be infinite relative to the leaf thickness and width, but leaf width can significantly affect the measurements of leaf reflectance and transmittance. For example, the needle width of black spruce, which is the dominant species in the boreal ecosystem, is only around two times the leaf thickness. Therefore, the amount of light that escapes from needle edges, that is, along the direction of the leaf width, could be large. This portion of light "loss" is not included in the PROSPECT model for the estimation of reflectance and transmittance.

Moorthy et al. [75] adapted PROSPECT for estimating the chlorophyll content of jack pine needles by using a simple geometric correction of needle form. With the consideration of the needle morphology as an equivalent flat plate, the accuracy of estimating chlorophyll content in jack pine needles can be improved, although a significant departure remained in the ability of PROSPECT to quantitatively predict the measured optical properties of the needles as a function of pigment content.

The edge effects of needleleaves on light transfer and measured spectra were considered by incorporating two morphology factors, needleleaf thickness and width, in PROSPECT [24]. First, laboratory experiments were conducted to investigate the effects of needle morphology and the needle-holding device on leaf spectral measurements. The effects induced by the leaf morphology and size were evident in leaf optical properties, which demonstrated that the PROSPECT model, as designed for broadleaf species, was not adequate to accurately represent the link between the optical properties and the pigment content of needleleaves. Second, to compensate for the portion of light "loss" to the needleleaf edge that is not considered in PROSPECT simulation of leaf reflectance and transmittance, two leaf biophysical parameters, needle width and thickness, were introduced in the model to take into account the effects of leaf morphology on chlorophyll content retrieval. Light scattering was assumed to be nonisotropic in all directions. The Henyey-Greenstein phase function, which characterizes the angular distribution of scattered light in layered materials such as biological tissues of leaves [76], was used to describe the directional scattering from the incoming direction to the outgoing direction of the light

$$p(\cos\theta) = \frac{1-g^2}{4(1+g^2-2g\cos\theta)^{3/2}} \cdot \frac{1}{4\pi} \qquad (9.4)$$

where θ is the angle between the incoming and outgoing direction, and g is the average cosine of the scattered angle. The phase function is characterized by the parameter g. Light scattering is symmetric about the incident direction. For different values of g, the phase function indicates different scattering:

$$\begin{cases} g=0, \text{isotropic scattering} \\ g>0, \text{predominantly forward scattering} \\ g<0, \text{predominantly backward scattering} \end{cases}$$

The value of g can be determined through an optimization procedure. When the incoming light penetrates through the upper leaf surface, it is scattered at point $P(x, y)$ (Figure 9.1b). Leaf width is denoted as w, leaf thickness as t, and the light scattered toward the leaf width (along 90° and 270°) and thickness (along 0° and 180°) direction as L_{width} (w) and L_{thick} (t), respectively. A_g is the global absorption coefficient by chlorophyll a and b, liquid water, and biochemicals. Based on Beer's law, L_{width} (w) and L_{thick} (t) can be calculated:

$$L_{width}(x) = \exp[-A_g P(\cos 90^0)x] + \exp[-A_g P(\cos 270^0)(w-x)] \qquad (9.5)$$

$$L_{thick}(y) = \exp[-A_g P(\cos 0^0)y] + \exp[-A_g P(\cos 180^0)(t-y)] \qquad (9.6)$$

Then the total light penetrating through the leaf width and thickness are:

$$L_{width} = \frac{1}{w} \int_0^w \{\exp[-A_g P(\cos 90°)x] + \exp[-A_g P(\cos 270°)(w-x)]\}dx \qquad (9.7)$$

$$L_{thick} = \frac{1}{t} \int_0^t \{\exp[-A_g P(\cos 0°)y] + \exp[-A_g P(\cos 180°)(t-y)]\}dy \qquad (9.8)$$

These two portions of light are not accounted for in the simulation of leaf reflectance and transmittance by PROSPECT. With the consideration of these two portions of light transfer, the corresponding reflectance R_{width} and transmittance T_{width} losses from the needle width direction can be simply inferred from the measurements:

$$R_{width} = \frac{L_{width}}{\frac{1}{t}\int_0^t \exp[-A_g P(\cos 0°)y]dy} \times R_{mes} \qquad (9.9)$$

$$T_{width} = \frac{L_{thick}}{\frac{1}{t}\int_0^t \exp[-A_g P(\cos 180°)y]dy} \times T_{mes} \qquad (9.10)$$

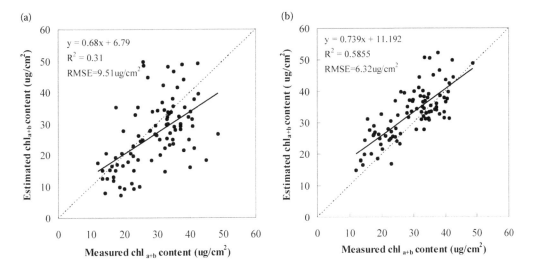

FIGURE 9.3 Relationship between the measured and estimated chlorophyll contents using (a) the original PROSPECT inversion model and (b) the modified PROSPECT model, which takes into account the effects of needle thickness and width. The dotted line is the 1:1 line.

Then the following adjustments were given to estimate needle reflectance and transmittance:

$$R_{mod} = R - R_{width} \qquad (9.11)$$

$$T_{mod} = T - T_{width} \qquad (9.12)$$

where R_{mod} and T_{mod} are the needle reflectance and transmittance after modifications; and R and T are the reflectance and transmittance estimated from the PROSPECT model. With the modifications to the model for reflectance and transmittance estimation, the chlorophyll content inversions were noticeably improved for small and short black spruce needles ($R^2 = 0.59$ and RMSE = 6.32 µg/cm^2) in comparison to the original model ($R^2 = 0.31$ and RMSE = 9.51 µg/cm^2), which tended to underestimate chlorophyll$_{a+b}$ content of black spruce needles (Figure 9.3).

The modified model captures the variation of needle chlorophyll content well from different sites, age classes, and branch orientations. As black spruce needles are significantly small and difficult to measure, the improvements achieved by this new approach are significant. Nevertheless, this improvement in PROSPECT applicability to needles did not appreciably reduce the departure of the slope of the estimated versus measured total needle chlorophyll content from the 1:1 line. This remaining issue calls for further model refinements specific to the needle internal structure.

9.3 ESTIMATING LEAF CHLOROPHYLL CONTENT FROM ABOVE-CANOPY HYPERSPECTRAL MEASUREMENTS

Hyperspectral images measure the confounding reflectance of ground surfaces. For vegetation canopies, the reflectance from the hyperspectral images is the canopy-level reflectance. Canopy biochemical composition depends strongly on leaf optical properties as well as on canopy structure. Leaf optical properties contribute directly to canopy-level reflectance. In closed canopies with high leaf area index (LAI), leaf optical properties strongly influence canopy-level reflectance [77]. The structure of forest canopies, such as the vertical and horizontal distribution, orientation, and density of foliage, determines light attenuation and thus influences canopy reflectance, distribution of photosynthesis, respiration, transpiration, and nutrient cycling in the canopy [78]. Photon scattering by leaves, stems, and soils also influences the radiation regime of forest canopies. LAI, leaf angle distribution (LAD), and foliage clumping represent the effects of leaf, stem, and soil optical properties on canopy reflectance and thus are dominant controllers of the relationship between leaf and canopy spectral characteristics [4]. Canopy reflectance is also a function of wavelength, soil reflectance, solar illumination conditions, and viewing geometry of the remote sensing instrument [79]. From the leaf to canopy scale, the complicated perturbations of canopy structure to light transfer need to be carefully considered. For complex forest canopies, it is impossible to derive leaf biochemical parameters from above-canopy hyperspectral measurements without consideration of canopy structural effects.

9.3.1 EMPIRICAL AND SEMI-EMPIRICAL METHODS

Statistical estimation of leaf biochemical contents from above-canopy reflectance is performed through different methods. The simplest way is to directly develop statistical relationships between ground-measured biochemical contents and canopy reflectance measured in the field or by airborne or satellite sensors [80–83]. Alternatively, some leaf-level relationships between optical indices and pigment content are directly applied to the canopy-level estimation [74–86].

Two chlorophyll spectral indices, the transformed chlorophyll absorption in reflectance index/optimized soil-adjusted vegetation index (TCARI/OSVAI) [87] and MERIS terrestrial chlorophyll index (MTCI) [88], were developed for chlorophyll content estimation. These two spectral indices were based on model spectra, field spectra, and airborne/spaceborne remote sensing data, so they may be viewed as semi-empirical indices sensitive to chlorophyll content. TCARI/OSVAI minimizes the effects of varying LAI and soil background using narrow-band hyperspectral data and has been reported to be suitable for crop, as well as open and structured, canopies [87,89,90]. MTCI relates a measure of the red-edge position with canopy chlorophyll content [88].

The application of these two indices to forest leaf chlorophyll content estimation was evaluated for nine black spruce sites that have open canopies [25]. It was found that TCARI/OSAVI correlated best with leaf chlorophyll a and b, but with its relative insensitivity to LAI variations, it showed only poor relationships with leaf chlorophyll a and b content (Table 9.4). For open forest canopies, when the forest background has a green and variable vegetation signature instead of bare soil, the relationship of TCARI/OSAVI to leaf chlorophyll a and b content was not as tight as those shown for crop canopies or open canopies [85,89,90]. MTCI showed a good relationship with leaf chlorophyll$_{a+b}$ content, as it is based on a measure of the red-edge position that was influenced by both leaf chlorophyll content and LAI. However, it showed no measurable correlation with leaf chlorophyll$_{a+b}$ content because of the confounding effect of LAI variations. For both indices, when applied to open boreal canopies, a strong influence of the vegetated understory remains as a perturbing parameter.

Confounding factors that influence the remotely sensed optical properties make it difficult to spatially and temporally extrapolate leaf-level relationships to the canopy level. Statistical relationships are often site and species specific and thus cannot be directly applied to other study sites since the canopy structure and viewing geometry may vary from different sites and species.

TABLE 9.4

Performance of TCARI/OSAVI and MTCI Index for Estimating Chlorophyll Content at Leaf and Canopy Scales

Index	Correlation to Leaf Chlorophyll Content	Correlation to Canopy Chlorophyll Content
TCARI/OSAVI	$y = -12.41\text{Ln}(x) + 8.5886,$ $R^2 = 0.4025$	$y = -81.998\text{Ln}(x) - 48.634,$ $R^2 = 0.1915$
MTCI	$y = 53.265x^2 - 204.51x + 226.05,$ $R^2 = 0.2005$	$y = 151.29x - 184.35,$ $R^2 = 0.6456$

Note: y denotes leaf/canopy chlorophyll content, x denotes index.

9.3.2   PHYSICALLY BASED MODELING METHOD

9.3.2.1   Modeling Method for Closed Forest Canopies

At the canopy level, physically based modeling approaches have been applied to estimate forest leaf chlorophyll content. To derive forest biochemical parameters accurately from hyperspectral imagery, one approach is to estimate leaf reflectance spectrum from canopy-level hyperspectral data. Retrieving leaf-level spectral information from canopy-level measurements requires comprehensive consideration of the interactions of radiation with plant canopies. Canopy RT models or tracing models are often coupled with leaf RT models to quantify leaf chlorophyll content from canopy-level remote sensing data [69,75,89,91–93]. Studies using coupled leaf and canopy RT models attempt to understand the effects of controlling factors on leaf reflectance properties at the canopy scale [93]. Coupled modeling methods are also used to scale up the leaf-level relationship between optical indices and pigment content. This method refines the development of spectral indices that are insensitive to factors such as canopy structure, illumination geometry, and background reflectance for estimating foliar chlorophyll concentrations from canopy reflectance [69,94,95].

Canopy RT models often assume that a canopy is composed of horizontal, homogeneous vegetation layers with infinite extent and Lambertian scatterers. Elements of the canopy are assumed to be randomly distributed in space as a turbid medium [96–98]. This type of model is suitable for close and dense canopies such as corn, soybean, and grass canopies, where the foliage spatial distribution is close to randomness. As canopy architecture is not considered in canopy RT models, the models are valid only for closed canopies with high LAI (LAI > 3) [99]. Zarco-Tejada et al. [69] estimated broadleaf chlorophyll content through coupled leaf and canopy-level RT models using the red-edge index (R750/R710) as a merit function to minimize the effects of forest canopy structure, shadows, and openings. This method was extended to coniferous forests for scaling leaf-level pigment estimation to canopy level using high spatial resolution airborne imagery to select only bright crown pixels in the scene for analysis [75,100]. Promising results for leaf chlorophyll content retrieval were obtained using this method. The bright crown methodology is successful in the case where a pixel is completely occupied by sunlit foliage and the shadow effects are small. However, remote sensing pixels, even at submeter resolutions, generally contain both sunlit and shaded fractions. The structural effects imposed by open forest canopies are not tackled using this method.

9.3.2.2   Modeling Methods for Open Forest Canopies

Open forest canopies, especially heterogeneous, open, and clumped forest canopies, present a big challenge for the retrieval of leaf biochemical parameters. This is due to the compound effects of distinct spatial distribution of trees, the geometry of tree crowns, the structure of tree elements, forest background, and leaf optical properties on canopy reflectance. Efforts have been made to deal with canopy structural effects on the retrieval of forest biochemical parameters. A methodology that is applicable to not only closed but also open canopies is highly desirable.

Models based on radiosity [101,102] and ray tracing [103,104] can simulate the complexity of multiple scattering, but simplifications of the mathematical and canopy architectural descriptions are inevitable due to computational limitations [104,105]. Geometrical optical (GO) models use geometrical shapes, such as cones for conifers and spheres or spheroids for deciduous trees, to simulate the angular and spatial distribution patterns of reflected solar radiance from forests [106–108]. GO models combine the advantages of simplicity, easy implementation, and capability of simulating the effects of canopy structure on the single and multiple scattering processes [109]. When considering complex canopy architectural conditions, geometrical optical-radiative transfer (GO-RT) models are a good solution to solving multiple scattering issues, in which geometrical optics are used to describe the shadowing (first-order scattering) effects and RT theories are adopted to estimate the second and higher order scattering effects.

Zhang et al. [25,110] developed a lookup-table (LUT) approach to separate structural effects of the canopy from leaf optical properties to estimate individual leaf reflectance spectra from hyperspectral

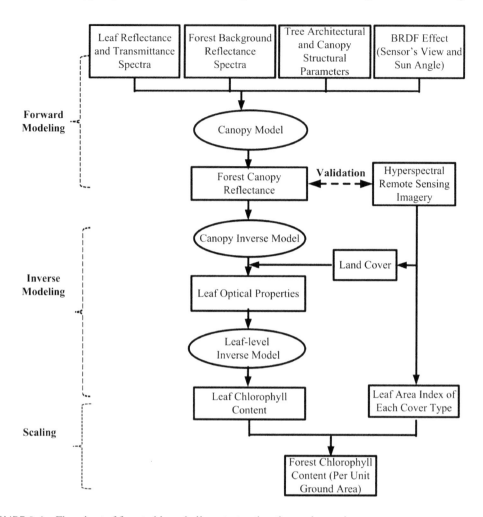

FIGURE 9.4 Flowchart of forest chlorophyll content estimation and mapping.

imagery, thereby deriving leaf chlorophyll content from estimated individual leaf reflectance. This method uses a combination of GO model 4-Scale and a leaf RT model PROSPECT. The method treats deciduous and coniferous trees separately, with the geometrical shapes of cones for conifers and spheres or spheroids for deciduous trees, to simulate the angular and spatial distribution patterns of reflected solar radiance from forests. The impacts of forest structural parameters, LAI, LAD, and clumping index on canopy reflectance were taken into account. The method is suitable for both closed and open forest canopies.

Figure 9.4 shows the procedures of the forest chlorophyll content estimation and mapping using this method. There are three main procedures to estimate leaf chlorophyll content from above-canopy hyperspectral remote sensing:

i. Canopy model forward simulation for canopy reflectance estimation. The forward modeling takes into account spectral scattering/absorbing properties of canopy components (leaf, forest background, canopy structure), canopy architecture, and directions of illumination and view. Forest canopy reflectance is calculated as a linear combination of four contributing components:

$$R = R_T P_T + R_{ZT} P_{ZT} + R_G P_G + R_{ZG} P_{ZG} \tag{9.13}$$

where R is the canopy reflectance; R_T, R_{ZT}, R_G, and R_{ZG} are the reflectivities of a sunlit tree crown, shaded tree crown, sunlit background, and shaded background, respectively; and P_T, P_{ZT}, P_G, and P_{ZG} are the probabilities of a sensor viewing sunlit and shaded tree crowns and sunlit and shaded background, respectively. Trees in forests are nonrandom discrete objects and spatially follow a Neyman distribution to simulate patchiness of a forest stand.

The forward modeling estimates the canopy reflectance and four scene components with inputs including leaf optical properties (reflectance and transmittance); forest background reflectance spectra; and general parameters of tree architecture for coniferous and deciduous canopies such as tree height, crown radius, forest density, and so on; canopy structure (leaf area index, clumping index); and image view geometry (solar zenith angle, view zenith angle, and azimuth angle). The estimated canopy reflectance is then compared with the reflectance measured by the hyperspectral sensor for validation.

ii. Canopy-level model inversion for estimating the contribution of an individual sunlit leaf to the measured canopy-level reflectance R across the spectrum.

To fulfill the canopy GO model inversion, a multiple scattering factor M was introduced to convert R_T, the reflectance of an assemblage of sunlit leaves, to individual leaf reflectance R_L. M includes the contributions of the two shaded components to canopy reflectance across the spectrum:

$$M = \frac{R - R_G \times P_G}{R_L \times P_T} \tag{9.14}$$

Given that canopy reflectance R can be remotely measured and forest background reflectivity R_G is known, the probabilities of observing the sunlit foliage (P_T) and background (P_G) components and the spectral multiple scattering factor (M) need to be estimated to invert the individual leaf reflectivity R_L.

At given solar zenith angle (SZA), view zenith angle (VZA), and azimuthal angle (PHI) between the sun and view point, variations of R, P_T, and P_G depend on stem density, LAI, tree height, and radius of tree crown. The sensitivity of R, P_T, and P_G to the model input parameters was investigated to develop the LUTs. For simplicity, two LUTs were developed as functions of viewing geometry (VZA, SZA, and PHI) and one dominant contributor LAI for both coniferous and deciduous species. One LUT provides the probabilities of viewing the sunlit foliage P_T and background P_G components, and the other LUT provides the spectral multiple scattering factors MS to convert the average reflectivity of sunlit leaves to the reflectivity of an individual sunlit leaf.

Combining the two LUTs, forest background reflectance measurements, and canopy reflectance derived from the hyperspectral image R_{image}, the sunlit leaf reflectivity R_L can be derived:

$$R_L = \frac{R_{image} - R_G \times P_G}{P_T \times M} \tag{9.15}$$

where R_{image} is the reflectance of the site derived from the hyperspectral image.

iii. With the retrieved individual leaf reflectance as input to the leaf-level inverse model, leaf chlorophyll content can be estimated. The retrieved individual leaf reflectance was input to the leaf-level inversion model PROSPECT to estimate leaf chlorophyll content. The original and modified PROSPECT models (as described in Section 9.2.2) were applied separately to deciduous, grass, and coniferous species (three vegetation types in the study area) for estimating leaf chlorophyll content of these cover types.

These procedures were performed for leaf chlorophyll content estimation and mapping over a study area in boreal forests near Sudbury, Ontario, Canada. Airborne hyperspectral data from the

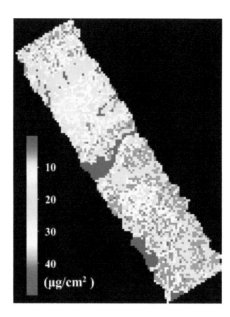

FIGURE 9.5   Leaf chlorophyll$_{a+b}$ content distribution. The image is produced based on the retrieved chlorophyll$_{a+b}$ content from 72-band CASI image for the three vegetated cover types grass, deciduous and coniferous forest species. The spatial resolution of the image is 20 × 20 m.

Compact Airborne Spectrographic Imager (CASI), with 72 bands and half bandwidth 4.25~4.36 nm in the visible and near-infrared region and a 2 m spatial resolution, was applied for retrieving forest leaf chlorophyll content. To meet the statistical tree distribution pattern defined in the 4-Scale model, the original 2 m resolution image was resampled to 20 m for LAI and forest chlorophyll content mapping. Figure 9.5 shows the spatially coarsened leaf chlorophyll content map using this method, with the spatial variability of leaf chlorophyll$_{a+b}$ content ranging from 16.2 to 43.6 µg/cm^2.

9.4   SCALING LEAF CHLOROPHYLL CONTENT ESTIMATES TO LANDSCAPE SCALE

Leaf chlorophyll content estimated from hyperspectral data can be further scaled to the landscape scale. This can be done using LAI, as it is an important forest structural variable. When the canopy is open, the forest background is often visually greener than the forest canopy and the effect of the forest background is large. Airborne hyperspectral CASI data with high spatial resolution can reproduce the heterogeneity of ground surface and is effective to derive forest biophysical parameters. Using vegetation indices such as the simple ratio for leaf area index mapping, the effect of the forest background could be subtracted from the canopy reflectance for the correlation of LAI to the vegetation index [25]. Combining leaf chlorophyll content (per unit leaf area) and LAI mapping, vegetation chlorophyll content (per unit ground area) can be generated. Figure 9.6 is the landscape chlorophyll content estimation using the leaf chlorophyll content retrieval and LAI mapping from the hyperspectral CASI image. The chlorophyll mapping shows the distribution of leaf chlorophyll content among different vegetation types, deciduous vs. coniferous forest vs. grass. The chlorophyll content ranges from 30 to 2170 mg/m^2 for vegetated pixels.

9.5   DISCUSSION

Hyperspectral data revealed the subtle spectral responses of leaves to leaf chlorophyll content, which facilitates the radiometric retrieval of leaf chlorophyll content from remotely sensed leaf and

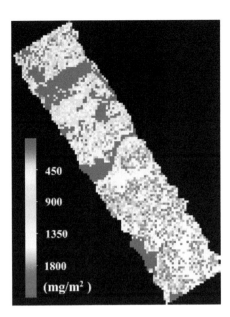

FIGURE 9.6 Landscape chlorophyll$_{a+b}$ content distribution per unit ground surface area. The image is derived based on the LAI image and retrieved chlorophyll$_{a+b}$ content from 72-band CASI image per unit leaf area of the three vegetated cover types grass, deciduous and coniferous forest species. The spatial resolution of the image is 20×20 m.

canopy optical properties. Leaf-level radiometric retrieval is a critical step for accurate mapping of canopy-level chlorophyll content using hyperspectral imagery. For broadleaf species, accurate estimation of leaf-level chlorophyll content should rely either on validated algorithms based on empirical indices or model inversions that take into account leaf thickness change. With leaf thickness as a surrogate to capture the seasonal and locational variations in leaf structure and nonchlorophyll light absorption, the improved PROSPECT model performed better than spectral indices and was capable of deriving the seasonal and canopy height variations in leaf chlorophyll content from leaf reflectance and transmittance spectra. For needleleaf species, the effects of needle width and thickness on light transfer need to be considered for chlorophyll retrieval. With the consideration of these two parameters, the model developed for deciduous leaves can be applied to needleleaf species with greater confidence. Needleleaf chlorophyll content was estimated with a reasonable accuracy. Nevertheless, this improvement to the broadleaf model PROSPECT did not appreciably reduce the departure of the slope of the estimated versus measured total needle chlorophyll content from the 1:1 line. This remaining issue calls for further model refinements specific to the needle internal structure. The current apparatus for measuring needle reflectance and transmittance spectra limits the leaf area exposed to the light source, which can result in a negative bias in the spectra measurements and accordingly an underestimation of chlorophyll content. A new apparatus that can fully expose the needle surface to the light source is desirable for accurate needle spectral measurements.

The transformation of chlorophyll content estimation from leaf scale to canopy scale is a very important issue in contemporary remote sensing. The lookup-table approach presented here successfully separated the leaf optical properties and canopy structural effect and links leaf-level optical properties to the canopy-level hyperspectral measurements. With lookup tables, the effects of canopy structure and forest background were removed for the retrieval of individual leaf reflectance and thereby leaf chlorophyll content. This physically based approach was developed for retrieving leaf reflectance and thereby leaf chlorophyll content from hyperspectral remote sensing imagery for both open and closed forests, which was in contrast to existing empirically based methods and some highly simplified methods for closed canopies. This approach produced systematically better estimates of chlorophyll content than a widely used spectral index approach.

This complex modeling process requires accurate measurements of leaf optical properties, forest background reflectance, and the forest structural parameter LAI as model inputs. For simplicity, LUTs were only developed as functions of LAI and view and sun geometry. The contribution of the two shaded components (shaded foliage and shaded background) was incorporated in the multiple scattering factor to reduce the unknown parameters in the model. In future studies, more comprehensive LUTs, which include LAI, sun and view geometry, stem density, and the two shaded components, can be generated to take into account the complexity of forests.

The probability of viewing forest background (such as soil, grass, and plant residue) by the sensor is large in open forests. The LUTs were developed under the assumption that forest background does not vary significantly across the landscape. Due to the heterogeneity and variability of the forest background, the effects of the forest background on the total remote sensing signals are uneven. To map chlorophyll content more accurately, the spatial variation of the forest background needs to be estimated. Multi-angle remote sensing data have demonstrated the capability for deriving background reflectance [111]. With the retrieval of spatially explicit forest background reflectance, forest chlorophyll content mapping will be improved.

It is noted that this method also uses a spatially coarsened LAI map for forest chlorophyll content mapping, as the GO model 4-Scale assumes trees in a forest have a certain patchiness. If a pixel is too small, one pixel might not even cover a complete tree crown. To meet the conditions for this statistical tree distribution pattern, the canopy-to-leaf inversion algorithm is applied to coarsened pixels. Further algorithm development is needed to retrieve chlorophyll content at the original CASI resolution. This algorithm would have to avoid using any tree distribution assumptions. This is one of the remaining challenges that have not been tackled.

9.5.1 APPLICATIONS

Leaf chlorophyll content is potentially one of the most important bio-indicators of vegetation stress. Long- or medium-range changes in canopy chlorophyll content can be related to photosynthetic capacity (thus productivity), developmental stage, vegetation phenology, and canopy stresses [112,113]. Chlorophyll content has been used for estimating gross primary productivity [114]. Studies found that leaf chlorophyll content provides an accurate, indirect estimate of plant nutrient status [47,115]. Canopy-level chlorophyll may appear to be most directly relevant for the prediction of productivity [116,117]. Forest chlorophyll content is a bio-indicator of forest health and physiological status, and estimates of forest chlorophyll content can reveal forest health status and serve forest management.

Detecting forest health requires a theoretical basis. The algorithms presented here quantitatively estimate leaf chlorophyll content across tree species and forest canopies from hyperspectral remote sensing. The algorithms would be valuable for various purposes, including ecosystem health assessment, natural resource management, and carbon cycle estimation. The spatial distribution of chlorophyll content can potentially be applied for the estimation of spatially explicit carbon budgets. Vegetation biochemical and biophysical parameters enable a linkage between optical remote sensing and the carbon cycle. Estimation of chlorophyll content provides useful insight into plant-environment interactions as chlorophylls change with light [118]. Leaf chlorophyll content can potentially serve as an input to photosynthesis and carbon cycle models.

9.6 CONCLUSION

Methods for retrieving leaf chlorophyll content from hyperspectral remote sensing at different scales for complex forest canopies were examined here. Leaf chlorophyll content of both broadleaf and needleleaf species was quantitatively estimated at leaf, canopy, and landscape scales through the combination of field measurements on forest structure and architecture, laboratory optical and chemical experiments, hysperspectral data, and statistical and modeling approaches. Some recent research has attempted to retrieve leaf chlorophyll distribution at the global scale [119]. The successful demonstration of the physically based approach to complex forests points to a future direction of retrieving vegetation information from hyperspectral imagery. The research here made an important step toward improving the accuracy of ecosystem models associated with carbon cycle modeling.

REFERENCES

1. Running, S.W. and Coughlan, J.C. 1988. A general model of forest ecosystem process for regional applications. 1. Hydrologic balance, canopy gas exchange and primary production processes. *Ecological Modelling* 42, 125–154.
2. Running, S.W. 1990. Estimating terrestrial primary productivity by combining remote sensing and ecosystem simulation. In: *Remote Sensing of Biosphere Functioning* (Eds: Hobbs, R.J., and Mooney, H.A.), Springer-Verlag, New York, pp. 65–86.
3. Goetz, S.L. and Prince, S.D. 1996. Remote sensing of net primary production in boreal forest stands. *Agricultural and Forest Meteorology* 78, 149–179.
4. Asner, G.P., Wessman, C.A., Schimel, D.S., and Archer, S. 1998a. Variability in leaf and litter optical properties: Implications for canopy BRDF model inversions using AVHRR, MODIS, and MISR. *Remote Sensing of Environment* 63, 243–257.
5. Blackburn, G.A. and Pitman, J.I. 1999. Biophysical controls on the directional spectral reflectance properties of bracken (*Pteridium aquilinum*) canopies: Results of a field experiment. *International Journal of Remote Sensing* 20(11), 2265–2282.
6. Baltzer, J.L. and Thomas, S.C. 2005. Leaf optical responses to light and soil nutrient availability in temperate deciduous trees. *American Journal of Botany* 92, 214–223.
7. Belward, A.S. 1991. Spectral characteristics of vegetation, soil and water invisible, near infrared and middle-infrared wavelengths. In: *Remote Sensing and Geographical Information Systems of Resource Management in Developing Countries* (Eds: Belward, and Valenzuela), ECSC, EEC, EAEC, Brussels and Luxembourg, pp. 31–53.
8. Rock, B.N., Hoshizaki, T., and Miller, J.R. 1988. Comparison of *in situ* and airborne spectral measurements of the blue shift associated with forest decline. *Remote Sensing of Environment* 24, 109–127.
9. Vogelmann, J.E., Rock, B.N., and Moss, D.M. 1993. Red-edge spectral measurements from sugar maple leaves. *International Journal of Remote Sensing* 14, 1563–1575.
10. Carter, G.A. 1994. Ratios of leaf reflectances in narrow wavebands as indicators of plant stress. *International Journal of Remote Sensing* 15, 697–704.
11. Gitelson, A.A., Merzlyak, M.N., and Lichtenthaler, H.K. 1996. Detection of red-edge position and chlorophyll content by reflectance measurements near 700 nm. *Journal of Plant Physiology* 148, 501–508.
12. Belanger, M.J., Miller, J.R., and Boyer, M.G. 1995. Comparative relationships between some red-edge parameters and seasonal leaf chlorophyll concentrations. *Canadian Journal of Remote Sensing* 21, 16–21.
13. Chang, S. and Collins, W. 1983. Confirmation of the airborne biogeophysical mineral exploration technique using laboratory methods. *Economic Geology* 1983, 723–736.
14. Horler, D.N.H., Barber, J., and Barringer, A.R. 1980. Effects of heavy metals on the absorbance and reflectance spectra of plants. *International Journal of Remote Sensing* 1, 121–136.
15. Horler, D.N.H., Dockray, M., and Barber, J. 1983. The red-edge of plant leaf reflectance. *International Journal of Remote Sensing* 4(2), 273–288.
16. Sampson, P.H., Mohammed, G.H., Zarco-Tejada, P.J., Miller, J.R., Noland, T.L., Irving, D., Treitze, P.M., Colombo, S.J., and Freemantle, J. 2000. The bioindicators of forest condition project: A physiological, remote sensing approach. *The Forestry Chronicle* 76(6), 941–952.
17. Sampson, P.H., Zarco-Tejada, P.J., Mohammed, G.H., Miller, J.R., and Noland, T.L. 2003. Hyperspectral remote sensing of forest condition: Estimating chlorophyll content in tolerant hardwoods. *Forest Science* 49(3), 381–391.
18. Thomas, S.C. 2005. Increased leaf reflectance in tropical trees under elevated CO_2. *Global Change Biology* 11, 197–202.
19. Curran, P.J. 1994. Imaging spectrometry. *Progress in Physical Geography* 18(2), 247–266.
20. Asner, G.P., Braswell, B.H., Schimel, D.S., and Wessman, C.A. 1998b. Ecological research needs from multi-angle remote sensing data. *Remote Sensing of Environment* 63, 155–165.

21. Jacquemoud, S., Ustin, S.L., Verdebout, J., Schmuck, G., Andreoli, G., and Hosgood, B. 1996. Estimating leaf biochemistry using the PROSPECT leaf optical properties model. *Remote Sensing of Environment* 56, 194–202.

22. Noland, T.L., Miller, J.R., Moorthy, I., Panigada, C., Zarco-Tejada, P.J., Mohammed, G.H., and Sampson, P.H. 2003. Bioindicators of forest sustainability: Using remote sensing to monitor forest condition. *Meeting Emerging Ecological, Economic, and Social Challenges in the Great Lakes Region: Popular Summaries.* Compiled by Buse, L.J. and A.H. Perera, Ontario Ministry of Natural Resources. Ontario Forest Research Institute. Forest Research Information. 155, pp. 75–77.

23. Zhang, Y., Chen, J.M., and Thomas, S.C. 2007. Retrieving seasonal variation in chlorophyll content of overstorey and understorey sugar maple leaves from leaf-level hyperspectral data. *Canadian Journal of Remote Sensing* 5, 406–415.

24. Zhang, Y., Chen, J.M., Miller, J.R., and Noland, T.L. 2008. Retrieving chlorophyll content in conifer needles from hyperspectral measurements. *Canadian Journal of Remote Sensing* 34(3), 296–310.

25. Zhang, Y., Chen, J.M., Miller, J.R., and Noland, T.L. 2008. Leaf chlorophyll content retrieval from airborne hyperspectral remote sensing imagery. *Remote Sensing of Environment* 112, 3234–3247.

26. Gates, D.M., Keegan, H.J.J., Schleter, C., and Wiedner, V.R. 1965. Spectral properties of plants. *Applied Optics* 4, 11–20.

27. Myneni, R.B., Ross, J.K., and Asrar, G. 1989. A review on the theory of photon transport in leaf canopies. *Agricultural Forest Meteorology* 45, 1–153.

28. Wessman, C.A. 1990. Evaluation of canopy biochemistry. In: *Remote Sensing of Biosphere Functioning* (Eds: Hobbs, R.J., and Mooney, H.A.), Springer-Verlag, New York, pp. 135–156.

29. Walter-Shea, E.A. and Norman, J.M. 1991. Leaf optical properties. In: *Photon Vegetation Interactions*, (Eds: Myneni, R.B., and Ross, J.), Springer-Verlag, New York, pp. 227–251.

30. Curran, P.J., Dungan, J.L., Macler, B.A., Plummer, S.E., and Peterson, D.L. 1992. Reflectance spectroscopy of fresh whole leaves for the estimation of chemical concentration. *Remote Sensing of Environment* 39, 153–166.

31. Fourty, T., Baret, F., Jacquemoud, S., Schmuck, G., and Verdebout, J. 1996. Leaf optical properties with explicit description of its biochemical composition: Direct and inverse problems. *Remote Sensing of Environment* 56, 104–117.

32. le Maire G., Francois, C., and Dufrene, E. 2004. Towards universal broad leaf chlorophyll indices using PROSPECT simulated database and hyperspectral reflectance measurements. *Remote Sensing of Environment* 89, 1–28.

33. Yang, X., Tang, J., Mustard, J.F., Wu, J., Zhao, K., Serbin, S., and Lee, J.-E. 2016. Seasonal variability of multiple leaf traits captured by leaf spectroscopy at two temperate deciduous forests. *Remote Sensing of Environment* 179, 1–12.

34. Kalacska, M., Lalonde, M., and Moore, T.R. 2015. Estimation of foliar chlorophyll and nitrogen content in an ombrotrophic bog from hyperspectral data: Scaling from leaf to image. *Remote Sensing of Environment* 169, 270–279.

35. Laurent, V.C.E., Verhoef, W., Damm, A., Schaepman, M.E., and Clevers, J.G.P.W. 2013. A Bayesian object-based approach for estimating vegetation biophysical and biochemical variables from APEX at-sensor radiance data. *Remote Sensing Of Environment* 139, 6–17.

36. Croft, H., Chen, J.M., and Zhang, Y. 2014. The applicability of empirical vegetation indices for determining leaf chlorophyll content over different leaf and canopy structures. *Ecological Complexity*, 17, 119–130.

37. Datt, B. 1999. Visible/near infrared reflectance and chlorophyll content in eucalyptus leaves. *International Journal of Remote Sensing* 20, 2741–2759.

38. Gamon, J.A., Serrano, L., and Surfus, J.S. 1997. The photochemical reflectance index: An optical indicator of photosynthetic radiation-use efficiency across species, functional types, and nutrient levels. *Oecologia* 112, 492–501.

39. Gitelson, A.A. and Merzlyak, M.N. 1997. Remote estimation of chlorophyll content in higher plant leaves. *International Journal of Remote Sensing* 18, 2691–2697.

40. Gitelson, A.A., Buschman, C., and Lichtenthaler, H.K. 1999. The chlorophyll fluorescence ratio F735/F700 as an accurate measure of chlorophyll content in plants. *Remote Sensing of Environment* 69, 296–302.

41. Maccioni, A., Agati, G., and Mazzinghi, P. 2001. New vegetation indices for remote measurement of chlorophylls based on leaf directional reflectance spectra. *Journal of Photochemistry and Photobiology, B: Biology* 61, 52–61.

42. Peñuelas, J., Filella, I., Llusia, J., Siscart, D., and Pinol, J. 1998. Comparative field study of spring and summer leaf gas exchange and photobiology of the Mediterranean trees *Quercus ilex* and *Phillyrea latifolia*. *Journal of Experimental Botany* 49, 229–238.

43. Schepers, J.S., Blackmer, T.M., Wilhelm, W.W., and Resende, M. 1996. Transmittance and reflectance measurements of corn leaves from plants with different nitrogen and water supply. *Journal of Plant Physiology* 148, 523–529.

44. Blackburn, G.A. 1998a. Spectral indices for estimating photosynthetic pigment concentrations: A test using senescent tree leaves. *International Journal of Remote Sensing* 19, 657–675.

45. Blackburn, G.A. 1998b. Quantifying chlorophylls and carotenoids from leaf to canopy scales: An evaluation of some hyperspectral approaches. *Remote Sensing of Environment* 66, 273–285.

46. Filella, I., Serrano, L., Serra, J., and Peñuelas, J. 1995. Evaluating wheat nitrogen status with canopy reflectance indices and discriminant analysis. *Crop Science* 35, 1400–1405.

47. Chappelle, E.W., Kim, M.S., and McMurtrey, J.E. 1992. Ratio analysis of reflectance spectra (RARS): An algorithm for the remote estimation of the concentrations of chlorophyll A, chlorophyll B and the carotenoids in soybean leaves. *Remote Sensing of Environment* 39, 239–247.

48. Sims, D.A. and Gamon, J.A. 2002. Relationship between pigment content and spectral reflectance across a wide range of species, leaf structures and developmental stages. *Remote Sensing of Environment* 81, 337–354.

49. Gitelson, A.A., Gritz, Y., and Merzlyak, M.N. 2003. Relationships between leaf chlorophyll content and spectral reflectance and algorithms for non-destructive chlorophyll assessment in higher plant leaves. *Journal of Plant Physiology* 160, 271–282.

50. Datt, B. 1998. Remote sensing of chlorophyll a, chlorophyll b, chlorophyll a+b and total carotenoid content in eucalyptus leaves. *Remote Sensing of Environment* 66, 111–121.

51. Yao, H.B. and Tian, L. 2003. A genetic algorithm-based selective principal component analysis (GA-SPCA) method for high-dimensional data feature extraction. *IEEE Transactions on Geoscience and Remote Sensing* 41, 1469–1478.

52. Coops, N., Drury, S., Smith, M.L., Martin, M., and Ollinger, S. 2002. Comparison of green leaf eucalypt spectra using spectral decomposition. *Australian Journal of Botany* 50, 567–576.

53. Tumbo, S.D., Wagner, D.G., and Heinemann, P.H. 2002. Hyperspectral-based neural network for predicting chlorophyll status in corn. *Transactions of the ASAE* 45, 825–832.

54. O'Neill, A.L., Kupiec, J.A., and Curran, P.J. 2002. Biochemical and reflectance variation throughout a Sitka spruce canopy. *Remote Sensing of Environment* 80, 134–142.

55. Osborne, S.L., Schepers, J.S., Francis, D.D., and Schlemmer, M.R. 2002. Use of spectral radiance to estimate in-season biomass and grain yield in nitrogen- and water-stressed corn. *Crop Science* 42, 165–171.

56. Allen, W.A., Gausman, H.W., Richardson, A.J., and Thomas, J.R. 1969. Interaction of isotropic light with a compact plant leaf. *Journal of the Optical Society of America* 59(10), 1376–1379.

57. Gamon, J. A., Serrano, L. and Surfus, J. S. 1997. The photochemical reflectance index: an optical indicator of photosynthetic radiation use efficiency across species, functional types, and nutrient levels. *Oecologia* 112 (4), 492–501.

58. Jacquemoud, S. and Baret, F. 1990. PROSPECT: A model of leaf optical properties spectra. *Remote Sensing of Environment* 34, 75–91.

59. Allen W.A. and Richardson A.J. 1968. Interaction of light with a plant canopy. *Journal of the Optical Society of America* 58(8), 1023–1028.

60. Yamada, N. and Fujimura, S. 1991. Nondestructive measurement of chlorophyll pigment content in plant leaves from three-color reflectance and transmittance. *Applied Optics* 30, 3964–3973.

61. Allen W.A., Gausman H.W., and Richardson A.J. 1973. Willstätter-Stoll theory of leaf reflectance evaluation by ray tracing. *Applied Optics* 12(10), 2448–2453.

62. Brakke, T.W. and Smith, J.A. 1987. A ray tracing model for leaf bidirectional scattering studies. In: *Proceedings of the 7th International Geoscience and Remote Sensing Symposium (IGARSS'87)*, May 18–21, Ann Arbor, MI, USA, pp. 643–648.

63. Govaerts, Y.M., Jacquemoud, S., Verstraete, M.M., and Ustin, S.L. 1996. Three-dimensional radiation transfer modeling in a dicotyledon leaf. *Applied Optics* 35(33),6585–6598.

64. Ustin S.L., Jacquemoud S., and Govaerts Y.M. 2001. Simulation of photon transport in a three-dimensional leaf: Implication for photosynthesis. *Plant Cell Environment* 24, 1095–1103.

65. Tucker, C.J. and Garratt, M.W. 1977. Leaf optical system modeled as a stochastic process. *Applied Optics* 16(3), 635–642.

66. Ganapol, B., Johnson, L., Hammer, P., Hlavka, C., and Peterson, D. 1998. LEAFMOD: A new within-leaf radiative transfer model. *Remote Sensing of Environment* 6, 182–193.

67. Maier, S.W., Lüdeker, W., and Günther, K.P. 1999. SLOP: A revised version of the stochastic model for leaf optical properties. *Remote Sensing of Environment* 68(3), 273–280.

68. Demarez, V., Gastellu-Etchegorry, J.P., Mougin, E., Marty, G., Proisy, C., Duferene, E., and Dantec, V. LE. 1999. Seasonal variation of leaf chlorophyll content of a temperate forest. Inversion of the PROSPECT model. *International Journal of Remote Sensing* 20(5), 879–894.

69. Zarco-Tejada, P.J., Miller, J.R., Noland, T.L., Mohammed, G.H., and Sampson, P.H. 2001. Scaling-up and model inversion methods with narrowband optical indices for chlorophyll content estimation in closed forest canopies with hyperspectral data. *IEEE Transactions in Geosciences and Remote Sensing* 39(7), 1491–1507.

70. Renzullo, L.J., Blanchfield, A.L., Guillermin, R., Powell, K.S., and Held, A.A. 2006. Comparison of PROSPECT and HPLC estimates of leaf chlorophyll contents in a grapevine stress study. *International Journal of Remote Sensing* 27(4), 817–823.

71. Forsythe, G.E., Malcolm, M.A., and Moler, C.B. 1976. *Computer Methods for Mathematical Computations*, Prentice-Hall, Englewood Cliffs, NJ.

72. Hosgood, B., Jacquemoud, S., Andreoli, G., Verdebout, J., Pedrini, G., and Schmuck, G. 1995. *Leaf Optical Properties EXperiment 93 (LOPEX93) Report EUR-16095-EN*, European Commission, Joint Research Centre, Institute for Remote Sensing Applications, Ispra, Italy.

73. Féret, J.B., François, C., Asner, G.P., Gitelson, A.A., Martin, R.E., Bidel, L.P.R., Ustin, S.L., le Maire, G., and Jacquemoud, S. 2008. PROSPECT-4 and 5: Advances in the leaf optical properties model separating photosynthetic pigments. *Remote Sensing of Environment* 112(6), 3030–3043.

74. Dawson, T.P., Curran, J.P., and Plummer, S.E. 1998. LIBERTY—Modeling the effects of leaf biochemical concentration on reflectance spectra. *Remote Sensing of Environment* 65(1), 50–60.

75. Moorthy, I., Miller, J.R., and Noland, T.L. 2008. Estimating chlorophyll concentration in conifer needles: An assessment at the needle and canopy level. *Remote Sensing of Environment* 12(6), 2824–2838.

76. Hanrahan, P. and Krueger, W. 1993. Reflection from layered surfaces due to subsurface scattering, *Proceedings of the 20th Annual Conference on Computer Graphics and Interactive Techniques*, pp. 165–174.

77. Myneni, R.B. and Asrar, G. 1993. Radiative transfer in three-dimensional atmosphere-vegetation media. *Journal of Quantitative Spectroscopy and Radiative Transfer* 49, 585–598.

78. Ross, J.K. 1981. *The Radiation Regime and Architecture of Plant Stands*. Kluwer Boston, Hingham, MA, USA.

79. Goel, N.S. 1988. Models of vegetation canopy reflectance and their use in estimation of biophysical parameters from reflectance data. *Remote Sensing Reviews* 4, 1–212.

80. Johnson, L.F., Hlavka, C.A., and Peterson, D.L. 1994. Multivariate analysis of AVIRIS data for canopy biochemical estimation along the Oregon transect. *Remote Sensing of Environment* 47, 216–230.

81. Matson, P., Johnson, L., Billow, C., Miller, J.R., and Pu, R. 1994. Seasonal patterns and remote spectral estimation of canopy chemistry across the Oregon transect. *Ecological Applications* 4, 280–298.

82. Curran, P.J., Kupiec, J.A., and Smith, G.M. 1997. Remote sensing the biochemical composition of a slash pine canopy. *IEEE Transactions on Geosciences and Remote Sensing* 35, 415–420.

83. Zarco-Tejada, P.J. and Miller, J.R. 1999. Land cover mapping at BOREAS using red-edge spectral parameters from CASI imagery. *Journal of Geophysics Research* 104(D22), 27921–27948.

84. Peterson, D.L., Aber, J.D., Matson, P.A., Card, D.H., Swanberg, N.A., Wessman, C.A., and Spanner, M.A. 1988. Remote sensing of forest canopy leaf biochemical contents. *Remote Sensing of Environment* 24, 85–108.

85. Yoder, B.J. and Pettigrew-Crosby, R.E. 1995. Predicting nitrogen and chlorophyll content and concentrations from reflectance spectra (400–2500 nm) at leaf and canopy scales. *Remote Sensing of Environment* 53(3), 199–211.

86. Zagolski, F., Pinel, V., Romier, J., Alcayde, D., Fotanari, J., Gastellu-Etchegorry, J.P., Giordano, G., Marty, G., and Joffre, R. 1996. Forest canopy chemistry with high spectral resolution remote sensing. *International Journal of Remote Sensing* 17, 1107–1128.

87. Haboudane, D., Miller, J.R., Tremblay, N., Zarco-Tejada, P.J., and Dextraze, L. 2002. Integrated narrow-band vegetation indices for prediction of crop chlorophyll content for application to precision agriculture. *Remote Sensing of Environment*, 81(2–3), 416–426.

88. Dash, J. and Curran, P.J. 2004. The MERIS terrestrial chlorophyll index. *International Journal of Remote Sensing* 25, 5403–5413.

89. Zarco-Tejada, P.J., Miller, J.R., Morales, A., Berjón, A. and Agüera, J. 2004a. Hyperspectral indices and model simulation for chlorophyll estimation in open-canopy tree crops. *Remote Sensing of Environment* 90(4), 463–476.

90. Zarco-Tejada, P.J., Berjón, A., López-Lozano, R., Miller, J.R., Martín, P., Cachorro, V., González, M.R., and Frutos, A. 2005. Assessing vineyard condition with hyperspectral indices: Leaf and canopy

reflectance simulation in a row-structured discontinuous canopy. *Remote Sensing of Environment* 99, 271–287.

91. Jacquemoud, S., Baret, F., Andrieu, B., Danson, F.M., and Jaggard, K. 1995. Extraction of vegetation biophysical parameters by inversion of the PROSPECT + SAIL models on sugar beet canopy reflectance data: application to TM and AVIRIS sensors. *Remote Sensing of Environment* 52, 163–172.

92. Demarez, V. and Gastellu-Etchegorry, J.P. 2000. A modeling approach for studying forest chlorophyll content. *Remote Sensing of Environment* 71, 226–238.

93. Dawson, T.P., Curran, P.J., North, P.R.J., and Plummer, S.E. 1997. The potential for understanding the biochemical signal in the spectra of forest canopies using a coupled leaf and canopy model. In: *Physical Measurements and Signatures in Remote Sensing*, 2 (Eds: Guyot, G., and Phulpin, T.), Balkema, Rotterdam, pp. 463–470.

94. Broge, N.H. and Leblanc, E. 2000. Comparing prediction power and stability of broadband and hyperspectral vegetation indices for estimation of green leaf area index and canopy chlorophyll density. *Remote Sensing of Environment* 76, 156–172.

95. Daughtry, C.S.T., Walthall, C.L., Kim, M.S., Brown, d.C.E., and McMurtrey, J.E. 2000. Estimating corn leaf chlorophyll concentration from leaf and canopy reflectance. *Remote Sensing of Environment* 74, 229–239.

96. Verhoef, W. 1984. Light scattering by leaf layers with application to canopy reflectance modelling: The SAIL model. *Remote Sensing of Environment* 16, 125–141.

97. Verstaete, M.M., Pinty, B., and Dickinson, R.E. 1990. A physical model of the bidirectional reflectance vegetation canopies, I Theory. *Journal of Geophysics Research* 95(D8), 11,755–11,765.

98. Liang, S. and Strahler, A.H. 1993. Calculation of the angular radiance distribution for a coupled atmosphere and canopy. *IEEE Transactions on Geoscience and Remote Sensing* 31(2), 491–502.

99. Zarco-Tejada, P.J. 2000. Hyperspectral remote sensing of closed forest canopies: Estimation of chlorophyll fluorescence and pigment content. PhD thesis, York University, Toronto, ON, Canada.

100. Zarco-Tejada, P.J., Miller, J.R., Harron, J., Hu, B., Noland, T.L., Goel, N., Mohammed, G.H., and Sampson, P. 2004b. Needle chlorophyll content estimation through model inversion using hyperspectral data from boreal conifer forest canopies. *Remote Sensing of Environment* 89, 189–199.

101. Borel, C.C., Gerstl, S.A.W., and Powers, B.J. 1991. The radiosity method in optical remote sensing of structured 3-D surfaces. *Remote Sensing of Environment* 36, 13–44.

102. Goel, N.S., Rozehnal, I., and Thompson, R.L. 1991. A computer graphics based model for scattering from objects of arbitrary shapes in the optical region. *Remote Sensing of Environment* 36(2), 73–104.

103. Myneni, R., Asrar, G., and Gerstl, S. 1990. Radiative transfer in three dimensional leaf canopies. *Transport Theory and Statistical Physics* 19, 205–250.

104. Gastellu-Etchegorry, J.P., Demarez, V., Pinel, V., and Zagolski, F. 1996. Modelling radiative transfer in heterogeneous 3-D vegetation canopies. *Remote Sensing of Environment* 58, 131–156.

105. Thompson, R.L. and Goel, N.S. 1999. SPRINT: A universal canopy reflectance model for kilometer level scene. In: *Abstract of the Second International Workshop on Multiple-Angle Measurements and Models*, September 15–17, Ispra, Italy.

106. Li, X. and Strahler, A.H. 1988. Modeling the gap probability of discontinuous vegetation canopy. *IEEE Transactions on Geoscience and Remote Sensing* 26, 161–170.

107. Li, X., Strahler, A.H., and Woodcock, C.E. 1995. A hybrid geometric optical-radiative transfer approach for modeling albedo and directional reflectance of discontinuous canopies. *IEEE Transactions on Geoscience and Remote Sensing* 33, 466–480.

108. Chen, J.M. and Leblanc, S. 1997. A 4-Scale bidirectional reflection model based on canopy architecture. *IEEE Transactions on Geoscience and Remote Sensing* 35, 1316–1337.

109. Chen, J.M. and Leblanc, S.G. 2001. Multiple-scattering scheme useful for hyperspectral geometrical optical modelling. *IEEE Transactions on Geoscience and Remote Sensing* 39(5), 1061–1071.

110. Zhang, Y., Hyperspectral remote sensing algorithms for retrieving forest chlorophyll content. PhD thesis, University of Toronto, Toronto, ON, Canada.

111. Canisius F. and Chen, J.M. 2007. Retrieving forest background reflectance in a boreal region from Multi-angle Imaging SpectroRadiometer (MISR) data. *Remote Sensing of Environment* 107, 312–321.

112. Ustin, S.L., Smith, M.O., Jacquemoud, S., Verstraete, M.M., and Govaerts, Y. 1998. GeoBotany: Vegetation mapping for earth sciences. In: *Manual of Remote Sensing: Remote Sensing for the Earth Sciences*, 3rd ed., vol. 3, (Eds: Rencz, A.N.), John Wiley, Hoboken, NJ, pp. 189–248.

113. Zarco-Tejada, P.J., Miller, J.R., Mohammed, G.H., Noland, T.L., and Sampson, P.H. 2002. Vegetation stress detection through chlorophyll a+b estimation and fluorescence effects on hyperspectral imagery. *Journal of Environmental Quality* 31, 1433–1441.

114. Gitelson, A.A., Vina, A., Verma, S.B., Rundquist, D.C., Arkebauer, T.J., Keydan, G., Leavitt, B., Ciganda, V., Burba, G.G., and Suyker, A.E. 2006. Relationship between gross primary production and chlorophyll content in crops: Implications for the synoptic monitoring of vegetation productivity. *Journal of Geophysical Research—Atmospheres* 111, D08S11, 2006, doi: 10.1029/2005JD006017.

115. Moran, J.A., Mitchell, A.K., Goodmanson, G., and Stockburger, K.A. 2000. Differentiation among effects of nitrogen fertilization treatments on conifer seedlings by foliar reflectance: A comparison of methods. *Tree Physiology* 20, 1113–1120.

116. Whittaker, R.H. and Marks, P.L. 1975. Methods of assessing terrestrial productivity. In: *Primary Productivity of the Biosphere, Ecological Studies*, vol. 14, (Eds: Lieth, H. and Whittaker, R.H.), Springer, New York, pp. 55–118.

117. Dawson, T.P., North, P.R.J., Plummer, S.E., and Curran, P.J. 2003. Forest ecosystem chlorophyll content: Implications for remotely sensed estimates of net primary productivity. *International Journal of Remote Sensing* 24, 611–617.

118. Fang, Z., Bouwkamp, J., and Solomos, T. 1998. Chlorophyllase activities and chlorophyll degradation during leaf senescence in nonyellowing mutant and wild type of *Phaseolus vulgaris* L. *Journal of Experimental Botany* 49, 503–510.

119. Croft, H., Chen, J.M., Mo, G., Luo, S., Luo, X., He, L., Gonsamo, A. et al. The global distribution of leaf chlorophyll content. *Remote Sensing of Environment*. (In press).

Section IV

Conclusions

10 Fifty Years of Advances in Hyperspectral Remote Sensing of Agriculture and Vegetation—Summary, Insights, and Highlights of Volume II

Hyperspectral Indices and Image Classifications for Agriculture and Vegetation

Prasad S. Thenkabail, John G. Lyon, and Alfredo Huete

CONTENTS

Hyperspectral point or imaging spectroscopy refers to a simultaneous acquisition of data by registered narrow spectral wavebands across the electro-magnetic spectrum. The key factor is the data are acquired in numerous unique spectral bands in the entire length of the spectrum (e.g., 400–2500 nm; 2500–14500 nm). Ideally, hyperspectral data are acquired continuously or near-continuously (e.g., every 1 nm) throughout the spectrum, resulting in a spectral signature (Figure 10.1; Hamzeh et al., 2016) from each picture element or pixel. In contrast, multispectral broadband sensors such as Landsat Enhanced Thematic Mapper Plus (ETM+) only capture data from specific broad spectral bands (e.g., Figure 10.1). In Figure 10.1, hyperspectral data of a sugarcane crop acquired from the Earth Observing-1 (EO-1) spaceborne Hyperion sensor consisting of 242 narrowbands (each

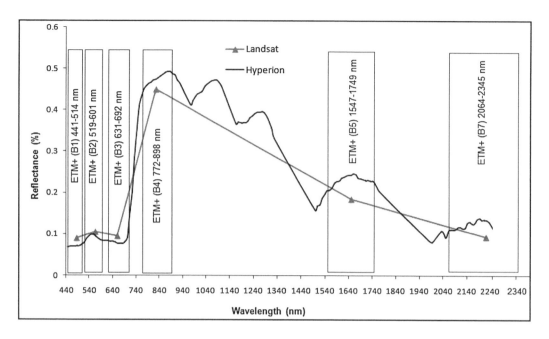

FIGURE 10.1 Sugarcane spectral profile from the hyperspectral Hyperion narrowband sensor versus the multispectral Landsat ETM+ broadband sensor. Sugarcane spectral curve, extracted from the hyperspectral narrowband Hyperion and multispectral broadband Landsat 7 ETM+ images with the overlap between the spectral wavelength coverage of these images. (Hamzeh, S. et al. 2016. *International Journal of Applied Earth Observation and Geoinformation*, Volume 52, Pages 412–421, ISSN 0303-2434, https://doi.org/10.1016/j.jag.2016.06.024.)

of 10-nm bandwidths across the 400–2500 nm spectral range) are compared with six broadbands of Landsat ETM+ across the optical spectral region. It is abundantly clear from Figure 10.1 that hyperspectral data capture spectral variability throughout the spectrum, whereas Landsat ETM+ averages the spectrum over its broad wavelength ranges. Hyperspectral data can, of course, be captured in much narrower wavelengths (e.g., 1 nm or even less) if point spectroradiometers are used (Thenkabail et al., 2000, 2004a,b) or even, recently, by using hyperspectral imaging data such as from the spaceborne Earth Observing-1 Hyperion, which captures data in 10-nm bandwidths in 400–2500 nm (Pearlman et al., 2003). In addition, hyperspectral data (point or imaging) do not necessarily need to cover the entire spectral range or have hundreds of bands. However, it is essential to have adequate numbers of narrowbands (<20 nm) spread across the electromagnetic spectrum, such as visible (VIS), near-infrared (NIR), short-wave infrared (SWIR), midwave infrared (MWIR), and long-wave infrared (LWIR) ranges. As needed, hyperspectral data can be acquired across these entire wavelength ranges or in selective regions, but always as continuous or near-continuous data in discrete narrow bands along the spectrum under consideration.

Hyperspectral data are gathered from various platforms: ground-based (lab and field), drone-flown, airborne (low and high altitude), and spaceborne platforms (Thenkabail, 2015). In each of these cases, data are acquired over a spatial dimension or pixel (e.g., Figures 10.2 and 10.3). To measure spectral properties of objects on the planet (e.g., Figure 10.2 showing crops versus bare soil), we not only need high spectral resolution, but we also need high spatial resolution. In reality, most every pixel in hyperspectral imaging data has spectral signatures of mixed objects. For example, EO-1 Hyperion data with 30 by 30 m pixels (1 pixel = 0.09 hectares) can capture pure signatures of crops when fields are larger than 0.09 hectares, which is often the case for an overwhelming proportion of farms in the world. So, crop types will have pure crop signatures (by ensuring edge pixels are ignored). However, in a 30 by 30 m area of a tropical forest, for example,

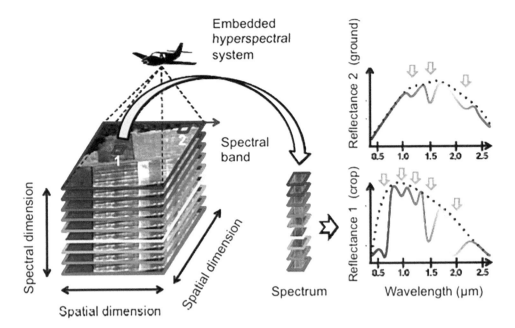

FIGURE 10.2   Diagram of a remote sensing hyperspectral image and acquiring a hyperspectral data bank of features or themes. The reflectance spectrum of each of the two materials (crops and bare ground) is shown with its continuum (dotted line) and its absorption bands (arrows). (Ceamanos, X., Valero, S. 2016. in: *Optical Remote Sensing of Land Surface*, edited by Baghdadi, N. and Zribi, M., Elsevier, Pages 163–200, ISBN 9781785481024, https://doi.org/10.1016/B978-1-78548–102-4.50004-1.)

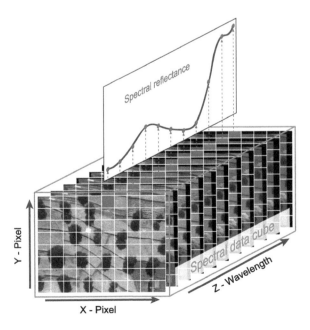

FIGURE 10.3   Hyperspectral data cube and gathering spectral profiles. Structure of the spectral data cube. In the field campaign, the Liquid Crystal Tunable Filter (LCTF) imager recurrently captured spectral images ranging from 460 to 780 nm at 10 nm intervals, which equates to 33 images per cycle. (Ishida, T. et al. 2018. *Computers and Electronics in Agriculture*, Volume 144, Pages 80–85, ISSN 0168-1699, https://doi.org/10.1016/j. compag.2017.11.027.)

one can have many species within a single pixel. So, the hyperspectral data of a 30-m pixel in the forest case will contain a mixture of tree species signatures. Ideally, we should be in a position to capture pure spectra of individual objects (e.g., a single plant). However, that is feasible only when high spatial and spectral resolution capabilities are available (e.g., a hyperspectral imaging sensor with spatial resolution of a few centimeters). For most studies pertaining to agriculture and vegetation, hyperspectral data captured at about 1-m spatial resolution will be quite adequate. By meeting these conditions, hyperspectral imaging spectroscopy data will allow us to capture unique signatures of various objects on the planet (e.g., Figure 10.2; Ceamanos and Valero, 2016), which are then stacked as a data cube (e.g., Figure 10.3; Ishida et al., 2018) for thematic data analysis and classification.

Many hyperspectral sensors are available today worldwide. Widely used operational airborne hyperspectral sensors include: (1) US National Aeronautics and Space Administration's (NASA) Airborne Visible/Infrared Imaging Spectrometer (AVIRIS), and AVIRIS-New Generation (AVIRIS NG), (2) the FENIX 1 K sensor from SPACIM (who have sold more than 120 airborne sensors at a variety of spectral ranges), (3) Australia's airborne hyperspectral imaging sensor (HyMap), and (4) many others (Panda et al., 2015). Readers are also encouraged to study Panda et al. (2015) for a more detailed listing and characterization of a multitude of hyperspectral sensors across platforms over the last 50 years.

10.1 HYPERSPECTRAL VEGETATION INDICES

It is feasible to compose a very large number of hyperspectral vegetation indices (HVIs) using hyperspectral narrowband (HNB) data. For example, 242 HNBs of Earth Observing-1 Hyperion data will allow 29,161 unique two-band HVIs relative to just 55 unique vegetation indices (VIs) possible from an 11-band Landsat Operational Land Imager (OLI) scene. That is a massive 29,102 more indices from EO-1 Hyperion relative to Landsat OLI. However, it is also true that an overwhelming proportion of the EO-1 Hyperion HVIs are redundant. At the same time, there are far more unique indices that are now available to study specific features using HVIs relative to broadband VIs. In a comprehensive study of agricultural crops and vegetation over the years, Thenkabail et al. (2014) proposed six distinct categories of HVIs (Table 10.1) based on their utility in studying various vegetation characteristics that were grouped as:

1. Hyperspectral vegetation indices for biophysical quantities (to best study biomass, LAI, plant height, and grain yield)
2. Hyperspectral vegetation indices for biochemical quantities (for study of pigments like carotenoids, anthocyanins, and chlorophyll, as well as nitrogen)
3. Hyperspectral vegetation indices for plant stress (to best study plant stress and drought)
4. Hyperspectral vegetation indices for plant water characteristics (to best study plant water and moisture)
5. Hyperspectral vegetation indices for light use efficiency (to best study plant light use efficiency)
6. Hyperspectral vegetation indices for lignin and cellulose (to best study plant lignin, cellulose, and plant residue)

Multispectral broadband remote sensing does not allow for such specific categorization since it consists of data averaged over broad spectral ranges. Indeed, other hyperspectral sensors that gather data over even finer spectral bandwidths such as 1 nm would have a far greater number of HVIs even when compared to the 29,161 unique two-band HVIs derived from the 242 HNBs of EO-1 Hyperion. Overall, HVIs allow us to capture specific vegetation characteristics (e.g., plant water, plant stress, carotenoids) with a greater degree of precision and accuracy as compared to traditional

TABLE 10.1

Two-Band Hyperspectral Vegetation Indices (HVIs) to Study Agricultural and Vegetation Biophysical and Biochemical Properties, Stress, Plant Water/Moisture, and Other Characteristics

Band Number (#)	Hyperspectral Narrowband (nanometers) (λ1)	Bandwidth (nanometers) ($\Delta\lambda$1)	Hyperspectral Narrowband (nanometers) (λ2)	Bandwidth (nanometers) ($\Delta\lambda$2)	Hyperspectral Vegetation Index (HVI) (nanometers)	Best Index Under Each Category
I. Hyperspectral Vegetation Indices for Biophysical Quantities (HVI-BIOP) (to Best Study Biomass, LAI, Plant Height, and Grain Yield)						
HVI-BIOP1	855	20	682	5	$(855 - 682)/$ $(855 + 682)$	HBSI: Hyperspectral biomass and structural index
HVI-BIOP2	910	20	682	5	$(910 - 682)/$ $(910 + 682)$	
HVI-BIOP3	550	5	682	5	$(550 - 682)/$ $(550 + 682)$	
HVI-BIOP4	1075	5	682	5	$(1075 - 682)/$ $(1075 + 682)$	
HVI-BIOP5	1245	5	682	5	$(1245 - 682)/$ $(1245 + 682)$	
HVI-BIOP6	1650	5	682	5	$(1650 - 682)/$ $(1650 + 682)$	
HVI-BIOP7	2205	5	682	5	$(2205 - 682)/$ $(2205 + 682)$	
II. Hyperspectral Vegetation Indices for Biochemical Quantities (HVI-BIOC) (Pigments Like Carotenoids, Anthocyanins as well as Nitrogen, Chlorophyll)						
HVI-BIOC1	550	5	515	5	$(550 - 515)/$ $(550 + 515)$	HBCI: Hyperspectral biochemical index
HVI-BIOC2	550	5	490	5	$(550 - 490)/$ $(550 + 490)$	
HVI-BIOC3	720	5	550	5	$(720 - 550)/$ $(720 + 550)$	
HVI-BIOC4	550	5	375	5	$(550 - 375)/$ $(550 + 375)$	
HVI-BIOC5	855	20	550	5	$(855 - 550)/$ $(855 + 550)$	
HVI-BIOC6	550	5	682	5	$(550 - 682)/$ $(550 + 682)$	
III. Hyperspectral Vegetation Indices for Plant Stress (HVI-PSTR) (to Best Study Plant Stress, Drought)						
HVI-PSTR1	700–740	40	First-order derivative integrated over red edge (700–740)			HREI: Hyperspectral red-edge index
HVI-PSTR2	855	5	720	5	$(855 - 720)/$ $(855 + 720)$	
HVI-PSTR3	910	5	705	5	$(910 - 705)/$ $(910 + 705)$	

(Continued)

TABLE 10.1 (*Continued*)

Two-Band Hyperspectral Vegetation Indices (HVIs) to Study Agricultural and Vegetation Biophysical and Biochemical Properties, Stress, Plant Water/Moisture, and Other Characteristics

Band Number (#)	Hyperspectral Narrowband (nanometers) ($\lambda 1$)	Bandwidth (nanometers) ($\Delta\lambda 1$)	Hyperspectral Narrowband (nanometers) ($\lambda 2$)	Bandwidth (nanometers) ($\Delta\lambda 2$)	Hyperspectral Vegetation Index (HVI) (nanometers)	Best Index Under Each Category
IV. Hyperspectral Vegetation Indices for Plant Water (HVI-PWTR) (to Best Study Plant Water and Moisture)						
HVI-PWTR1	855	20	970	10	$(855 - 970)/$ $(855 + 970)$	HWMI: Hyperspectral water and moisture index
HVI-PWTR2	1075	5	970	10	$(1075 - 970)/$ $(1075 + 970)$	
HVI-PWTR3	1075	5	1180	5	$(1075 - 1180)/$ $(1075 + 1180)$	
HVI-PWTR4	1245	5	1180	5	$(1245 - 1180)/$ $(1245 + 1180)$	
HVI-PWTR5	1650	5	1450	5	$(1650 - 1450)/$ $(1650 + 1450)$	
HVI-PWTR6	2205	5	1450	5	$(2205 - 1450)/$ $(2205 + 1450)$	
HVI-PWTR7	2205	5	1950	5	$(2205 - 1950)/$ $(2205 + 1950)$	
V. Hyperspectral Vegetation Indices for Light Use Efficiency (HVI-LUEF) (to Best Study Light Use Efficiency or LUE)						
HYVI-LUEF1	570	5	531	1	$(570 - 531)/$ $(570 + 531)$	HLEI: Hyperspectral light-use efficiency index
VI. Hyperspectral Vegetation Indices for Lignin and Cellulose (HVI-LICE) (to Best Study Plant Lignin, Cellulose, and Residue)						
HYVI-LICE1	2205	5	2025	1	$(2205 - 2025)/$ $(2205 + 2025)$	HLCI: Hyperspectral legnin cellulose index

Source: Adopted with modifications from comprehensive research reported in Thenkabail (2015) and Thenkabail et al. (2000, 2002, 2004a,b, 2012a,b, 2014).

Note: There are six distinct categories of two-band HVIs that are best to study agriculture and vegetation biophysical and biochemical properties, stress, plant water/moisture, and other characteristics.

　　1. Physical relevance of these two-band hyperspectral indices are presented and discussed in Table 10.4 in Thenkabail et al. (2012a) and various chapters in the book by Thenkabail et al. (2012b).

　　2. The first index under each of the six categories performs the best, but further research needs to confirm this.

　　3. For extensive research on hyperspectral wavebands and vegetation indices, refer to papers by Thenkabail et al. (2002, 2004a,b,c, 2012a,b, 2013, 2014).

　　4. Under each of the six categories (I to VI), you may select the best index (mentioned first in the category and highlighted).

TABLE 10.2

Best Hyperspectral Narrowband (HNB) Combinations to Study and Classify Agricultural and Vegetation Characteristics[a]

Best 5 bands	550, 682, 855, 970, 1450
Best 10 bands	550, 682, 720, 855, 970, 1245, 1450, 1650, 1950, 2205
Best 15 bands	490, 515, 550, 570, 682, 720, 855, 970, 1075, 1180, 1245, 1450, 1650, 1950, 2205
Best 20 bands	490, 515, 531, 550, 570, 682, 720, 855, 910, 970, 1075, 1180, 1245, 1450, 1650, 1725, 1950, 2205, 2260, 2359

Source: Thenkabail, P.S., Lyon, J. Huete, A. (Editors) 2012b. *Hyperspectral Remote Sensing of Vegetation*, CRC Press–Taylor and Francis group, Boca Raton, Page 781 (80+ pages in color).

Note: Best 5-, 10-, 15-, and 20-band HNBs to study agriculture and vegetation.

[a] Gathered from research by Thenkabail et al. (2012a,b).

---

methods. Just as hyperspatial data (e.g., 1 or 5 m) provides finer information on objects or themes, hyperspectral data provides greater details of objects or themes. For example, hyperspatial data helps in detecting a building or a tree as opposed to coarser-resolution (e.g., 250-m) data, which only provide broad land cover classes or themes (e.g., forest, agriculture). Similarly, hyperspectral data such as a 10-nm-wide hyperspectral narrowband centered at 970 nm help in the study of plant or leaf water conditions, while a 140-nm-wide broadband at 760–900 nm helps model biomass but does not help in determining plant water content or stress with the same degree of precision or accuracy as hyperspectral data. The HVIs in Table 10.1 help us address numerous plant biophysical, biochemical, stress, and water characteristics. These indices were computed using 10-nm bandwidth hyperspectral narrowbands such as the ones shown in Table 10.2. HNBs are used to compute many two-band, multiband, and derivative HVIs (Tables 10.3 and 10.4). The indices shown in Tables 10.1 through 10.4 are mainly focused on agricultural crops and\or other vegetation types. These indices were also re-affirmed in several other studies. A recent example estimated percent green vegetation cover (PVC; Liu et al., 2017, and highlighted the use of several valuable HNBs centered at important absorption features: 681.20 nm (pigment absorption); 721.90 and 732.07 nm (red edge); 1174.77 and 1184.87 nm (water absorption); and 1447.14, 1457.23, 2072.65, and 2102.94 nm (cellulose and lignin absorption). Most of these wavebands also appear in Table 10.1 within ± 5 nm.

In Chapter 1, Dr. Dar Roberts et al. described and discussed HVIs by categorizing them into the following groups: (1) vegetation structure, (2) biochemistry (pigments, canopy moisture, lignin and cellulose, nitrogen), and (3) plant physiology. An excellent list of HVIs was summarized in tabular form in Chapter 1 and, along with those mentioned in this Chapter (Tables 10.1 through 10.4), provide comprehensive coverage of the HVIs pertaining to agriculture and vegetation studies. Chapter 1 also illustrated the applications of HVIs in estimating leaf area index (LAI) of hybrid poplars (*Populus trichocarpa*) and in the study of two invasive species: *Brachypodium distachyon* (Purple false brome) and *Carduus pycnocephalus* (Italian thistle). They further extended the analysis by evaluating performance of HVIs in the study of potential (PET) and actual (AET) evapotranspiration. They concluded that HVIs, in many cases, are less sensitive to atmospheric contamination. They emphasized the advanced capabilities of HVIs in measuring specific vegetation traits such as the photochemical reflectance index (PRI), which is a proxy to light use efficiency, and the red-edge vegetation stress index (RVSI).

It also must be noted that the availability of hundreds or thousands of hyperspectral narrowbands facilitates computation of HVIs that involve more than two bands. A classic example is the optimum multiple narrow-band reflectance (OMNBR) first proposed by Thenkabail et al. (2000). This was also discussed in Chapter 1 by Dr. Roberts et al. Many (Yu et al., 2013; Gnyp et al., 2014) have shown the great strengths of OMNBR-type indices relative to other HVIs in modeling crop and vegetation

TABLE 10.3

Hyperspectral Vegetation Indices (HVIs) to Study Agricultural and Vegetation Pigments and Stress

Name	Related to	Equation	Reference
Normalized Difference Vegetation Index (NDVI)	Chlorophyll	$\dfrac{R_{800} R_{R670}}{R_{800} + R_{670}}$	Rouse et al. (1974)
Simple Ratio Index (SR)	Chlorophyll	$\dfrac{R_{800}}{R_{670}}$	Rouse et al. (1974)
Enhanced Vegetation Index (EVI)	Chlorophyll	$2.5 \dfrac{R_{800} R_{670}}{R_{800} + 6R_{670} 7.5R_{490} + 1}$	Huete et al. (1997)
Atmospherically Resistant Vegetation Index (ARVI)	Chlorophyll	$\dfrac{R_{800} 2(R_{670} R_{490})}{R_{800} + 2(R_{670} R_{490})}$	Kaufman and Tanre (1996)
Sum Green Index (SG)	Greenness	$\dfrac{1}{n} \sum_{i-500}^{599} R_i$	Gamon and Surfus (1999)
Red Edge NDVI (RENDVI)	Chlorophyll	$\dfrac{R_{750} R_{705}}{R_{750} + R_{705}}$	Gitelson and Merzlyak (1994)
Modified Red Edge Simple Ratio Index (mRESR)	Chlorophyll	$\dfrac{R_{750} R_{445}}{R_{705} R_{445}}$	Sims and Gamon (2002)
Modified Red Edge NDVI (rnRENDVI)	Stress Senescence	$\dfrac{R_{750} R_{705}}{R_{750} + R_{705} 2R_{445}}$	Sims and Gamon (2002)
Vogelmann Red Edge Index 1 (VOG1)	Chlorophyll, Water	$\dfrac{R_{740}}{R_{720}}$	Vogelmann et al. (1993)
Vogelmann Red Edge Index 2 (VOG2)	Chlorophyll, Water	$\dfrac{R_{734} R_{747}}{R_{715} + R_{726}}$	Vogelmann et al. (1993)
Vogelmann Red Edge Index 3 (VOG3)	Chlorophyll, Water	$\dfrac{R_{734} R_{747}}{R_{715} + R_{720}}$	Vogelmann et al. (1993)
Red Edge Position Index (REP)	Chlorophyll	$\arg\max_{wl}(R_{690\,740})'$	Curran et al. (1995)
Photochemical Reflectance Index (PRI)	Photosynthesis, Carot.	$\dfrac{R_{531} R_{570}}{R_{531} + R_{570}}$	Gamon et al. (1992)
Structure Insensitive Pigment Index (SIPI)	Carot.-Chl.-Ratio	$\dfrac{R_{800} R_{445}}{R_{800} + R_{680}}$	Peñuelas et al. (1995)
Red Green Ratio Index (RGRI)	Anth.-Chl.-Ratio	$\dfrac{\text{Mean}(R_{800\,800})}{\text{Mean}(R_{800\,700})}$	Gamon and Surfus (1999)
Plant Senescence Reflectance Index (PSRI)	Stress, Senescence	$\dfrac{R_{680} R_{500}}{R_{750}}$	Merzlyak et al. (1999)
Carotenoid Reflectance Index 1 (CAR1)	Carotenoids	$\dfrac{1}{R_{510}} - \dfrac{1}{R_{550}}$	Gitelson et al. (2002)
Carotenoid Reflectance Index 2 (CAR2)	Carotenoids	$\dfrac{1}{R_{510}} - \dfrac{1}{R_{700}}$	Gitelson et al. (2003)
Anthocyanin Reflectance Index (ANTH1)	Anthocyanins	$\dfrac{1}{R_{550}} - \dfrac{1}{R_{700}}$	Gitelson et al. (2001)
Anthocyanin Reflectance Index (ANTH2)	Anthocyanins	$R_{800}\left[\dfrac{1}{R_{550}} - \dfrac{1}{R_{700}}\right]$	Gitelson et al. (2001)

Source: Behmann, J. et al. 2014. *ISPRS Journal of Photogrammetry and Remote Sensing*, Volume 93, Pages 98–111, ISSN 0924-2716, https://doi.org/10.1016/j.isprsjprs.2014.03.016.

Note: HVIs to best study agricultural and vegetation pigments (carotenoids, anthocyanins, and chlorophyll) and stress.

TABLE 10.4

Hyperspectral Vegetation Indices (HVIs) to Study Agricultural and Vegetation Characteristics Such as Leaf Water, Stress, Disease, Chlorophyll, and Photochemical Reflectance Index

Variables	Equation/Method	Reference
1. Depth of the 671 nm absorption band (chlorophyll)	Continuum removal method (edges at 569 and 763 nm)	Clark and Roush (1984)
2. Depth of the 983 nm and 1205 nm absorption bands (leaf liquid water)	Continuum removal method (edges at 933 and 1094, 1094 and 1286 nm)	Clark and Roush (1984)
3. Depth of the 2103 and 2304 nm absorption bands (lignin-cellulose)	Continuum removal method (edges at 2052 and 2214 nm, 2214 and 2385 nm)	Clark and Roush (1984)
4. Disease Water Stress Index (DSWI)	$(\rho_{803\,nm} + \rho_{549\,nm})/(\rho_{1659\,nm} + \rho_{681\,nm})$	Apan et al. (2004)
5. First derivative and red-edge position	Savitzky-Golay smoothing method (691–763 nm)	Tsai and Philpot (1998)
6. Liquid water content	ACORN-derived method	Imspec (2001)
7. Modified Chlorophyll Absorption in Reflectance Index (MCARI)	$[(\rho_{701\,nm} - \rho_{671\,nm}) - 0.2(\rho_{701\,nm} - \rho_{549\,nm})]$ $(\rho_{701\,nm}/\rho_{671\,nm})$	Dauglitry et al. (2000)
8. Normalized Difference Vegetation Index (NDVI)	$(\rho_{864\,nm} - \rho_{671\,nm})/(\rho_{864\,nm} + \rho_{671\,nm})$	Rouse et al. (1973)
9. Normalized Difference Water Index (NDWI)	$(\rho_{864\,nm} - \rho_{1245\,nm})/(\rho_{864\,nm} + \rho_{1245\,nm})$	Gao (1996)
10. Leaf Water Vegetation Index (LWVI-1)	$(\rho_{1094\,nm} - \rho_{893\,nm})/(\rho_{1094\,nm} + \rho_{983\,nm})$	This study, an NDWI variant
11. Leaf Water Vegetation Index (LWVI-2)	$(\rho_{1094\,nm} - \rho_{1205\,nm})/(\rho_{1094\,nm} + \rho_{1205\,nm})$	This study, an NDWI variant
12. Photochemical Reflectance Index (PRI)	$(\rho_{529\,nm} - \rho_{569\,nm})/(\rho_{529\,nm} + \rho_{569\,nm})$	Gamon et al. (1992)
13. Red-Edge Vegetation Stress Index (RVSI)	$((\rho_{712\,nm} + \rho_{752\,nm})/2) - \rho_{732\,nm}$	Merton and Huntington (1999)

Source: Galvão, L.S. et al. 2005. *Remote Sensing of Environment*, Volume 94, Issue 4, Pages 523–534, ISSN 0034-4257, https://doi.org/10.1016/j.rse.2004.11.012.

Note: HVIs to best study agricultural and vegetation pigments (carotenoids, anthocyanins, and chlorophyll) and stress. ρ is the reflectance. In the equations, the wavelengths indicated are from Hyperion bands.

quantities. Yu et al. (2013) showed that OMNBR models significantly increase the accuracy for estimating leaf and plant nitrogen content when six HNBs are used.

10.2   DERIVATIVE HYPERSPECTRAL VEGETATION INDICES

The derivative of a spectrum is its rate of change with respect to wavelength (Demetriades-Shah et al., 1990). There are two main reasons to use derivative hyperspectral vegetation indices (dHVIs) rather than just hyperspectral vegetation indices. First, hyperspectral data are continuous or nearly continuous over a wavelength range (e.g., 400–2500 nm). So, sometimes the best approach for hyperspectral data analysis is to use these continuous spectra, often drawing from different portions of the spectrum (e.g., 680–760 nm, 626–795 nm; Elvidge and Chen, 1995) rather than using selective hyperspectral vegetation indices. Treating hyperspectral data as truly continuous allows access to information that is often suppressed by the standard analysis methods (Tsai and Philpot, 1998). Second, derivatives are relatively insensitive to variations in illumination intensity whether caused by changes in sun angle, cloud cover, or topography, and also relatively less sensitive to the spectral variations of sunlight and skylight (Tsai and Philpot, 1998). However, derivative indices have not been widely used so far, probably due to reasons including the data redundancy issue that we are familiar with from some of the previous chapters of this volume and dHVIs working best only when the original spectra are smoothed or appear in high-quality form, as they are very sensitive to noise.

Derivative indices capture slope or curvature of reflectance rather than reflectance values (Tsai and Philpot, 1998). Derivative indices are especially useful in the red-edge region where the change in spectral reflectivity to unit change in spectral wavelength is quite dramatic (Datt and Paterson, 2000). The derivative of the spectrum is defined as the change in reflectivity of the spectrum between two hyperspectral narrowbands (e.g., 680–690 nm) divided by the change in wavelength (10 nm; a.k.a. slope of the spectrum). Elvidge and Chen (1995) computed first-order derivative spectra over a continuous range (e.g., 626 to 795 nm) and then integrated them. Transformation of original reflectances of plants to their first-order derivatives (e.g., Figure 10.4) and second-order derivatives

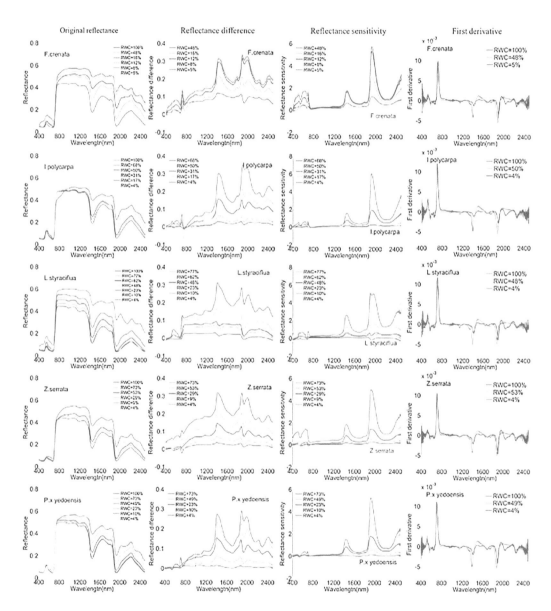

FIGURE 10.4   Hyperspectral spectra of plant species and their transformations including first-order derivative versions. Changes of reflectance spectra, reflectance differences, reflectance sensitivities, and first-derivative reflectance spectra for different plant species attributed to progressive relative water content (RWC). (Cao, Z. et al. 2015. *Ecological Indicators*, Volume 54, Pages 96–107, ISSN 1470-160X, https://doi.org/10.1016/j. ecolind.2015.02.027.)

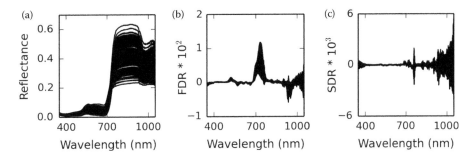

FIGURE 10.5 Hyperspectral spectra of plant species and their transformations including first- and second-order derivative versions. Spectral measurements of durum wheat on 111 days after planting in the 2011–2012 growing season (March 31, 2012), including (a) canopy spectral reflectance, (b) the first derivative of canopy spectral reflectance (FDR), and (c) the second derivative of canopy spectral reflectance (SDR). (Thorp, K.R. et al. 2013. *Remote Sensing of Environment*, Volume 132, Pages 120–130, ISSN 0034-4257, https://doi.org/10.1016/j.rse.2013.01.008.)

(Figure 10.5) has shown significant advantages over normal HVIs in several instances by several authors (Li and Wand, 2016; Cao et al., 2015; Thorp et al., 2017), for example, showed that the plant relative water content (RWC) and equivalent water thickness (EWT) were best modeled using the first derivative reflectance based on the normalized difference type derivative index or dND (1415 nm, 1530 nm) and a simple ratio-type derivative index or dSR (1530 nm, 1895 nm). Thorp et al. (2017) demonstrated improved wheat crop trait establishment using derivative indices. Jin and Wang (2016) determined that dSR (660 nm, 1040 nm) based on the first-derivative spectra provided the best estimate of canopy transpiration in desert plants. Sonobe and Wang (2018) found that dND (516 nm, 744 nm), a normalized differences type index using reflectance derivatives at 516 and 744 nm [(D516 – D744)/(D516 + D744)] performed the best in estimating carotenoid content of broadleaved plant species.

In Chapter 2, Dr. Quan Wang et al. explored numerous derivative indices (see the table in the appendix of chapter 2 of this volume). They computed first- to sixth-order derivative hyperspectral vegetation indices along with normal hyperspectral vegetation indices and developed relationships to estimate various forest biophysical and biochemical quantities such as chlorophyll a, chlorophyll b, and carotenoids. They concluded that dHVIs performed significantly better than other HVIs. However, they recognized that derivative indices were very sensitive to spectral noise and needed to be used only after smoothing the spectra. This is a fact generally recognized by a wide range of researchers. We would like to caution readers that derivative indices require far more rigorous study than that found in the existing literature to determine their real strengths and limitations. We especially propose studies over a wide range of crops and other vegetation and in varying conditions before a definitive determination can be made on the strengths and limitations of dHVIs. In this context, Chapter 2 by Dr. Quan Wang is a very good starting point that not only captures the past literature well, but also provides a good comprehensive set of research of its own.

10.3 HYPERSPECTRAL IMAGE CLASSIFICATION METHODS IN VEGETATION AND AGRICULTURAL CROPLAND STUDIES

Hyperspectral image classification methods, to a great extent, are also used in multispectral broadband remote sensing image classifications. However, hyperspectral sensors increase data volume by several magnitudes due to the large number of spectral bands. This requires high computing capacity such as cloud computing or supercomputers backed by machine learning algorithms (MLAs) and artificial intelligence systems when classifying images of large areas. This is especially true when hyperspectral data are also multitemporal or the analysis integrates additional co-registered datasets such as LiDAR.

Some of the well-known image classification methods that are used with any type of imagery, hyperspectral or multispectral, include (Jia and Richards, 1994; Lu and Weng, 2007): (1) per-pixel classifiers that are either parametric (e.g., maximum likelihood) or nonparametric (neural networks, decision trees, support vector machines, and expert systems), (2) subpixel classification (e.g., spectral mixture analysis), (3) per-field classification (e.g., object-based classifications), (4) contextual classification (e.g., Markov random field-based contextual classifiers), (5) knowledge-based classification (e.g., decision trees that use data from a number of secondary sources such as soils, elevation), and (6) hybrid classifications (e.g., combining parametric and nonparametric approaches). When hyperspectral data were integrated with other data sources such as hyperspatial, temporal, or LiDAR data or other datasets such as elevation, there was often an increase in classification accuracies as well as an increase in the number of classes that could be discerned.

The most prominent pixel-based machine learning algorithms to classify images include the: (1) random forest (Gislason et al., 2006; Tatsumi et al., 2015); (2) automated cropland mapping algorithm (ACMA) (Xiong et al., 2017); (3) automated cropland classification algorithm (ACCA) (Thenkabail et al., 2010; Teluguntla et al., 2017); (4) spectral matching techniques (Thenkabail et al., 2007; Teluguntla et al., 2017); (5) support vector machines (Mountrakis et al., 2011); (6) decision trees (Friedl and Brodley, 1997; DeFries and Chan, 2000; Waldner et al., 2015); (7) linear discriminant analysis (Imani and Ghassemian, 2015); (8) principal component analysis (Jensen, 2009); (9) k-means, Isoclass clustering (Jensen, 2009; Duveiller et al., 2015); (10) classification and regression tree (CART) (Deng and Wu, 2013; Egorov et al., 2015); (11) tree-based regression algorithm (Ozdogan and Gutman, 2008); (12) phenology-based methods (Dong et al., 2015); and (13) Fourier harmonic analysis (Geerken et al., 2005; Zhang, et al., 2015). The most prominent object-based algorithm approach is the recursive hierarchical image segmentation (RHSEG) (Tilton et al., 2012).

However, hyperspectral remote sensing is not a panacea and has its own challenges (Thenkabail et al., 2012a, 2013, 2014; Thenkabail, 2015). In order to attain a high degree of accuracy, there is a need for adequate training samples (He et al., 2018). The requirement for the number of training samples per class increases with data dimensionality. As a result, it is important to reduce the data dimensionality (by removing redundant bands) before applying classifiers. For example, George et al. (2014) reduced the Hyperion data from its initial 242 hyperspectral narrowbands to 29 nonredundant HNBs. Then they applied support vector machines (SVMs) and a spectral angle mapper (SAM) approach using the 29 HNBs to classify forest vegetation and compared the results with those obtained from classifying Landsat TM data (Figure 10.6). The 29-band HNBs provided greater precision as well as accuracy in mapping plant species as opposed to broadband Landsat TM data (Figure 10.6). The performance of pixel-based supervised classifiers such as support vector machines is dependent on robust, accurate, and large sample size training and validation datasets. Machine learning, deep learning, and artificial intelligence methods are expected to advance hyperspectral image classification in terms of computation speed, accuracies, and the ability to self-learn based on training data.

In Chapter 3, Dr. Edoardo Pasolli et al. provided comprehensive discussions on various widely used as well as emerging hyperspectral classification methods and approaches. First, feature selection methods for dimensionality reduction were addressed including: (1) parametric and nonparametric supervised and unsupervised feature selection methods, (2) spectral indices, (3) linear transformations, and (4) manifold learning. In all these methods, redundant hyperspectral narrowbands are removed and the best nonredundant HNBs retained. This process is important in any hyperspectral data analysis to reduce data volumes and ensure computing efficiency and resource optimization. However, it must be noted that HNBs that are redundant in one application (e.g., agriculture) may be very useful in another application (e.g., land cover or minerals). So, hyperspectral dimensionality reduction should take the application into consideration, which results in determining the best nonredundant optimal bands for a given application (Thenkabail et al., 2000, 2004a,b, 2013; Thenkabail, 2015). Second, machine learning image classification methods and approaches were examined that included: (1) pixel-based classifications such as support vector machines; (2) Bayesian parametric and nonparametric classification such as Gaussian mixture models; and (3) deep learning

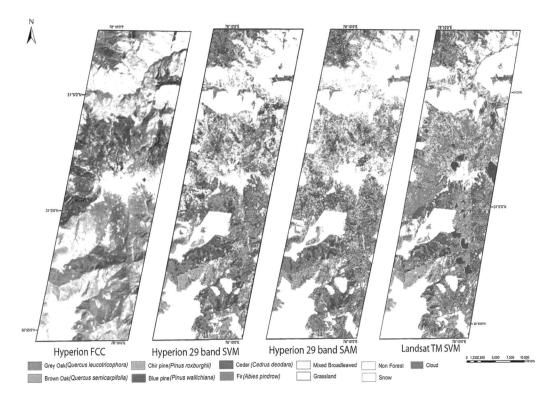

FIGURE 10.6   False color composite (FCC; NIR: 853, Red: 671, Green: 508), SVM, and SAM classified images of Hyperion, and SVM classified image of Landsat TM. (For interpretation of the references to color in this figure legend, the reader is referred to the web version of this article.) (George, R. et al. 2014. *International Journal of Applied Earth Observation and Geoinformation*, Volume 28, Pages 140–149, ISSN 0303-2434, https://doi.org/10.1016/j.jag.2013.11.011.)

algorithms like deep convolutional neural networks (CNNs), recurrent neural networks (RNNs), and belief networks (DBNs). They also discussed segmentation methods such as recursive hierarchical segmentation (RHSeG).

10.4   MASSIVELY LARGE HYPERSPECTRAL REMOTE SENSING DATA PROCESSING ON CLOUD COMPUTING ARCHITECTURES

This is a new era of remote sensing in the twenty-first century. In this era, there is a clear paradigm shift in how remote sensing science is conducted. The number of Earth observation satellites has multiplied severalfold (Panda et al., 2015) even since the past decade. Temporal, spatial, spectral, and radiometric resolutions have all increased in the new generation of satellite sensors (Table 10.5). For example, PlanetScope (Table 10.5) has 175+ mini-satellites that gather data of the entire planet daily at 3–4-meter resolution and in four spectral bands (Table 10.5). A number of other Cubesats and Smallsats are also planning to carry hyperspectral sensors. Data from these satellites, acquired in hyperspectral and hyperspatial mode, result in massively large datasets that require cloud computing, machine learning, and artificial intelligence to manage-process-analyze these data. Powerful cloud computing platforms such as Google Earth Engine (GEE) enable global-level data processing (Gorelick et al., 2017) with the availability offered by image collection from multiple satellites over time and space resulting in petabytes (1 petabyte = 1 billion gigabytes) of data. Further, processing such massively large datasets for information of various kinds requires us to train, code, classify, model, map, and validate these massively large datasets through various machine learning algorithms and artificial intelligence.

TABLE 10.5

Characteristics of Some of the Twenty-First Century Satellites[a]

	PlanetScope	RapidEye	SkySat	Landsat 8	Sentinel-2
Number of satellites	175+	5	13	1	2
Spectral resolution	4 bands (R, G, B, NIR)	5 bands (R, G, B, RE, NIR)	5 bands (R, G, B, NIR, pan)	11 bands (VNIR, SWIR, TIR, pan)	13 bands (VNIR, SWIR)
Spatial resolution	3–4 m	6.5 m	0.8–1 m	30 m multispectral, 15 m pan	10, 20, or 60 m depending on band
Temporal resolution	Daily at nadir	Daily off-nadir, 5.5 days at nadir	4–5 days	16 days	5 days
Image dimension	24 by 7 km	77 km by up to 300 km	3.2 by 1.3 km	185 by 180 km	290 by 300 km
Radiometric resolution	8 to 16 bit	8 to 16 bit	8 to 16 bit	12 to 16 bit	12 bit

[a] This table was compiled by Dr. Itiya Aneece, Mendenhall postdoctoral researcher at USGS.

Thenkabail and his team recently produced a global 30-m cropland extent product based on Landsat Operational Land Imager 16-day time series data of the 2013–2015 time period. These products involved use of massively big data, cloud computing, and machine learning algorithms. MLAs require robust and carefully collected training data as they are used to generate knowledge, write scripts for MLAs, and perform classifications. The accuracy of the product will depend on the purity and correctness of the reference training data. Classification accuracies were established using reference validation data. One such product, a 30-m Landsat-derived cropland extent image (Figure 10.7) can be viewed and downloaded from the following links (Teluguntla et al., 2015; Xiong et al., 2017b):

> www.croplands.org
> https://lpdaac.usgs.gov/about/news_archive/release_gfsad_30_meter_cropland_extent_
> products

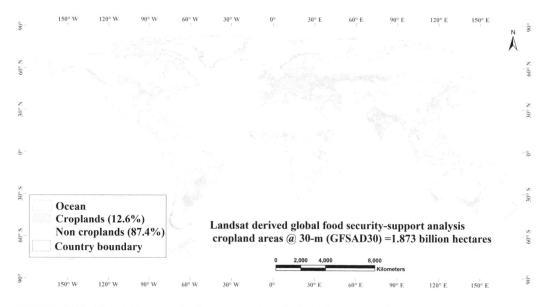

FIGURE 10.7   Global 30-m cropland extent product. Global 30-m cropland extent product derived using Landsat 16-day time-series data for the nominal year 2015.

Obtaining high-quality reference training data, generating knowledge, and implementing that knowledge to classify or analyze image collections on cloud computing platforms such as GEE will result in products such as the global cropland extent product at 30 m resolution (Figure 10.7; http://www.croplands.org) derived from Landsat time-series images. By having appropriate reference training and validation data, for example, in an agricultural application, numerous other cropland products can be generated, such as crop types, watering methods (irrigation or rainfed), cropping intensities, and crop productivities. The same "image collection" in GEE used to produce the global cropland extent outcome (Figure 10.7) can also be used to produce a multitude of other applications such as in forests, wetlands, and water conditions such as droughts. Other applications can be made as defined by the user community, provided there are appropriate training and validation data and relevant algorithms for those applications. In each case, appropriate reference training and validation data are required to train the MLAs. Once the MLAs start being robust and mature based on the best available reference training data to generate the best possible knowledge, the MLAs will be able to "self-train" themselves without human interaction, leading to refinement of products that are increasingly accurate. This process is commonly referred to as artificial intelligence, where machines can self-learn based on knowledge generated from reference training data.

In Chapter 4, Dr. Wu Zebin et al. started by defining cloud computing and explaining why it is crucial in hyperspectral data analysis. Indeed, they clearly demonstrated the need for cloud computing in every step of hyperspectral data analysis such as data preprocessing and data analysis for applications. The authors highlighted the fact that to implement parallel and distributed processing of massively large hyperspectral datasets, three issues need to be well understood and developed including the: (1) distributed programming model, (2) underlying engine that supports parallel computing, and (3) dynamic storage of distributed remote sensing data. Preprocessing involves steps to get the data ready for use after going through processes like atmospheric correction, converting raw data to radiance, and then converting to the top of the atmosphere radiance and ground reflectance. Processing of hyperspectral data will lead to various applications such as deriving a land use/land cover map or biomass map. The authors described the most well-known cloud computing model, MapReduce (MR), in detail. This was followed by distributed computing engines such as Apache Hadoop, public clouds like GEE, and dynamic storage mechanisms. Frameworks for big data processing were described for hyperspectral data. A case study on hyperspectral image processing was illustrated by a theoretical framework, followed by a practical application in the study of agricultural crops. Airborne Visible/Infrared Imaging Spectrometer (AVIRIS) hyperspectral image processing in the cloud for the study of agricultural crops was described and illustrated. The authors suggested that one model to process large hyperspectral datasets efficiently was to utilize Hadoop distributed file systems (HDFSs) for scalable and reliable data storage with Apache Spark as the computing engine, and parallel algorithms developed based on mapReduce.

10.5 NONDESTRUCTIVE ESTIMATION OF FOLIAR PIGMENT (CHLOROPHYLL, CAROTENOID, AND ANTHOCYANIN CONTENTS)

Hyperspectral remote sensing is a powerful tool in the nondestructive retrieval of biophysical and biochemical quantities due to the availability of narrow (10 nm or less) spectral waveband data throughout the electromagnetic spectrum (e.g., 400–2500 nm). Chapter 5 deals with the study of leaf pigments such as chlorophylls (Chl), carotenoids (Car), anthocyanins (Anth), xanthophylls, and flavonoids, which play a key role in plant photosynthesis.

Studies to determine plant pigments (Chl, Car, Anth, xanthophylls, and flavonoids) using remote sensing have been widely reported over the years, especially using appropriate hyperspectral vegetation indices (e.g., Table 10.6). For example, chlorophyll and carotenoid contents of three unrelated tree species were accurately estimated based on hyperspectral narrowbands established by vegetation indices, neural networks (NNs), and partial least-squares (PLS) regression (Kira et al., 2015). However, Kalacska et al. (2015) showed that HVIs did not predict chlorophyll in bog

TABLE 10.6

Leaf/Canopy Biochemical Hyperspectral Vegetation Indices (HVIs)

Index	Index ID	Formula	Reference	Scale
Leaf Area Index				
Normalized difference vegetation index	NDVI	$(R_{800} - R_{670})/(R_{800} + R_{670})$	Rouse et al. (1974)	Leaf/canopy
Chlorophyll Estimation				
Structure insensitive pigment index	SIPI	$(R_{800} - R_{445})/(R_{800} + R_{680})$	Peñuelas et al. (1995)	Leaf
Chlorophyll index red edge	$CI_{red\ edge}$	R_{750}/R_{710}	Gitelson and Merzlyak (1996); Zarco-Tejada et al. (2004)	Leaf/canopy
Transformed cab absorption in reflectance index	TCARI	$3 * [(R_{700} - R_{670}) - 0.2 * (R_{700} - R_{550}) * (R_{700}/R_{670})]$	Haboudane et al. (2002); Meggio et al. (2010)	Leaf/canopy
Optimized soil-adjusted vegetation index	OSAVI	$(1 + 0.16) * (R_{800} - R_{670})/(R_{800} + R_{670} + 0.16)$	Rondeaux et al. (1996); Meggio et al. (2010)	Leaf/canopy
	TCARI/OSAVI	$3 * [(R_{700} - R_{670}) - 0.2 * (R_{700} - R_{550}) * (R_{700}/R_{670})]/(1.16 * (R_{800} - R_{670})/(R_{800} + R_{670} + 0.16))$	Haboudane et al. (2002)	Leaf/canopy
			Meggio et al. (2010)	Leaf/canopy
Carotenoid Concentration				
Ratio analysis of reflectance spectra	RARS	R_{746}/R_{513}	Chappelle et al. (1992)	Leaf
Pigment-specific simple ratio	PSSRa	R_{800}/R_{680}	Blackburn (1998)	Leaf/canopy
Pigment-specific simple ratio	PSSRb	R_{800}/R_{635}	Blackburn (1998)	Leaf/canopy
Pigment-specific simple ratio	PSSRc	R_{800}/R_{470}	Blackburn (1998)	Leaf/canopy
Pigment-specific normalized difference	PSNDc	$(R_{800} - R_{470})/(R_{800} + R_{470})$	Blackburn (1998)	Leaf/canopy
Carotenoid concentration index	CRI_{550}	$(1/R_{515}) - (1/R_{550})$	Gitelson et al. (2003, 2006)	Leaf
Carotenoid concentration index	CRI_{700}	$(1/R_{515}) - (1/R_{700})$	Gitelson et al. (2003, 2006)	Leaf
Carotenoid concentration index	RNIR * CRI_{550}	$(1/R_{510}) - (1/R_{550})*R_{770}$	Gitelson et al. (2003, 2006)	Leaf

(Continued)

TABLE 10.6 (Continued)
Leaf/Canopy Biochemical Hyperspectral Vegetation Indices (HVIs)

Index	Index ID	Formula	Reference	Scale
Carotenoid concentration index	RNIR * CRI$_{700}$	$(1/R_{510}) - (1/R_{700}) * R_{770}$	Gitelson et al. (2003, 2006)	Leaf
Modified photochemical reflectance index	PRIm1	$(R_{515} - R_{530})/(R_{515} + R_{530})$	Hernández-Clemente et al. (2011)	Leaf/canopy
Photochemical reflectance index	PRI	$(R_{570} - R_{530})/(R_{570} + R_{530})$	Gamon et al. (1992)	Leaf
Carotenoid/chlorophyll ratio index	PRI * CI	$((R_{570} - R_{530})/(R_{570} + R_{530})) * ((R_{760}/R_{700}) - 1)$	Garrity et al. (2011)	Leaf
Plant senescencing reflectance index	PSRI	$(R_{680} - R_{500})/R_{750}$	Merzlyak et al. (1999)	Leaf
Reflectance band ratio index	Datt-CabCx+c	$R_{672}/(R_{550} * 3R_{708})$	Datt (1999a)	Leaf
Reflectance band ratio index	DattNIRCabCx+c	$R_{860}/(R_{550} * R_{708})$	Datt (1999b)	Leaf

Source: Hernández-Clemente, R. et al. 2012. *Remote Sensing of Environment*, Volume 127, Pages 298–315, ISSN 0034-4257, https://doi.org/10.1016/j.rse.2012.09.014.

Note: Some key leaf/plant biochemical hyperspectral vegetation indices.

vegetation, whereas continuous wavelet transforms predicted it with a high degree of accuracy. The strengths of wavelet analysis in leaf chlorophyll retrieval were affirmed by Blackburn and Ferwerda (2008), and this was furthered by Ju et al. (2010). They used red-edge boundary wavebands at 675 and 755 nm (R_{675} and R_{755}) and reflectance of the red-edge center wavelength at 718 nm (R718), with the equation RES = ($R_{718} - R_{675}$)/($R_{755} + R_{675}$). Tong and He (2017) found that broadband indices were as powerful as HVIs in deriving chlorophyll. However, the strengths of hyperspectral narrowbands and HVIs over multispectral broadbands (MBBs) are well established (Thenkabail et al., 2000, 2002, 2004a,b, 2013, 2014; Mariotto et al., 2013; Marshall and Thenkabail, 2014; Thenkabail, 2015). Carotenoids have been studied using other indices and Zarco-Tejada et al. (2013) suggested a ratio of R_{515}/R_{570} for vineyards, which was also used as an index by Hernández-Clemente et al. (2012) studying conifer forests. Anthocyanin content has been estimated by many such indices in the study of lychee (Yang et al., 2015). Sonobe and Wang (2018) investigated 13 hyperspectral vegetation indices in the study of plant species and determined that dND (516,744), a normalized differences type index using reflectance derivatives at 516 and 744 nm [(D516 – D744)/(D516 + D744)], performed the best.

Chapter 5 by Drs. Anatoly Gitelson and Alexei Solovchenko highlighted the fact that a nondestructive approach to determining leaf and plant pigments depends on correctly selecting informative spectral bands. The chapter provides a strong theoretical study of spectral characteristics of leaves, specifically very detailed synthesis of plant pigments including chlorophylls, carotenoids, anthocyanins, xanthophylls, and flavonoids. Implementing a knowledge-based approach, spectral bands were selected (see several tables in Chapter 5) based on three different models (NN, PLS, and VI), leding authors to recommend: (1) two hyperspectral narrowbands in the red edge and near-infrared to estimate chlorophyll and (2) four hyperspectral narrowbands to estimate carotenoids.

10.6 ESTIMATING LEAF NITROGEN CONCENTRATION OF CEREAL CROPS WITH HYPERSPECTRAL DATA

Nitrogen (N) is a major component of chlorophyll and its availability is important for crop growth, health, and productivity. Indeed, it is considered the most important crop limiting factor (Vigneau et al., 2011). Lack of N will result in crop stress, stunted growth, and reduced productivity (e.g., biomass, yield). How much N to apply will depend on soil fertility and crop types. For example, C4 crops (e.g., corn, sorghum, sugarcane) use N, CO_2, and water much more efficiently than C3 crops (e.g., rice, wheat, beans). However, heavy N application also pollutes ground and surface water through runoff and discharge back to streams; hence, N needs to be applied carefully to optimize crop yields in a sustainable way. The direct implication of N application of plant growth and yield is illustrated for a rice crop (Figure 10.8; Tan et al., 2018). Four different applications of N (N0: 0 kg/hm^2, N1: 60 kg/hm^2, N2: 90 kg/hm^2, N3: 120 kg/hm^2) in the tillering growth stage resulted in four distinct productivities, as illustrated by hyperspectral measurements in the 400–1100 nm range shown in Figure 10.8 (Tan et al., 2018).

To measure N in crops, various hyperspectral vegetation indices are recommended (Table 10.7; Chen et al., 2010). Yao et al. (2010) demonstrated that leaf N accumulation per unit soil area (LNA, g N m^{-2}) in winter wheat (*Triticum aestivum* L.) and hyperspectral reflectance at various wavelengths changed with N application rates. Indeed, numerous studies (Nigon et al., 2015; Ji-Yong et al., 2012; Feng et al., 2015) have indicated that hyperspectral data were probably the most sensitive data to detect, model, and map N-status of crops using one or more of the indices suggested in Table 10.7 (Chen et al., 2010).

In Chapter 6, Dr. Tao Cheng et al. focused on leaf nitrogen content (LNC) estimations of cereal crops (e.g., wheat, rice) using hyperspectral data. They discussed LNC estimation methods such as vegetation indices, stepwise multiple linear regression (SMLR), partial least-squares regression (PLSR), and continuous wavelet analysis (CWA). They referred to Yao et al. (2010) for other methods like continuum removal (CR), artificial neural networks (ANNs), and support vector machines. They

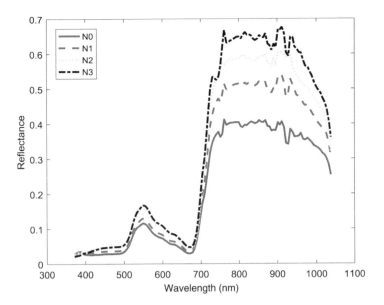

FIGURE 10.8 Spectral response to plant nitrogen (N). Spectral reflectances of different nitrogen levels in the tillering stage. (Tan, K. et al. 2018. *Chemometrics and Intelligent Laboratory Systems*, Volume 172, Pages 68–79, ISSN 0169-7439, https://doi.org/10.1016/j.chemolab.2017.11.014.)

also provided an excellent summary of various hyperspectral vegetation indices in a table, as well as outlining the advances that still need to be made for better understanding, modeling, and mapping of leaf N content. These advances required: a need for better understanding of LNC variations during various growth stages for developing more robust models, addressing a lack of appropriate physical models for LNC inversion, and a better understanding of confounding contributions of canopy structure. They also recommend increased use of short-wave infrared bands (in addition to visible and near infrared), multi-angle observations, and a need for regional mapping.

10.7 OPTICAL REMOTE SENSING OF VEGETATION ("LIQUID") WATER CONTENT

The ability to detect plant and leaf water content throughout the growing period is key for studying plant health, vigor, stress, drought, fire susceptibility, and productivity (for example, biomass, yield). Leaf and plant water content are two of the most common physiological parameters limiting efficiency of photosynthesis and biomass productivity in plants (Jin et al., 2017). In agriculture, early detection of crop water stress is key to efficient crop management and optimizing crop productivity. Also, optimizing water use in agriculture requires innovations in detection of plant water stress at various stages of the growing season to minimize crop physiological damage and yield loss (Ihuoma and Madramootoo, 2017). Daily water consumption (or evapotranspiration) is largely determined by water treatment, whereas biomass accumulation per unit of water used (WUE) is clearly determined by genotype, indicating a strong genetic control of WUE (Ge et al., 2016).

Leaf and plant water are often quantified in terms of: (1) leaf water content (LWC) (percent) and (2) canopy equivalent water thickness (g cm^{-2}).

LWC is defined as (Jin et al., 2017):

$$\text{Water content (\%)} = (W_f - W_d)/W_f * 100$$

where W_f, fresh weight and W_d, dry weight.

TABLE 10.7

Various Hyperspectral Vegetation Indices (HVIs) for Studying Nitrogen (N) in Crops and Vegetation

Index	Name	Formula	Developed by
Nitrogen Indices			
Vi_{opt}	Optimal vegetation index	$(1 + 0.45)((R_{800})^2 + 1)/(R_{670} + 0.45)$	Reyniers and Vrindts (2006)
$NDVI_{g-b}$[#]	Normalized difference vegetation index green-blue[#]	$(R_{573} - R_{440})/(R_{573} + R_{440})$	Hansen and Schjoerring (2003)
RVI I[#]	Ratio vegetation index I[#]	R_{810}/R_{660}	Zhu et al. (2008)
RVIII[#]	Ratio vegetation index II[#]	R_{810}/R_{560}	Xue et al. (2004)
NRI	Nitrogen reflectance index	$(R_{800}/R_{550})_{target}/(R_{800}/R_{550})_{reference}$	Bausch and Duke (1996)
MCARI/ MTV12	Combined index	MCARI/MTVI2 MCARI: $(R_{700} - R_{670} - 0.2(R_{700} - R_{550}))$ (R_{700}/R_{670}) MTVI2: $1.5(1.2(R_{800} - R_{550}) - 2.5(R_{670} - R_{550}))/$ $sqrt((2R_{800} + 1)^2 - (6R_{800} - 5sqrt(R_{670})) - 0.5)$	Eitel et al. (2007)
REP-LE[#]	Red edge position: linear extrapolation method[#]	Based on linear extrapolation of two straight lines through two points on the far-red and two points on the NIR flanks of the first derivative reflectance spectrum of the red-edge region. REP is defined by the wavelength value at the intersection of the straight lines.	Cho and Skidmore (2006)
DCNI[#]	Double-peak canopy nitrogen index[#]	$(R_{720} - R_{700})/(R_{700} - R_{670})/(R_{720} - R_{670} + 0.03)$	This study
Chlorophyll Indices			
MCARI	Modified chlorophyll absorption ratio index	$(R_{700} - R_{670} - 0.2(R_{700} - R_{550}))(R_{700}/R_{670})$	Daughtry et al. (2000)
TCARI	Transformed chlorophyll absorption in reflectance index	$3((R_{700} - R_{670}) - 0.2(R_{700} - R_{550})(R_{700}/R_{670}))$	Haboudane et al. (2002)
TCARI/ OSAVI	Combined Index II[#]	TCARI/OSAVI TCARI: $3((R_{700} - R_{670}) - 0.2(R_{700} - R_{550})$ $(R_{700}/R_{670}))$ OSAVI: $1.16(R_{800} - R_{670})/(R_{800} + R_{670} + 0.16)$	Haboudane et al. (2002)
MTCI	MERIS terrestrial chlorophyll index	$(R_{750} - R_{710})/(R_{710} - R_{680})$	Dash and Curran (2004)
R-M[#]	Red model[#]	$R_{750}/R_{720} - 1$	Gitelson et al. (2005)
CCI	Canopy chlorophyll index	D_{720}/D_{700}	Sims et al. (2006)
REP-LI[#]	Red-edge position: linear interpolation method[#]	$700 + 40 (R_{re} - R_{700})/(R_{740} - R_{700})$ $R_{re}: (R_{670} + R_{780})/2$	Guyot et al. (1988)
Other Indices			
NDVI	Normalized difference vegetation index	$(R_{800} - R_{670})/(R_{800} + R_{670})$	Rouse et al. (1974)
RVI	Ratio vegetation index	R_{800}/R_{670}	Pearson and Miller (1972)

Source: Chen, P. et al. 2010. *Remote Sensing of Environment*, Volume 114, Issue 9, Pages 1987–1997, ISSN 0034-4257, https://doi.org/10.1016/j.rse.2010.04.006.

The fresh leaves of each sample were weighed and recorded as W_f, then dried at 104°C for 2 hours and at 80°C for 72 hours. The dry matter weight was recorded as W_d.

EWT is defined as mass of leaf water per unit leaf area (cm, 1 g/cm^2 = 1 cm^3/cm^2 = 1 cm) (Fang et al., 2017). EWT relates to plant biomass and leaf area index and is defined as (Champagne et al., 2003):

$$EWT_{biomass} = (FM - DM) * LA$$

where LA is the leaf area expressed as an area of leaf matter (cm^2), and FM and DM are the fresh and dry masses expressed as a volume (cm^3). The LA for this study was not measured directly, but was estimated from plant dry matter as (Champagne et al., 2003):

$$LA = DM \times SLA$$

where SLA is the specific leaf area. SLA is defined as the area of leaf per unit of dry leaf matter and is a crop-specific parameter that quantifies the internal structure of plant leaves and is more or less constant for nonsenescent leaves (Champagne et al., 2003).

Hyperspectral remote sensing has demonstrated great potential for accurate retrieval of canopy water content (CWC) (Clevers et al., 2010). CWC is defined by the product of the leaf equivalent water thickness and the leaf area index (Clevers et al., 2010). Vegetation water content is measured in terms of: (1) fuel moisture content (FMC, %, mass based) and (2) equivalent water thickness (g cm^{-2}, area based). Fuel moisture content refers to the mass of water per mass of dry matter in the vegetation (Al-Moustafa et al., 2012). FMC is expressed here as (Al-Moustafa et al., 2012):

$$FMC(\%) = EWT/DM * 100.$$

In field sampling, FMC can easily be measured based on the fresh and dry weight of a vegetation sample:

$$FMC(\%) = W_f - W_d/W_d$$

where W_f refers to the fresh weight of a vegetation sample and W_d refers to the dry weight of the same sample.

Literature shows three broad methods for estimating EWT using hyperspectral remote sensing (Fang et al., 2017):

1. Methods based on hyperspectral vegetation indices (Table 10.8; Fang et al., 2017)
2. Methods based on the inversion of the PROSPECT radiative transfer model
3. Methods based on wavelet analysis and intelligent algorithms

Ge et al. (2016) illustrated the procedure for measuring plant LWC in Figure 10.9. Yi et al. (2013) found two first derivative reflectance (DR)–based indices, DR1647/DR1133 and DR1653/DR1687, to be the optimal indices for EWT. Table 10.8 provides some of the commonly used indices. Clevers et al. (2008, 2010), Peñuelas et al. (1993a), Curran (1989), and others (Champagne et al., 2003; Claudio et al., 2006; Jacquemoud et al., 2009; Al-Moustafa et al., 2012; Fang et al., 2017) have suggested the importance of water absorption bands centered at 970, 1450, and 1950 nm, among others, to best study crop water.

In Chapter 7, Dr. Roberto Colombo et al. defined EWT, FMC, and relative water content of leaf and canopy and discussed the influences of water content in various portions of the electromagnetic spectrum based on spectroradiometer measurements and PROSAIL radiative transfer model simulations. Here they discussed the importance of water absorption bands and concluded that the spectral regions that are most suitable for vegetation water content studies

TABLE 10.8

Hyperspectral Moisture and Water Indices

Spectral Index	Acronym	Formula	Source
Moisture Stress Index	MSI	R_{1600}/R_{820}	Hunt and Rock (1989), Ceccato et al. (2001)
Normalized Difference Water Index	NDWI	$(R_{860} - R_{1240})/(R_{860} + R_{1240})$	Gao (1996), Stimson et al. (2005)
Normalized Difference Infrared Index	NDII	$(R_{820} - R_{1600})/(R_{820} + R_{1600})$	Hardisky et al. (1983), Maki et al. (2004), Yilmaz et al. (2008)
Maximum Difference Water Index	MDWI	$(R_{max} - R_{min})/(R_{max} + R_{min})_{1500-1750}$	Eitel et al. (2007)
Shortwave Angle Normalized Index	SANI	$\beta_{1240} * (R_{1640} - R_{858})/(R_{1640} + R_{858})$	Palacios-Orueta et al. (2006)
Shortwave Angle Slope Index	SASI	$\beta_{1240} * (R_{858} - R_{1640})$	Khanna et al. (2007)

Source: Fang, M. et al. 2017. *Remote Sensing of Environment*, Volume 196, Pages 13–27, ISSN 0034-4257, https://doi.org/10.1016/j.rse.2017.04.029.

Note: Spectral indices for predicting leaf equivalent water thickness or EWT. Here R_λ = reflectance at wavelength λ; R_{max} and R_{min} are the maximum and minimum reflectance values between 1500 and 1750 nm, respectively; β_{1240} is the angle formed at vertex 1240 nm by the 858-1240-1640 nm reflectance.

are regions showing: (1) intermediate absorption in short-wave infrared (around 1520–1540, 1650, and 2130–2200 nm) and (2) weak absorption in near-infrared (near 975 nm and 1150–1260 nm). The methods discussed to study vegetation water content using hyperspectral data were comprehensive and began by listing the hyperspectral vegetation indices that are most sensitive to plant water content studies. These include: (1) near-infrared–based indices, (2) SWIR-based indices, and (3) greenness indices with bands taken from the visible portion of the spectrum. They offered other methods like continuum-removed indices, derivative indices, and radiative transfer models. Extensive discussions of retrieving leaf and canopy EWT through the inversion of physically based models like PROSAIL followed. They also suggested other methods like artificial neural networks.

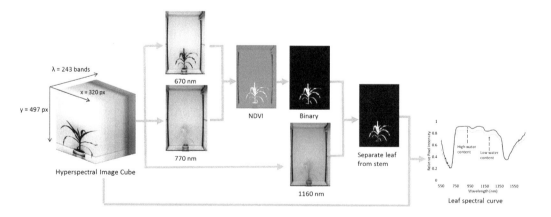

FIGURE 10.9 Procedures of hyperspectral image analysis to extract leaf pixels and the average leaf reflectance (pixel intensity) information. Average leaf reflectance spectra were used to predict plant leaf water content. (Ge, Y. et al. 2016. *Computers and Electronics in Agriculture*, Volume 127, Pages 625–632, ISSN 0168-1699, https://doi.org/10.1016/j.compag.2016.07.028.)

10.8 ESTIMATION OF NITROGEN CONTENT IN CROPS AND PASTURES USING HYPERSPECTRAL VEGETATION INDICES

Plant nitrogen has a direct impact on productivity. How much N is present in plants also helps us determine how much N fertilizer to add. This also depends on a host of factors such as available soil N, type of crops or vegetation, and cultivars. Without adequate N, plants fail in optimal productivity. However, N application should always be done while minimizing or avoiding polluting ground and surface water resources. For agricultural crops, sustainable application of N requires continuous monitoring throughout the plant growing cycle.

Numerous researchers (Feng et al., 2015; Zhao et al., 2018) have shown tremendous advantages of nondestructive approaches to monitoring N in crops using hyperspectral remote sensing. Wang et al. (2012) showed the power of hyperspectral vegetation indices in studying rice and wheat N. They recommended a linear model based on HVI:

$$(R_{924} - R_{703} + 2 \times R_{423})/(R_{924} + R_{703} - 2 \times R_{423})$$

where R is reflectance (%). The wavebands (with bandwidths within brackets) are: 924 nm (36 nm), 703 nm (15 nm), and 423 nm (21 nm). Yao et al. (2010) suggested the best HVIs for estimating leaf N accumulation per unit soil area (LNA, g N m^{-2}) in winter wheat (*Triticum aestivum* L.) were normalized reflectance-based HVIs involving wavebands (R_{860}, and R_{720}) and ratio-based HVIs involving wavebands (R_{990}, and R_{720}). Tian et al. (2011) (Figure 10.10) showed the best hyperspectral narrowbands in the visible and near-infrared range of the spectrum for modeling N in rice crops based on ratio-based HVIs, derivative HVIs, and normalized difference HVIs. The "bulls-eye" areas in the λ versus λ plot (Figure 10.10) where R-square values are the highest and were deemed the best HNBs from which HVIs need to be computed for N modeling.

In Chapter 8, Dr. Daniela Stroppiana et al. summarized extensive literature on the study of crop N using HVIs, looking at leading world crops such as wheat, rice, and corn as well as natural vegetation like grasslands, mixed savannas, and pasture. They also reviewed various HVIs and listed a number of them as espoused by various researchers. They suggested that even though N retrievals using hyperspectral remote sensing are generally performed by simple linear or nonlinear models, slightly improved results are obtained by using multiple linear regression (MLR), principal component regression (PCR), and partial least-squares regression. They also reviewed the strengths of classification approaches such as artificial neural networks and decision trees (DTs) in crop N assessment. A normalized difference red-edge index ($NDRE = \frac{\rho_{790} - \rho_{720}}{\rho_{790} + \rho_{720}}$, where ρ represents reflectance at 790 and 720 nm) was found to be highly correlated with wheat crop N. Numerous other simple ratio and normalized difference HVIs to model and map N in crops and vegetation are summarized in a table and discussed extensively.

10.9 FOREST LEAF CHLOROPHYLL CONTENT ESTIMATION USING HYPERSPECTRAL REMOTE SENSING

Forests of all kinds play an immense role on Earth in terms of biodiversity; sustaining livelihoods of humans, animals, and other species; providing wood for fuel, furniture, and paper; sequestering carbon; harboring a richness of flora and fauna; as well as other values like ecosystem services and hydrological cycle. For example, tropical forests host the largest biodiversity of terrestrial ecosystems and have a fundamental role in the carbon cycle (Laurin et al., 2016).

Hyperspectral data have the ability to quantify a wide variety of forest characteristics such as forest species, biomass, productivity, biodiversity, nutrient cycling, and carbon flux, and they do so with much greater accuracies than multispectral broadband data (Thenkabail et al., 2004a,b). Early hyperspectral studies of forest leaf chlorophyll were done using AVIRIS data (Curran et al., 1991; Green et al., 1998), supplying leaf biochemical parameters like chlorophyll content (μg cm^{-2}), leaf

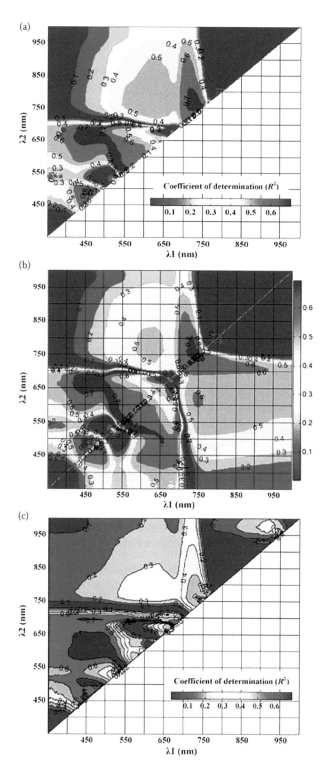

FIGURE 10.10 Contour maps for coefficients of determination (R^2) between rice leaf nitrogen concentrations and normalized hyperspectral: (a) ratio indices, (b) difference indices, (c) two-band ($\lambda 1$ by $\lambda 2$) spectral indices. (Note: n = 312). (Tian, Y.C. et al. 2011. *Field Crops Research*, Volume 120, Issue 2, Pages 299–310, ISSN 0378-4290, https://doi.org/10.1016/j.fcr.2010.11.002.)

TABLE 10.9

Hyperspectral Vegetation Indices (HVIs) in Forest Vegetation Studies

Vegetation Index	Equation Structure	Reference
Normalized Difference Vegetation Index (NDVI)	$(\rho_{864} - \rho_{660})/(\rho_{864} + \rho_{660})$	Rouse et al. (1973)
Enhanced Vegetation Index (EVI)	$2.5 \times ((\rho_{864} - \rho_{660})/$ $(\rho_{864} + 6 \times \rho_{660} - 7.5 \times \rho_{487} + 1))$	Huete et al. (2002)
Visible Index Green (VIg)	$(\rho_{559} - \rho_{660})/(\rho_{559} + \rho_{660})$	Gitelson et al. (2002)
Visible Atmospherically Resistant Index (VARI)	$(\rho_{559} - \rho_{660})/(\rho_{559} + \rho_{660} - \rho_{487})$	Gitelson et al. (2002)
Biochemistry		
Moisture Stress Index (MSI)	$(\rho_{1598})/(\rho_{823})$	Hunt and Rock (1989)
Normalized Difference Infrared Index (NDII)	$(\rho_{823} - \rho_{1649})/(\rho_{823} + \rho_{1649})$	Hunt and Rock (1989)
Structure Insensitive Pigment Index (SIPI)	$(\rho_{803} - \rho_{467})/(\rho_{803} + \rho_{681})$	Peñuelas et al. (1993b)
Normalized Difference Water Index (NDWI)	$(\rho_{854} - \rho_{1245})/(\rho_{854} + \rho_{1245})$	Gao (1996)
Water Band Index (WBI)	ρ_{905}/ρ_{972}	Peñuelas et al. (1997)
Pigment-Specific Simple Ratio (QPSSR)	ρ_{803}/ρ_{671}	Blackburn (1998)
Plant Senescence Reflectance Index (PSRI)	$(\rho_{681} - \rho_{498})/\rho_{752}$	Merzlyak et al. (1999)
Modified Chlorophyll Absorption Ratio Index (MCARI)	$[(\rho_{701} - \rho_{671}) - 0.2 * (\rho_{701} - \rho_{549})]$ $* (\rho_{701}/\rho_{671})$	Daughtry (2001)
Anthocyanin Reflectance Index (ARI)	$(1/\rho_{559}) - (1/\rho_{721})$	Gitelson et al. (2002)
Cellulose Absorption Index (CAI)	$100 * [0.5 * (\rho_{2032} + \rho_{2213}) - \rho_{2102}]$	Daughtry (2001)
Normalized Difference Lignin Index (NDLI)	$[\log(1/\rho_{1754}) - \log(1/\rho_{1680})]/$ $[\log(1/\rho_{1754}) + \log(1/\rho_{1680})]$	Serrano et al. (2002)
Carotenoid Reflectance Index 1 (CRI1)	$(1/\rho_{510}) - (1/\rho_{550})$	Gitelson et al. (2002)
Leaf Water Vegetation Index 2 (LWVI2)	$(\rho_{1094} - \rho_{1205})/(\rho_{1094} + \rho_{1205})$	Galvão et al. (2005)
Physiology		
Red-Edge Position Index (REPI)	Wavelength of maximum derivative $(690 - 740 \text{ nm})$	Horler et al. (1983)
Vogelmann Red-Edge Index (VOG)	ρ_{742}/ρ_{722}	Vogelmann et al. (1993)
Red-Edge NDVI (RENDVI)	$(\rho_{752} - \rho_{701})/(\rho_{752} + \rho_{701})$	Gitelson et al. (1996)
Photochemical Reflectance Index (PRI)	$(\rho_{529} - \rho_{569})/(\rho_{529} + \rho_{569})$	Gamon et al. (1997)
Red-Edge Vegetation Stress Index (RVSI)	$[(\rho_{712} + \rho_{752})/2] - \rho_{732}$	Merton and Huntington (1999)

Source: Toniol et al., 2017. *Remote Sensing Applications: Society and Environment,* Volume 8, Pages 20–29, ISSN 2352-9385, https://doi.org/10.1016/j.rsase.2017.07.004.

Note: Narrowband vegetation indices, associated with vegetation structure, biochemistry, and physiology, selected in this study for the Hyperion/EO-1 data analysis.

water thickness (g cm^{-2}), and leaf mass per area (LMA, g cm^{-2}) required by most biogeochemical models that describe ecosystem functions (Wang and Li, 2012). Leaf chlorophyll content is a good indicator of photosynthesis activity, light harvesting potential, mutations, stress, and nutritional state (Wu et al., 2008). Forest leaf chlorophyll studies using hyperspectral data have been conducted by numerous researchers (Datt, 1999a; le Marie et al., 2004; Zhang et al., 2008). In addition, a number of researchers (le Maire et al., 2004; Wang and Li, 2012) have found strong relationships between hyperspectral vegetation indices (e.g., Table 10.9; Toniol et al., 2017) and forest biochemical parameters including forest leaf chlorophyll, vegetation structure, biochemistry, and physiology. le Maire et al. (2008) found that broadleaved forest chlorophyll was best estimated by a normalized index involving narrowbands at 710 nm and 925 nm. They also suggested that the broadband in the near-infrared does as well as narrowbands, but a narrowband at 710 nm was most appropriate.

Leaf-level Chlorophyll
Content Estimation

Canopy-level Chlorophyll
Content Estimation

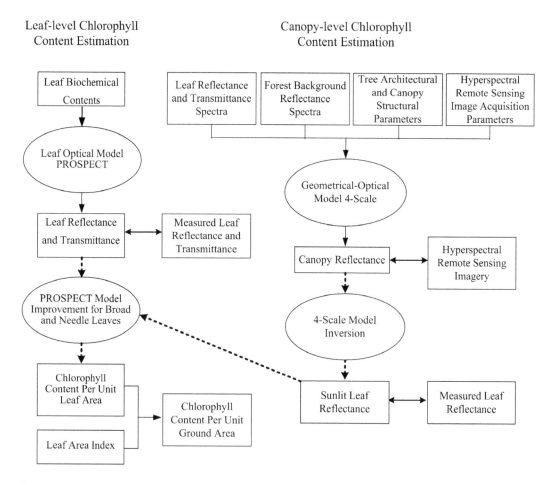

FIGURE 10.11 Flowchart for leaf- and canopy-level chlorophyll retrieval. In this chart, dashed lines represent inversion processes. The left and right facing arrows represent the comparison for modeled and measured parameters. (Zhang, Y. et al. 2008. *Remote Sensing of Environment*, Volume 112, Issue 7, Pages 3234–3247, ISSN 0034-4257, https://doi.org/10.1016/j.rse.2008.04.005.)

Croft et al. (2014) studied 47 hyperspectral vegetation indices across coniferous and deciduous forest leaf types and a range of canopy structures and concluded: (1) a number of HVIs provided strong relationships, (2) the best-performing HVIs varied with observational scale and plant type, and (3) background composition had a significant impact on performance of the HVIs.

A flowchart depicting the process of leaf- and canopy-level forest chlorophyll retrieval is shown in Figure 10.11 (Zhang et al., 2008). This study used Compact Airborne Spectrographic Imager (CASI)-derived hyperspectral data of open spruce and closed forests and showed that a process-based approach produced systematically better estimates of chlorophyll a + b content than a widely used spectral index (Table 10.9) approach. The authors mapped forest chlorophyll based on the retrieval of leaf chlorophyll content and a LAI map derived from hyperspectral remote sensing imagery. Kalacska et al. (2015) studied 19 forest species in Ottawa, Canada, and concluded that HVIs did not predict chlorophyll, whereas continuous wavelet transformation improved prediction of chlorophyll. Sonobe and Wang (2017) studied 86 hyperspectral vegetation indices, 71 reflectance-based and 10 derivative-based, and established that leaf-type information was critical in applicability of these indices in quantifying leaf chlorophyll. They found that 680 nm performed best for sunlit leaf orientation, whereas 710 nm performed best for shaded leaf orientation. Moorthy et al. (2008) mapped leaf chlorophyll of *Pinus banksiana* and grouped them into six distinct groups (Figure 10.12).

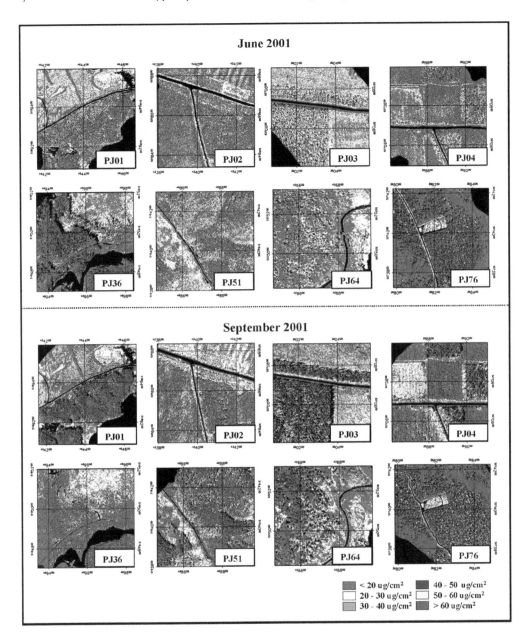

FIGURE 10.12   Chlorophyll a and b (Chl a + b) mapping using hyperspectral data. Chl a + b estimation using mapping mission CASI data (seven spectral channels and 1×1 m spatial resolution) over all eight *Pinus banksiana* for June and September. Shadowed areas and nonvegetated targets were masked out (in black) using multithresholding techniques. PROSPECT coupled with the SAILH model while accounting for a normalization factor (0.76) and site age proportionality (0.75) was used to generate chlorophyll estimates (µg/cm^2) which are grouped into six classes. (Moorthy, I. et al. 2008. *Remote Sensing of Environment*, Volume 112, Issue 6, Pages 2824–2838, ISSN 0034-4257, https://doi.org/10.1016/j.rse.2008.01.013.)

In Chapter 9, Dr. Yongquin Zhang et al. provided methods for estimating leaf chlorophyll from leaf-level hyperspectral measurements including: (1) Empirically- and (2) physically-based methods. Empirical methods use hyperspectral vegetation indices such as those listed in Table 10.9 as well as those discussed in the tables in Chapter 9. PROSPECT is the most widely used physically based approach in the literature and has been presented in detail in the chapter for modeling and estimating broadleaf chlorophyll content. Then they presented LIBERTY, a radiative transfer (RT) model, for estimating needle leaf chlorophyll content. This is followed by physically based modeling methods for closed forests using RT models and open forests using geometrical optical (GO) models. Finally, they presented methods for scaling leaf chlorophyll content estimated from hyperspectral data to the landscape scale by using leaf area index as a forest structural variable. They discussed estimating LAI through HVIs derived from the 72-band CASI hyperspectral sensor that acquires data at 20 m spatial resolution. They showed lead chlorophyll distribution among various vegetation types such as deciduous forests, coniferous forests, and grasses. Chlorophyll content ranged from 30 mg/m^2 to 2170 mg/m^2 for vegetated pixels. Their chapter showed the possibilities of modeling and mapping biochemistry parameters such as leaf chlorophyll of complex forests with increased accuracies using hyperspectral data.

REFERENCES

Al-Moustafa, T., Armitage, R.P., Danson, F.M. 2012. Mapping fuel moisture content in upland vegetation using airborne hyperspectral imagery, *Remote Sensing of Environment*, Volume 127, Pages 74–83, ISSN 0034-4257, https://doi.org/10.1016/j.rse.2012.08.034.

Apan, A., Held, A., Phinn, S., Markley, J. 2004. Detecting sugarcane 'orange rust' disease using EO-1 Hyperion hyperspectral imagery, *International Journal of Remote Sensing*, Volume 25, Pages 489–498.

Bareth, G., Waldhoff, G. 2018. 2.01–GIS for mapping vegetation, in: *Comprehensive Geographic Information Systems*, edited by Huang, B., Elsevier, Oxford, Pages 1–27, ISBN 9780128047934, https://doi.org/10.1016/B978-0-12-409548-9.09636-6.

Bausch, W.C., Duke, H.R. 1996. Remote sensing of plant nitrogen status in corn, *Transactions of the ASAE*, Volume 39, Pages 1869–1875.

Behmann, J., Steinrücken, J., Plümer, L. 2014. Detection of early plant stress responses in hyperspectral images, *ISPRS Journal of Photogrammetry and Remote Sensing*, Volume 93, Pages 98–111, ISSN 0924-2716, https://doi.org/10.1016/j.isprsjprs.2014.03.016.

Blackburn, G.A. 1998. Spectral indices for estimating photosynthetic pigment concentrations: A test using senescent tree leaves. *International Journal of Remote Sensing*, Volume 19, Pages 657–675.

Blackburn, G.A., Ferwerda, J.G. 2008. Retrieval of chlorophyll concentration from leaf reflectance spectra using wavelet analysis, *Remote Sensing of Environment*, Volume 112, Issue 4, Pages 1614–1632, ISSN 0034-4257, https://doi.org/10.1016/j.rse.2007.08.005.

Cao, Z., Wang, Q., Zheng, C. 2015. Best hyperspectral indices for tracing leaf water status as determined from leaf dehydration experiments, *Ecological Indicators*, Volume 54, Pages 96–107, ISSN 1470-160X, https://doi.org/10.1016/j.ecolind.2015.02.027.

Ceamanos, X., Valero, S. 2016. 4—Processing hyperspectral images, in: *Optical Remote Sensing of Land Surface*, edited by Baghdadi, N. and Zribi, M., Elsevier, Pages 163–200, ISBN 9781785481024, https://doi.org/10.1016/B978-1-78548-102-4.50004-1.

Ceccato, P., Flasse, S., Tarantola, S., Jacquemoud, S., Grégoire, J.-M. 2001. Detecting vegetation leaf water content using reflectance in the optical domain, *Remote Sensing of Environment*, Volume 77, Pages 22–33.

Champagne, C.M., Staenz, K., Bannari, A., McNairn, H., Deguise, J.C. 2003. Validation of a hyperspectral curve-fitting model for the estimation of plant water content of agricultural canopies, *Remote Sensing of Environment*, Volume 87, Issues 2–3, Pages 148–160, ISSN 0034-4257, https://doi.org/10.1016/S0034-4257(03)00137-8.

Chappelle, E.W., Kim, M.S., McMurtrey III, J.E. 1992. Ratio analysis of reflectance spectra (RARS): An algorithm for the remote estimation of the concentrations of chlorophyll a, chlorophyll b, and carotenoids in soybean leaves, *Remote Sensing of Environment*, Volume 39, Pages 239–247.

Chen, P., Haboudane, D., Tremblay, N., Wang, J., Vigneault, P., Li, B. 2010. New spectral indicator assessing the efficiency of crop nitrogen treatment in corn and wheat, *Remote Sensing of Environment*, Volume 114, Issue 9, Pages 1987–1997, ISSN 0034-4257, https://doi.org/10.1016/j.rse.2010.04.006.

Cho, M.A., Skidmore, A.K. 2006. A new technique for extracting the red edge position from hyperspectral data: The linear extrapolation method, *Remote Sensing of Environment*, Volume 101, Pages 181–193.

Clark, R.N., Roush, T.L. 1984. Reflectance spectroscopy: Quantitative analysis techniques for remote sensing applications, *Journal of Geophysical Research*, Volume 89, Pages 6329–6340.

Claudio, H.C., Cheng, Y., Fuentes, D.A., Gamon, J.A., Luo, H., Oechel, W., Qiu, H.L., Rahman, A.F., Sims, D.A. 2006. Monitoring drought effects on vegetation water content and fluxes in chaparral with the

970 nm water band index, *Remote Sensing of Environment*, Volume 103, Issue 3, Pages 304–311, ISSN 0034-4257, https://doi.org/10.1016/j.rse.2005.07.015.

Clevers, G.P.W., Kooistra, L., Schaepman, M.E. 2010. Estimating canopy water content using hyperspectral remote sensing data, *International Journal of Applied Earth Observation and Geoinformation*, Volume 12, Issue 2, Pages 119–125, ISSN 0303-2434, https://doi.org/10.1016/j.jag.2010.01.007.

Clevers, J.G.P.W., Kooistra, L., Schaepman, M.E. 2008. Using spectral information from the NIR water absorption features for the retrieval of canopy water content, *International Journal of Applied Earth Observatory Geoinf*, Volume 10, Pages 388–397.9.

Croft, H., Chen, J.M., Zhang, Y. 2014. The applicability of empirical vegetation indices for determining leaf chlorophyll content over different leaf and canopy structures, *Ecological Complexity*, Volume 17, Pages 119–130, ISSN 1476-945X, https://doi.org/10.1016/j.ecocom.2013.11.005.

Curran, P.J. 1989. Remote sensing of foliar chemistry. *Remote Sensing of Environment*, Volume 30, Pages 271–278.

Curran, P.J., Dungan, J.L., Macler, B.A., Plummer, S.E. 1991. The effect of a red leaf pigment on the relationship between red edge and chlorophyll concentration, *Remote Sensing of Environment*, Volume 35, Issue 1, Pages 69–76, ISSN 0034-4257, https://doi.org/10.1016/0034-4257(91)90066-F.

Curran, P., Windham, W., Gholz, H. 1995. Exploring the relationship between reflectance red edge and chlorophyll concentration in slash pine leaves, *Tree Physiology*, Volume 15, Issue 3, Pages 203–206.

Datt, B. 1999a. A new reflectance index for remote sensing of chlorophyll content in higher plants: Tests using eucalyptus leaves, *Journal of Plant Physiology*, Volume 154, Pages 30–36.

Datt, B. 1999b. Remote sensing of water content in Eucalyptus leaves. *Australian Journal of Botany*, Volume 47, Pages 909–923.

Datt, B., Paterson, M. 2000. Vegetation-soil spectral mixture analysis. *Proc. IEEE International Geoscience and Remote Sensing Symposium (IGARSS)*, Volume 5, Pages 1936–1938.

Dash, J., Curran, P.J. 2004. The MERIS terrestrial chlorophyll index, *International Journal of Remote Sensing*, Volume 25, Pages 5003–5013.

Daughtry, C.S.T. 2001. Agroclimatology: Discriminating crop residues from soil by shortwave infrared reflectance, *Agronomy Journal*, Volume 93, Pages 125–131.

Daughtry, C.S.T., Walthall, C.L., Kim, M.S., de Colstoun, E.B., McMurtrey III, J.E. 2000. Estimating corn leaf chlorophyll concentration from leaf and canopy reflectance. *Remote Sensing of Environment*, Volume 74, Pages 229–239.

DeFries, R.S., Chan, J.C.-W. 2000. Multiple criteria for evaluating machine learning algorithms for land cover classification from satellite data, *Remote Sensing of Environment*, Volume 74, Issue 3, Pages 503–515.

Demetriades-Shah, T.H., Steven, M.D., Clark, J.A. 1990. High resolution derivative spectra in remote sensing, *Remote Sensing of Environment*, Volume 33, Issue 1, Pages 55–64, ISSN 0034-4257, https://doi.org/10.1016/0034-4257(90)90055-Q.

Deng, C., Wu, C. 2013. The use of single-date MODIS imagery for estimating large-scale urban impervious surface fraction with spectral mixture analysis and machine learning techniques, *ISPRS Journal of Photogrammetry and Remote Sensing*, Volume 86, Pages 100–110.

Dong, J., Xiao, X., Kou, W., Qin, Y., Zhang, G., Li, L., Jin, C., Zhou, Y., Wang, J., Biradar, C. 2015. Tracking the dynamics of paddy rice planting area in 1986–2010 through time series Landsat images and phenology-based algorithms, *Remote Sensing of Environment*, Volume 160, Pages 99–113.

Duveiller, G., Lopez-Lozano, R., Cescatti, A. 2015. Exploiting the multi-angularity of the MODIS temporal signal to identify spatially homogeneous vegetation cover: A demonstration for agricultural monitoring applications, *Remote Sensing of Environment*, Volume 166, Pages 61–77.

Egorov, A., Hansen, M., Roy, D., Kommareddy, A., Potapov, P. 2015. Image interpretation-guided supervised classification using nested segmentation, *Remote Sensing of Environment*, Volume 165, Pages 135–147.

Eitel, J.U.H., Long, D.S., Gessler, P.E., Smith, A.M.S. 2007. Using in-situ measurements to evaluate the new RapidEye™ satellite series for prediction of wheat nitrogen status, *International Journal of Remote Sensing*, Volume 28, Pages 4183–4190.

Elvidge, C.D., Chen, Z.K. 1995. Comparison of broad-band and narrow-band red and near-infrared vegetation indices. *Remote Sensing of Environment*, Volume 54, Issue 1, Pages 38–48.

Fang, M., Ju, W., Zhan, W., Cheng, T., Qiu, F., Wang, J. 2017. A new spectral similarity water index for the estimation of leaf water content from hyperspectral data of leaves, *Remote Sensing of Environment*, Volume 196, Pages 13–27, ISSN 0034-4257, https://doi.org/10.1016/j.rse.2017.04.029.

Feng, W., Guo, B.-B., Zhang, H.Y., He, L., Zhang, Y.S., Wang, Y.H., Zhu, Y.J., Guo, T.C., 2015. Remote estimation of above ground nitrogen uptake during vegetative growth in winter wheat using hyperspectral red-edge ratio data, *Field Crops Research*, Volume 180, Pages 197–206, ISSN 0378-4290, https://doi.org/10.1016/j.fcr.2015.05.020.

Friedl, M.A., Brodley, C.E. 1997. Decision tree classification of land cover from remotely sensed data. *Remote Sensing of Environment*, Volume 61, Issue 3, Pages 399–409.

Galvão, L.S., Formaggio, A.R., Tisot, D.A. 2005. Discrimination of sugarcane varieties in Southeastern Brazil with EO-1 Hyperion data, *Remote Sensing of Environment*, Volume 94, Issue 4, Pages 523–534, ISSN 0034-4257, https://doi.org/10.1016/j.rse.2004.11.012.

Gamon, J., Penuelas, J., Field, C. 1992. A narrow-waveband spectral index that tracks diurnal changes in photosynthetic efficiency, *Remote Sensing of Environment*, Volume 41, Issue 1, Pages 35–44.

Gamon, J.A., Serrano, L., Surfus, J.S. 1997. The photochemical reflectance index: An optical indicator of photosynthetic radiation-use efficiency across species, functional types, and nutrient levels, *Oecologia*, Volume 112, Pages 492–501.

Gamon, J., Surfus, J. 1999. Assessing leaf pigment content and activity with a reflectometer, *New Phytologist*, Volume 143, Issue 1, Pages 105–117.

Gao, B.C. 1996. NDWI—a normalized difference water index for remote sensing of vegetation liquid water from space, *Remote Sensing of Environment*, Volume 58, Pages 257–266.

Garrity, S.R., Eitel, J.U.H., Vierling, L.A. 2011. Disentangling the relationships between plant pigments and the photochemical reflectance index reveals a new approach for remote estimation of carotenoid content, *Remote Sensing of Environment*, Volume 115, Pages 628–635.

Ge, Y., Bai, G., Stoerger, V., Schnable, J.C. 2016. Temporal dynamics of maize plant growth, water use, and leaf water content using automated high throughput RGB and hyperspectral imaging, *Computers and Electronics in Agriculture*, Volume 127, Pages 625–632, ISSN 0168-1699, https://doi.org/10.1016/j.compag.2016.07.028.

Geerken, R., Zaitchik, B., Evans, J.P. 2005. Classifying rangeland vegetation type and coverage from NDVI time series using Fourier filtered cycle similarity, *International Journal of Remote Sensing*, Volume 26, Issue 24, Pages 5535–5554.

George, R., Padalia, H., Kushwaha, S.P.S. 2014. Forest tree species discrimination in western Himalaya using EO-1 Hyperion, *International Journal of Applied Earth Observation and Geoinformation*, Volume 28, Pages 140–149, ISSN 0303-2434, https://doi.org/10.1016/j.jag.2013.11.011.

Gislason, P.O., Benediktsson, J.A., Sveinsson, J.R. 2006. Random forests for land cover classification, *Pattern Recognition Letters*, Volume 27, Issue 4, Pages 294–300.

Gitelson, A., Merzlyak, M. 1994. Spectral reflectance changes associated with autumn senescence of Aesculus hippocastanum L. and Acer platanoides L. leaves. Spectral features and relation to chlorophyll estimation, *Journal of Plant Physiology*, Volume 143, Issue 3, Pages 286–292.

Gitelson, A., Merzlyak, M., Chivkunova, O. 2001. Optical properties and nondestructive estimation of anthocyanin content in plant leaves, *Photochemistry and Photobiology*, Volume 74, Issue 1, Pages 38–45.

Gitelson, A.A., Kaufman, Y.J., Stark, R., Rundquist, D. 2002. Novel algorithms for remote estimation of vegetation fraction. *Remote Sensing of Environment*, Volume 80, Issue 2002, Pages 76–87.

Gitelson, A.A., Keydan, G.P., Merzlyak, M.N. 2006. Three-band model for noninvasive estimation of chlorophyll carotenoids and anthocyanin contents in higher plant leaves. Papers in Natural Resources. 258. http://digitalcommons.unl.edu/natrespapers/258.

Gitelson, A.A., Merzlyak, M.N. 1996. Signature analysis of leaf reflectance spectra: Algorithm development for remote sensing of chlorophyll, *Journal of Plant Physiology*, Volume 148, Pages 494–500.

Gitelson, A.A., Merzlyak, M.N., Lichtenthaler, H.K. 1996. Detection of red edge position and chlorophyll content by reflectance measurements near 700 nm, *Journal of Plant Physiology*, Volume 148, Pages 501–508.

Gitelson, A.A., Viña, A., Arkebauer, T.J., Rundquist, D.C., Keydan, G., Leavitt, B. 2003. Remote estimation of leaf area index and green leaf biomass in maize canopies. First published: 13 March, https://doi.org/10.1029/2002GL016450.

Gitelson, A.A., Viña, A., Ciganda, V., Rundquist, D.C. 2005. Remote estimation of canopy chlorophyll content in crops, *Geophysical Research Letter*, Volume 32, Page L08403.

Gnyp, M.L., Miao, Y., Yuan, F., Ustin, S.L., Yu, K., Yao, Y., Huang, S., Bareth, G. 2014. Hyperspectral canopy sensing of paddy rice aboveground biomass at different growth stages, *Field Crops Research*, Volume 155, Pages 42–55, ISSN 0378-4290, https://doi.org/10.1016/j.fcr.2013.09.023.

Gorelick, N., Hancher, M., Dixon, M., Ilyushchenko, S., Thau, D., Moore, R. 2017. Google Earth Engine: Planetary-scale geospatial analysis for everyone, *Remote Sensing of Environment*, Volume 202, Pages 18–27, ISSN 0034-4257, https://doi.org/10.1016/j.rse.2017.06.031.

Green, R.O., Eastwood, M.L., Sarture, C.M., Chrien, T.G., Aronsson, M., Chippendale, B.J. et al. 1998. Imaging spectroscopy and the Airborne Visible/Infrared Imaging Spectrometer (AVIRIS), *Remote Sensing of Environment*, Volume 65, Issue 3, Pages 227–248, ISSN 0034-4257, https://doi.org/10.1016/S0034-4257(98)00064-9.

Guyot, G., Baret, F., Major, D.J. 1988. High spectral resolution: Determination of specral shifts between the red and the near infrared, *International Archives of Photogrammetry and Remote Sensing*, Volume 11, Pages 740–760.

Haboudane, D., Miller, J.R., Tremblay, N., Zarco-Tejada, P.J., Dextraze, L. 2002. Integrated narrow-band vegetation indices for prediction of crop chlorophyll content for application to precision agriculture, *Remote Sensing of Environment*, Volume 81, Pages 416–426.

Hamzeh, S., Naseri, A.A., Kazem, S., Panah, A., Bartholomeus, H., Herold, M. 2016. Assessing the accuracy of hyperspectral and multispectral satellite imagery for categorical and quantitative mapping of salinity stress in sugarcane fields, *International Journal of Applied Earth Observation and Geoinformation*, Volume 52, Pages 412–421, ISSN 0303-2434, https://doi.org/10.1016/j.jag.2016.06.024.

Hansen, P.M., Schjoerring, J.K. 2003. Reflectance measurement of canopy biomass and nitrogen status in wheat crops using normalized difference vegetation indices and partial least squares regression, *Remote Sensing of Environment*, Volume 86, Pages 542–553.

Hardisky, M., Klemas, V., Smart, M. 1983. The influence of soil salinity, growth form, and leaf moisture on the spectral radiance of *Spartina alterniflora* canopies, *Photogramm. Eng. Remote. Sens.*, Volume 49, Pages 77–83.

He, Z., Hu, J., Wang, Y. 2018. Low-rank tensor learning for classification of hyperspectral image with limited labeled samples, *Signal Processing*, Volume 145, Pages 12–25, ISSN 0165-1684, https://doi.org/10.1016/j.sigpro.2017.11.007.

Hernández-Clemente, R., Navarro-Cerrillo, R.M., Suárez, F., Morales, L., Zarco-Tejada, P.J. 2011. Assessing structural effects on PRI for stress detection in conifer forests, *Remote Sensing of Environment*, Volume 115, Pages 2360–2375.

Hernández-Clemente, R., Navarro-Cerrillo, R.M., Zarco-Tejada, P.J. 2012. Carotenoid content estimation in a heterogeneous conifer forest using narrow-band indices and PROSPECT+DART simulations, *Remote Sensing of Environment*, Volume 127, Pages 298–315, ISSN 0034-4257, https://doi.org/10.1016/j.rse.2012.09.014.

Horler, D.N.H., Dockray, M., Barber, J. 1983. The red-edge of plant leaf reflectance, *International Journal of Remote Sensing*, Volume 4, Pages 273–288.

Huete, A., Liu, H., Batchily, K., Van Leeuwen, W. 1997. A comparison of vegetation indices over a global set of TM images for EOS-MODIS, *Remote Sensing of Environment*, Volume 59, Issue 3, Pages 440–451.

Huete, A.R., Didan, K., Miura, T., Rodriguez, E.P., Gao, X., Ferreira, L.G. 2002. Overview of the radiometric and biophysical performance of the MODIS vegetation indices, *Remote Sensing of Environment*, Volume 83, Pages 195–213.

Huete, A.R., Liu, H.Q., Batchily, K., van Leeuwen, W. 1997. A comparison of vegetation indices over a global set of TM images for EOS-MODIS, *Remote Sensing of Environment*, Volume 59, Issue 3, Pages 440–451, ISSN: 0034-4257, https://doi.org/10.1016/S0034-4257(96)00112-5.

Hughes, G.F. 1968. On the mean accuracy of statistical pattern recognizers, *IEEE Transactions on Information Theory*, Volume 14, Pages 55–63. [Google Scholar] [CrossRef].

Hunt, E.R., Rock, B.N. 1989. Detection of changes in leaf-water content using nearinfrared and middle-infrared reflectances, *Remote Sensing of Environment* , Volume 30, Pages 43–54.

Ihuoma, S.O., Madramootoo, C.A. 2017. Recent advances in crop water stress detection, *Computers and Electronics in Agriculture*, Volume 141, Pages 267–275, ISSN 0168-1699, https://doi.org/10.1016/j.compag.2017.07.026.

Imani, M., Ghassemian, H. 2015. Two dimensional linear discriminant analyses for hyperspectral data, Volume 81, Issue 10, Pages 777–786.

Imspec. 2001. *ACORN™ User's Guide*. Analytical Imaging and Geophysics, Boulder, Colorado, USA, 64 pp.

Ishida, T., Kurihara, J., Viray, F.A., Namuco, S.B., Paringit, E.C., Perez, G.J., Takahashi, Y., Marciano, J.J. 2018. A novel approach for vegetation classification using UAV-based hyperspectral imaging, *Computers and Electronics in Agriculture*, Volume 144, Pages 80–85, ISSN 0168-1699, https://doi.org/10.1016/j.compag.2017.11.027.

Jacquemoud, S., Verhoef, W., Baret, F., Bacour, C., Zarco-Tejada, P.J., Asner, G.P., François, C., Ustin, S.L. 2009. PROSPECT+SAIL models: A review of use for vegetation characterization, *Remote Sensing of Environment*, Volume 113, Supplement 1, Pages S56-S66, ISSN 0034-4257, https://doi.org/10.1016/j.rse.2008.01.026.

Jensen, J.R. 2009. Remote sensing of the environment: An earth resource perspective 2/e: Pearson Education, London, UK

Jia, X., Richards, J. 1994. Efficient maximum likelihood classification for imaging spectrometer data sets. *IEEE Transactions on Geoscience and Remote Sensing*, Volume 41, Issue 5, Pages 1129–1131.

Jin, J., Wang, Q. 2016. Hyperspectral indices based on first derivative spectra closely trace canopy transpiration in a desert plant, *Ecological Informatics*, Volume 35, Pages 1–8, ISSN 1574-9541, https://doi.org/10.1016/j.ecoinf.2016.06.004.

Jin, X., Shi, C., Yu, C.Y., Yamada, T., Sacks, E.J. 2017. Determination of leaf water content by visible and near-infrared spectrometry and multivariate calibration in Miscanthus. *Frontiers in Plant Science*, Volume 8, Page 721, doi:10.3389/fpls.2017.00721.

Ji-Yong, S., Xiao-Bo, Z., Jie-Wen, Z., Kai-Liang, W., Zheng-Wei, C., Xiao-Wei, H., De-Tao, Z., Holmes, M. 2012. Nondestructive diagnostics of nitrogen deficiency by cucumber leaf chlorophyll distribution map based on near infrared hyperspectral imaging, *Scientia Horticulturae*, Volume 138, Pages 190–197, ISSN 0304-4238, https://doi.org/10.1016/j.scienta.2012.02.023.

Ju, C. H., Tian, Y. C., Yao, X., Cao, W. X., Zhu, Y., Hannaway, D. 2010. Estimating leaf chlorophyll content using red edge parameters. *Pedosphere*. 20(5): 633-644.

Kalacska, M., Lalonde, M., Moore, T.R. 2015. Estimation of foliar chlorophyll and nitrogen content in an ombrotrophic bog from hyperspectral data: Scaling from leaf to image, *Remote Sensing of Environment*, Volume 169, Pages 270–279, ISSN 0034-4257, https://doi.org/10.1016/j.rse.2015.08.012.

Kaufman, Y., Tanré, D. 1996. Strategy for direct and indirect methods for correcting the aerosol effect on remote sensing: From AVHRR to EOS-MODIS, *Remote Sensing of Environment*, Volume 55, Issue 1, Pages 65–79.

Khanna, S., Palacios-Orueta, A., Whiting, M.L., Ustin, S.L., Riaño, D., Litago, J. 2007. Development of angle indexes for soil moisture estimation, dry matter detection and land-cover discrimination, *Remote Sensing of Environment*, Volume 109, Pages 154–165.

Kira, O., Linker, R., Gitelson, A. 2015. Non-destructive estimation of foliar chlorophyll and carotenoid contents: Focus on informative spectral bands, *International Journal of Applied Earth Observation and Geoinformation*, Volume 38, Pages 251–260, ISSN 0303-2434, https://doi.org/10.1016/j.jag.2015.01.003.

Laurin, G.V., Puletti, N., Hawthorne, W., Liesenberg, V., Corona, P., Papale, D., Chen, Q., Valentini, R. 2016. Discrimination of tropical forest types, dominant species, and mapping of functional guilds by hyperspectral and simulated multispectral Sentinel-2 data. *Remote Sensing of Environment*, Volume 176, Pages 163–176, ISSN 0034-4257, https://doi.org/10.1016/j.rse.2016.01.017.

le Maire, G., Francois, C., Dufrene, E. 2004. Towards universal broad leaf chlorophyll indices using PROSPECT simulated database and hyperspectral reflectance measurements, *Remote Sensing of Environment*, Volume 89, Pages 1–28.

Li, C., Wand, M. 2016. Precomputed real-time texture synthesis with Markovian Generative Adversarial Networks. In Leibe, B., Matas, J., Sebe, N., Welling, M. editors, *Computer Vision – ECCV 2016: 14th European Conference*, Amsterdam, The Netherlands, October 11-14, Proceedings, Part III, pages 702–716. Springer International Publishing.

Liu, N., Budkewitsch, P., Treitz, P. 2017. Examining spectral reflectance features related to Arctic percent vegetation cover: Implications for hyperspectral remote sensing of Arctic tundra, *Remote Sensing of Environment*, Volume 192, Pages 58–72, ISSN 0034-4257, https://doi.org/10.1016/j.rse.2017.02.002.

Lu, D., Weng, Q. 2007. A survey of image classification methods and techniques for improving classification performance, *International Journal of Remote Sensing*, Volume 28, Issue 5, Pages 823–870, https://doi.org/10.1080/01431160600746456.

Maki, M., Ishiahra, M., Tamura, M. 2004. Estimation of leaf water status to monitor the risk of forest fires by using remotely sensed data, *Remote Sensing of Environment*, Volume 90, Pages 441–450.

Mariotto, I., Thenkabail, P.S., Huete, H., Slonecker, T., Platonov, A. 2013. Hyperspectral versus multispectral crop- biophysical modeling and type discrimination for the HyspIRI mission. *Remote Sensing of Environment*, Volume 139, Pages 291–305. IP-049224.

Marshall, M.T., Thenkabail, P.S. 2014. Biomass modeling of four leading world crops using hyperspectral narrowbands in support of HyspIRI mission, *Photogrammetric Engineering and Remote Sensing*, Volume 80, Issue 4, Pages 757–772. IP-052043.

Meggio, F., Zarco-Tejada, P.J., Núñez, L.C., Sepulcre-Cantó, G., Gonzalez, M.R., Martin, P. 2010. Grape quality assessment in vineyards affected by iron deficiency chlorosis using narrow-band physiological remote sensing indices, *Remote Sensing of Environment*, Volume 114, Pages 1968–1986.

Merton, R., Huntington, J. 1999. Early simulation of the ARIES-1 satellite sensor for multi-temporal vegetation research derived from AVIRIS. Paper presented at the JPL Airborne Earth Science Workshop 8, Pasadena, USA, February 9–11, pp. 299–307.

Merzlyak, M.N., Gitelson, A.A., Chivkunova, O.B., Rakitin, V.Y.U. 1999. Non-destructive optical detection of pigment changes during leaf senescence and fruit ripening, *Physiologia Plantarum*, Volume 106, Issue 1, Pages 135–141.

Moorthy, I., Miller, J.R., Noland, T.L. 2008. Estimating chlorophyll concentration in conifer needles with hyperspectral data: An assessment at the needle and canopy level, *Remote Sensing of Environment*, Volume 112, Issue 6, Pages 2824–2838, ISSN 0034-4257, https://doi.org/10.1016/j.rse.2008.01.013.

Mountrakis, G., Im, J., Ogole, C. 2011. Support vector machines in remote sensing: A review, *ISPRS Journal of Photogrammetry and Remote Sensing*, Volume 66, Issue 3, Pages 247–259.

Nigon, T.J., Mulla, D.J., Rosen, C.J., Cohen, Y., Alchanatis, V., Knight, J., Rud, R. 2015. Hyperspectral aerial imagery for detecting nitrogen stress in two potato cultivars. *Computers and Electronics in Agriculture*, Volume 112, Pages 36–46. DOI: 10.1016/j.compag.2014.12.018.

Ozdogan, M., Gutman, G. 2008. A new methodology to map irrigated areas using multi-temporal MODIS and ancillary data: An application example in the continental US, *Remote Sensing of Environment*, Volume 112, Issue 9, Pages 3520–3537.

Palacios-Orueta, A., Khanna, S., Litago, J., Whiting, M.L., Ustin, S.L. 2006. Assessment of NDVI and NDWI spectral indices using MODIS time series analysis and development of a new spectral index based on MODIS shortwave infrared bands. *Proceedings of the 1st International Conference of Remote Sensing and Geoinformation Processing*, Trier, Germany, Pages 207–209.

Panda, S.S., Rao, M.N., Thenkabail, P.S., Fitzerald, J.E. 2015. Remote Sensing Systems—Platforms and Sensors: Aerial, Satellites, UAVs, Optical, Radar, LiDAR, Chapter 1, in: Thenkabail, P.S., (Editor-in-Chief), 2015. *Remote Sensing Handbook (Volume I): Remotely Sensed Data Characterization, Classification, and Accuracies*. ISBN 9781482217865—CAT# K22125, Taylor and Francis Inc./CRC Press, Boca Raton, Pages 3–60. IP-060641.

Pearlman J.S., Barry P.S., Segal C.C., Shepanski J., Beiso D., Carman S.L. 2003. Hyperion, a space-based imaging spectrometer, *IEEE Transactions on Geoscience and Remote Sensing*, Volume 41, Pages 1160–1173.

Pearson, R.L., Miller, L.D. 1972. Remote mapping of standing crop biomass for estimation of the productivity of the short-grass prairie. Pawnee National Grasslands, Colorado, ERIM, Ann Arbor, MI, Pages 1357–1381.

Peñuelas, J., Filella, I., Biel, C., Serrano, L., Save, R. 1993a. The reflectance at the 950–970 nm region as an indicator of plant water status, *International Journal of Remote Sensing*, Volume 14, Pages 1887–1905.

Peñuelas, J., Filella, I., Elvira, S., Inclán, R. 1995. Reflectance assessment of summer ozone fumigated Mediterranean white pine seedlings. *Environmental and Experimental Botany*, Volume 35, Issue 3, Pages 299–307. ISSN 0098-8472, https://doi.org/10.1016/0098-8472(95)00019-0.

Peñuelas, J., Gamon, J.A., Griffin, K.L., Field, C.B. 1993b. Remote Sensing of Environment Assessing community type, plant biomass, pigment composition, and photosynthetic efficiency of aquatic vegetation from spectral reflectance. *Remote Sensing of Environment*, Volume 46, Issue 2, Pages 110–118. https://doi.org/10.1016/0034-4257(93)90088-F.

Peñuelas, J., Pinol, J., Ogaya, R., Lilella, I. 1997. Estimation of plant water content by the reflectance water index WI (R900/R970), *International Journal of Remote Sensing*, Volume 18, Pages 2869–2875.

Reyniers, M., Vrindts, E. 2006. Measuring wheat nitrogen status from space and ground-based platform, *International Journal of Remote Sensing*, Volume 27, Pages 549–567.

Rondeaux, G., Steven, M., Baret, F. 1996. Optimization of soil-adjusted vegetation indices, *Remote Sensing of Environment*, Volume 55, Pages 95–107.

Rouse, J.W., Haas, R.H., Schell, J.A., Deering, D.W. 1973. Monitoring vegetation systems in the Great Plains with ERTS. Paper presented at the *Third ERTS-1 Symposium*, December 10–14, Washington, DC, NASA SP-351, Volume 1, Pages 309–317.

Rouse, J.W., Haas, R.H., Schell, J.A., Deering, D.W., and Harlan, J.C. 1974. Monitoring the vernal advancement of retrogradation (green wave effect) of natural vegetation. NASA/GSFC, Type III, Final Report, Greenbelt, Maryland, USA, pp. 1–371.

Serrano, L., Peñuelas, J., Ustin, S.L. 2002. Remote sensing of nitrogen and lignin in Mediterranean vegetation from AVIRIS data: Decomposing biochemical from structural signals, *Remote Sensing of Environment*, Volume 81, Pages 355–364.

Sims, D., Gamon, J. 2002. Relationships between leaf pigment content and spectral reflectance across a wide range of species, leaf structures and developmental stages, *Remote Sensing of Environment*, Volume 81, Issue 2-3, Pages 337–354.

Sims, D.A., Luo, H.Y., Hastings, S., Oechel, W.C., Rahman, A.F., Gamon, J.A. 2006. Parallel adjustments in vegetation greenness and ecosystem CO_2 exchange in response to drought in a southern California chaparral ecosystem, *Remote Sensing of Environment*, Volume 103, Pages 289–303.

Sonobe, R., Wang, Q. 2017. Hyperspectral indices for quantifying leaf chlorophyll concentrations performed differently with different leaf types in deciduous forests, *Ecological Informatics*, Volume 37, Pages 1–9, ISSN 1574-9541, https://doi.org/10.1016/j.ecoinf.2016.11.007.

Sonobe, R., Wang, Q. 2018. Nondestructive assessments of carotenoids content of broadleaved plant species using hyperspectral indices, *Computers and Electronics in Agriculture*, Volume 145, Pages 18–26, ISSN 0168-1699, https://doi.org/10.1016/j.compag.2017.12.022.

Stimson, H.C., Breshears, D.D., Ustin, S.L., Kefauver, S.C. 2005. Spectral sensing of foliar water conditions in two co-occurring conifer species: *Pinus edulis* and *Juniperus monosperma*, *Remote Sensing of Environment*, Volume 96, Pages 108–118.

Tan, K., Wang, S., Song, Y., Liu, Y., Gong, Z. 2018. Estimating nitrogen status of rice canopy using hyperspectral reflectance combined with BPSO-SVR in cold region, *Chemometrics and Intelligent Laboratory Systems*, Volume 172, Pages 68–79, ISSN 0169-7439, https://doi.org/10.1016/j.chemolab.2017.11.014.

Tatsumi, K., Yamashiki, Y., Torres, M.A.C., Taipe, C.L.R. 2015. Crop classification of upland fields using random forest of time-series Landsat 7 ETM+ data., *Computers and Electronics in Agriculture*, Volume 115, Pages 171–179.

Teluguntla, P., Thenkabail, P.S., Xiong, J., Gumma, M.K., Congalton, R.G., Oliphant, A., Poehnelt, J., Yadav, K., Rao, M., Massey, R. 2017. Spectral matching techniques (SMTs) and automated cropland classification algorithms (ACCAs) for mapping croplands of Australia using MODIS 250-m time-series (2000–2015) data, *International Journal of Digital Earth*, doi:10.1080/17538947.2016.1267269.

Teluguntla, P., Thenkabail, P.S., Xiong, J., Gumma, M.K., Giri, C., Milesi, C. et al. 2015. Global Food Security Support Analysis Data at Nominal 1 km (GFSAD1km) Derived from Remote Sensing in Support of Food Security in the Twenty-First Century: Current Achievements and Future Possibilities, Chapter 6. In Thenkabail, P.S., (Editor-in-Chief), 2015. *"Remote Sensing Handbook" (Volume II): Land Resources Monitoring, Modeling, and Mapping with Remote Sensing*. Taylor and Francis Inc.\CRC Press, Boca Raton, London, New York. ISBN 9781482217957 - CAT# K22130. Pp. 131–160. IP-054785.

Thenkabail, P.S. 2015. Hyperspectral remote sensing for terrestrial applications, Chapter 9, in: Thenkabail, P.S., (Editor-in-Chief), 2015. *Remote Sensing Handbook (Volume II): Land Resources Monitoring, Modeling, and Mapping with Remote Sensing*. Taylor and Francis Inc./CRC Press, Boca Raton. ISBN 9781482217957—CAT# K22130. Pages 201–236. IP-0606312.

Thenkabail, P.S., Enclona, E.A., Ashton, M.S., Legg, C., Jean De Dieu, M. 2004a. Hyperion, IKONOS, ALI, and ETM+ sensors in the study of African rainforests, *Remote Sensing of Environment*, Volume 90, Pages 23–43.

Thenkabail, P.S., Enclona, E.A., Ashton, M.S., Van Der Meer, V. 2004b. Accuracy assessments of hyperspectral waveband performance for vegetation analysis applications, *Remote Sensing of Environment*, Volume 91, Pages 2–3. 354–376.

Thenkabail, P.S., Gamage, N., Smakhin, V. 2004c. The use of remote sensing data for drought assessment and monitoring in south west Asia. IWMI Research report # 85. Page 25. IWMI, Colombo, Sri Lanka.

Thenkabail, P., GangadharaRao, P., Biggs, T., Krishna, M., Turral, H. 2007. Spectral matching techniques to determine historical land-use/land-cover (LULC) and irrigated areas using time-series 0.1-degree AVHRR Pathfinder datasets, *Photogrammetric Engineering & Remote Sensing*, Volume 73, Issue 10, Pages 1029–1040.

Thenkabail, P.S., Gumma, M.K., Teluguntla, P., Mohammed, I.A. 2014. Hyperspectral remote sensing of vegetation and agricultural crops. highlight article, *Photogrammetric Engineering and Remote Sensing*, Volume 80, Issue 4, Pages 697–709. IP-052042.

Thenkabail, P.S., Hanjra, M.A., Dheeravath, V., Gumma, M. 2010. A holistic view of global croplands and their water use for ensuring global food security in the 21st century through advanced remote sensing and non-remote sensing approaches, *Remote Sensing*, Volume 2, Pages 211–261.

Thenkabail, P.S., Knox, J.W., Ozdogan, M., Gumma, M.K., Congalton, R.G., Wu, Z., Milesi, C., Finkral, A., Marshall, M., Mariotto, I. 2012a. Assessing future risks to agricultural productivity, water resources and food security: How can remote sensing help? *Photogrammetric Engineering and Remote Sensing*, Volume 78, Issue 8, Pages 773–782.

Thenkabail, P.S., Lyon, J., Huete, A. (Editors) 2012b. *Hyperspectral Remote Sensing of Vegetation*, CRC Press–Taylor & Francis group, Boca Raton, Page 781 (80+ pages in color).

Thenkabail, P.S., Mariotto, I., Gumma, M.K., Middleton, E.M., Landis, D.R., Huemmrich, F.K. 2013. Selection of hyperspectral narrowbands (HNBs) and composition of hyperspectral twoband vegetation indices (HVIs) for biophysical characterization and discrimination of crop types using field reflectance and Hyperion/EO-1 data. *IEEE Journal of Selected Topics in Applied Earth Observations and Remote Sensing*, Volume 6, Issue 2, Pages 427–439. APRIL 2013.doi:10.1109/JSTARS.2013.2252601. (80%). IP-037139.

Thenkabail, P.S., Smith, R.B., De-Pauw, E. 2002. Evaluation of narrowband and broadband vegetation indices for determining optimal hyperspectral wavebands for agricultural crop characterization. *Photogrammetric Engineering and Remote Sensing*, Volume 68, Issue 6, Pages 607–621.

Thenkabail, P.S., Smith, R.B., Pauw, D.P. 2000. Hyperspectral vegetation indices and their relationships with agricultural crop characteristics. *Remote Sensing of Environment*, Volume 71, Issue 2, Pages 158–182. ISSN 0034-4257, https://doi.org/10.1016/S0034-4257(99)00067-X.

Thorp, K.R., French, A.N., Rango, A. 2013. Effect of image spatial and spectral characteristics on mapping semi-arid rangeland vegetation using multiple endmember spectral mixture analysis (MESMA), *Remote Sensing of Environment*, Volume 132, Pages 120–130, ISSN 0034-4257, https://doi.org/10.1016/j. rse.2013.01.008.

Thorp, K.R., Wang, G., Bronson, K.F., Badaruddin, M., Mon, J. 2017. Hyperspectral data mining to identify relevant canopy spectral features for estimating durum wheat growth, nitrogen status, and grain yield, *Computers and Electronics in Agriculture*, Volume 136, Pages 1–12, ISSN 0168-1699, https://doi. org/10.1016/j.compag.2017.02.024.

Tian, Y.C., Yao, X., Yang, J., Cao, W.X., Hannaway, D.B., Zhu, Y. 2011. Assessing newly developed and published vegetation indices for estimating rice leaf nitrogen concentration with ground- and space-based hyperspectral reflectance, *Field Crops Research*, Volume 120, Issue 2, Pages 299–310, ISSN 0378-4290, https://doi.org/10.1016/j.fcr.2010.11.002.

Tilton, J.C., Tarabalka, Y., Montesano, P.M., Gofman, E. 2012. Best merge region-growing segmentation with integrated nonadjacent region object aggregation. Geoscience and remote sensing, *IEEE Transactions on Geoscience and Remote Sensing*, Volume 50, Pages 4454–4467.

Toniol, A.C., Galvão, L.S., Ponzoni, F.J., Sano, E.E., Amore, D.D.J. 2017. Potential of hyperspectral metrics and classifiers for mapping Brazilian savannas in the rainy and dry seasons, *Remote Sensing Applications: Society and Environment*, Volume 8, Pages 20–29, ISSN 2352-9385, https://doi.org/10.1016/j. rsase.2017.07.004.

Tsai, F., Philpot, W. 1998. Derivative analysis of hyperspectral data, *Remote Sensing of Environment*, Volume 66, Issue 1, Pages 41–51.

Vigneau, N., Ecarnot, M., Rabatel, G., Roumet, P. 2011. Potential of field hyperspectral imaging as a non destructive method to assess leaf nitrogen content in wheat, *Field Crops Research*, Volume 122, Issue 1, Pages 25–31, ISSN 0378-4290, https://doi.org/10.1016/j.fcr.2011.02.003.

Vogelmann, J.E., Rock, B.N., Moss, D.M. 1993. Red edge spectral measurements from sugar maple leave. *International Journal of Remote Sensing*, Volume 14, Issue 8, Pages 1563–1575.

Waldner, F., Canto, G.S., Defourny, P. 2015. Automated annual cropland mapping using knowledge-based temporal features, *ISPRS Journal of Photo and Remote Sensing*, Volume 110, Pages 1–13.

Wang, Q., Li, P. 2012. Hyperspectral indices for estimating leaf biochemical properties in temperate deciduous forests: Comparison of simulated and measured reflectance data sets, *Ecological Indicators*, Volume 14, Issue 1, Pages 56–65, ISSN 1470-160X, https://doi.org/10.1016/j.ecolind.2011.08.021.

Wang, W., Yao, X., Yao, X.F., Tian, Y.C., Liu, X.J., Ni, J., Cao, W.X., Zhu, Y. 2012. Estimating leaf nitrogen concentration with three-band vegetation indices in rice and wheat, *Field Crops Research*, Volume 129, Pages 90–98, ISSN 0378-4290, https://doi.org/10.1016/j.fcr.2012.01.014.

Wu, C., Niu, Z., Tang, Q., Huang, W. 2008. Estimating chlorophyll content from hyperspectral vegetation indices: Modeling and validation, *Agricultural and Forest Meteorology*, Volume 148, Issue 8–9, Pages 1230–1241, ISSN 0168-1923, https://doi.org/10.1016/j.agrformet.2008.03.005.

Xiong, J., Thenkabail, P.S., Gumma, M., Teluguntla, P., Poehnelt, J., Congalton, R., Yadav, K. 2017. Automated cropland mapping of continental Africa using Google Earth Engine cloud computing, *The ISPRS Journal of Photogrammetry and Remote Sensing (P&RS)*, Volume 126, Pages 225–244. doi:10.1016/j. isprsjprs.2017.01.019.

Xiong, J., Thenkabail, P.S., Tilton, J.C., Gumma, M.K., Teluguntla, P., Oliphant, A., Congalton, R.G., Yadav, K., Gorelick, N. 2017b. Nominal 30-m cropland extent map of continental Africa by integrating pixel-based and object-based algorithms using Sentinel-2 and Landsat-8 data on Google Earth Engine, *Remote Sensing*, Volume 9, Issue 10, Page 1065; doi:10.3390/rs9101065, http://www.mdpi.com/2072-4292/9/10/1065.

Xue, L.H., Cao, W.X., Luo, W.H., Dai, T.B., Zhu, Y. 2004. Monitoring leaf nitrogen status in rice with canopy spectral reflectance, *Agronomy Journal*, Volume 96, Pages 135–142.

Yang, Y.-C., Sun, D.-W., Pu, H., Wang, N.-N., Zhu, Z. 2015. Rapid detection of anthocyanin content in lychee pericarp during storage using hyperspectral imaging coupled with model fusion, *Postharvest Biology and Technology*, Volume 103, Pages 55–65, ISSN 0925-5214, https://doi.org/10.1016/j. postharvbio.2015.02.008.

Yao, X., Zhu, Y., Tian, Y.C., Feng, W., Cao, W.X. 2010. Exploring hyperspectral bands and estimation indices for leaf nitrogen accumulation in wheat, *International Journal of Applied Earth Observation and Geoinformation*, Volume 12, Issue 2, Pages 89–100, ISSN 0303-2434, https://doi.org/10.1016/j. jag.2009.11.008.

Yi, Q.X., Bao, A.M., Wang, Q., Zhao, J. 2013. Estimation of leaf water content in cotton by means of hyperspectral indices, *Computers and Electronics in Agriculture*, Volume 90, Pages 144–151, ISSN 0168-1699, https://doi.org/10.1016/j.compag.2012.09.011.

Yilmaz, M.T., Hunt, E.R., Jackson, T.J. 2008. Remote sensing of vegetation water content from equivalent water thickness using satellite imagery, *Remote Sensing of Environment*, Volume 112, Pages 2514–2522.

Yu, K., Li, F., Gnyp, M.L., Miao, Y., Bareth, G., Chen, X. 2013. Remotely detecting canopy nitrogen concentration and uptake of paddy rice in the Northeast China Plain, *ISPRS Journal of Photogrammetry and Remote Sensing*, Volume 78, Pages 102–115, ISSN 0924-2716, https://doi.org/10.1016/j.isprsjprs.2013.01.008.

Zarco-Tejada, P.J., Guillén-Climent, M.L., Hernández-Clemente, R., Catalina, A., González, M.R., Martín, P. 2013. Estimating leaf carotenoid content in vineyards using high resolution hyperspectral imagery acquired from an unmanned aerial vehicle (UAV), *Agricultural and Forest Meteorology*, Volumes 171–172, Pages 281–294, ISSN 0168-1923, https://doi.org/10.1016/j.agrformet.2012.12.013.

Zarco-Tejada, P.J., Miller, J.R., Harron, J., Hu, B., Noland, T.L., Goel, N. et al. 2004. Needle chlorophyll content estimation through model inversion using hyperspectral data from boreal conifer forest canopies, *Remote Sensing of Environment*, Volume 89, Pages 189–199.

Zhang, G., Xiao, X., Dong, J., Kou, W., Jin, C., Qin, Y., Zhou, Y., Wang, J., Menarguez, M.A., Biradar, C. 2015. Mapping paddy rice planting areas through time series analysis of MODIS land surface temperature and vegetation index data. *ISPRS Journal of Photogrammetry and Remote Sensing*, Volume 106, Pages 157–171.

Zhang, Y., Chen, J.M., Miller, J.R., Noland, T.L. 2008. Leaf chlorophyll content retrieval from airborne hyperspectral remote sensing imagery, *Remote Sensing of Environment*, Volume 112, Issue 7, Pages 3234–3247, ISSN 0034-4257, https://doi.org/10.1016/j.rse.2008.04.005.

Zhao, B., Duan, A., Ata-Ul-Karim, S.T., Liu, Z., Chen, Z., Gong, Z. et al. 2018. Exploring new spectral bands and vegetation indices for estimating nitrogen nutrition index of summer maize, *European Journal of Agronomy*, Volume 93, Pages 113–125, ISSN 1161-0301, https://doi.org/10.1016/j.eja.2017.12.006.

Zhu, Y., Yao, X., Tian, Y.C., Liu, X.J., Cao, W.X. 2008. Analysis of common canopy vegetation indices for indicating leaf nitrogen accumulations in wheat and rice, *International Journal of Applied Earth Observation and Geoinformation*, Volume 10, Pages 1–10.

Index